MUSCLES
Testing and Function

THIRD EDITION

Henry Otis Kendall, P.T. (1898-1979)

Co-author of first and second editions

Former Director of Physical Therapy Department, Children's Hospital, Baltimore, Maryland; Supervisor of Physical Therapy, Baltimore Board of Education; Instructor in Body Mechanics, Johns Hopkins School of Nursing; private practice.

MUSCLES
Testing and Function

THIRD EDITION

Florence Peterson Kendall, P.T.

Lecturer; Consultant to the Surgeon General, U.S. Army; Consultant to, and former member of, the Maryland State Board of Physical Therapy Examiners. Formerly, physical therapist, Children's Hospital, Baltimore, Maryland; faculty member, School of Medicine, Department of Physical Therapy, University of Maryland; instructor in Body Mechanics, Johns Hopkins Hospital, School of Nursing.

Elizabeth Kendall McCreary, B.A.

Classified graduate student in Biomedical Sciences (Physiology), University of Hawaii.

Illustrations by   Ranice W. Crosby ▪ Diane K. Abeloff
Marjorie B. Gregerman ▪ William E. Loechel
Photographs by Charles C. Krause, Jr.

WILLIAMS & WILKINS
Baltimore • Hong Kong • London • Sydney

Copyright ©, 1983
Williams & Wilkins
428 E. Preston Street
Baltimore, Md. 21202, U.S.A.

Made in the United States of America

First edition, 1949
Second edition, 1971
Reprinted 1972, 1973, 1974, 1975, 1976, 1977, 1978, 1979, 1980, 1981, 1982
Third edition, 1983

Library of Congress Cataloging in Publication Data

Kendall, Florence Peterson, 1910–
    Muscles, testing and function.

    Rev. ed. of: Muscles, testing and function/Henry Otis Kendall, Florence Peterson Kendall, Gladys Elizabeth Wadsworth, 2nd ed. 1971.
    Bibliography: p.
    Includes index.
    1. Muscles—Examination. 2. Physical therapy. I. McCreary, Elizabeth Kendall. II. Kendall, Henry Otis, 1898–1979. Muscles, testing and function. III. Title. [DNLM: 1. Muscles—Physiology. WE 500 K326m]
    RM701.K46 1983   616.7′40754   82-20166
    ISBN 0-683-04575-X

15 14 13 12 11

Dedicated to
The Kendall Grandchildren

Foreword to the First Edition

For over twenty-five years we have been fortunate in having Henry O. Kendall in charge of the Physiotherapy Department of the Children's Hospital School. During that time not only has a great volume of clinical work been splendidly carried on, but also due to the teamwork of Henry Kendall and his wife, Florence Peterson Kendall, a valuable series of original scientific papers and treatises has appeared in which we were proud to share by observation, criticism, and the writing of brief forewords. So it was that when they proposed to concentrate their wide experience, their technical skill, and their critical judgment in one big volume to give a thorough exposition of the methods and results of neuromuscular physical examination we were delighted. Now that it is completed we feel they have succeeded admirably in their difficult task and that the book with its text, charts, and illustrations has made the difficult techniques clear and shown the way to correlate and evaluate the information obtained. Although a textbook containing necessarily established facts and methods, it presents them more clearly and graphically than ever heretofore and, in addition, it contains much that is entirely new and extremely valuable, especially the diagnostic nerve charts for the peripheral nerves. We take pleasure and pride in commending to the medical world this book, entitled *MUSCLES, Testing and Function*, by Henry O. and Florence P. Kendall.

GEORGE E. BENNETT
Emeritus Adjunct Professor of Orthopedic Surgery
ROBERT W. JOHNSON
Adjunct Professor of Orthopedic Surgery

Johns Hopkins Medical School

Preface to the Third Edition

The third edition of *MUSCLES, Testing and Function* preserves all the valuable and sound information that has characterized the first and second editions, and adds to this a good deal of textual material as well as many new photographs and drawings. A new chapter, "Muscle Function in Relation to Posture", has been added; an outline of chapter content appears at the beginning of each chapter; and material in several chapters has been reorganized for more logical sequence of ideas, easier reading, and better comprehension of the concepts presented.

Since muscle origins and insertions do not change, descriptions of these do not require "updating" in new editions. Muscle action is also a very constant factor. Among anatomy books there are some variations in the specific actions ascribed to muscles but the differences are not substantial. With the advent of electronic testing instruments, additional information about muscle actions, ascertained under closely controlled conditions, continues to become available. To date, information obtained serves to confirm the validity of the tests portrayed by the photographs used in the first and second editions. Because of the quality of these original photographs and the accuracy with which they portray the tests, they continue to be used in the third edition.

The drawings which have been added to this edition have been made by two of the three artists who worked on the second edition. For this reason, there is continuity in the type of drawing, the manner of presentation, and in understanding the authors' ideas that the artists have portrayed so effectively. By reducing the size of many of the photographs and some of the drawings, it has been possible to add the new material without greatly increasing the number of pages in the book.

In Chapter 1, some important definitions and concepts relating to muscle length and strength have been added. Chapter 2 includes several new paragraphs defining circumduction of various joints, and additional text and illustrations explaining movements of the spine. The material on posture, which constituted about one-half of Chapter

2 in the second edition, now appears in Chapter 8.

Chapter 3 has been reorganized and some material has been rewritten for clarity. After careful review of the data about spinal segment innervation to muscles, it became evident that there was no need to change any of this information except for two minor deletions. The S3 innervation attributed to Gemellis superior and Obturator internus was found to have been an addition error in the second edition and has been removed from the compilation and from the Spinal Nerve and Muscle Chart. The second edition contained a drawing of the brachial plexus. That drawing has been replaced by a new one to conform with three new drawings of the cervical, lumbar, and sacral plexuses that have been added to the third edition.

In Chapter 4, new text material describes, and photographs illustrate, the test position and fixation necessary for each of the palmar and dorsal Interossei muscles. A test differentiating the Extensor carpi radialis brevis from the longus is described, and illustrated by photographs. There are new photographs for the tests of upper and lower Trapezius, and an accompanying drawing for the upper Trapezius. The second edition full-page drawing of several shoulder and shoulder girdle muscles has been eliminated and the individual drawings have been put with the corresponding photographs of the tests. There are new photographs showing range-of-motion tests for shoulder joint rotation, and for combined shoulder joint and shoulder girdle movements involving rotation.

In Chapter 5, two new drawings of the Lumbricales and Interossei have been added. The drawings of the Hamstring and hip flexor length tests have been improved in order to show more precisely the analysis and interpretation of test findings. Numerous photographs have been added to illustrate significant variations in tests of hamstring and hip flexor length.

A revision of Chapter 6, which deals with trunk muscles, is the result of a determined effort to put much of the material in simple language and make it more understandable for the layman. Definitions have been added and illustrations have been placed in immediate proximity to the related test. Drawings which illustrate the joint movements that occur during sit-ups are followed by comparable drawings which show the muscle actions that occur. The detailed descriptions and the accompanying illustrations should dispel some of the erroneous concepts that continue to persist regarding the actions of abdominal and hip flexor muscles.

In Chapter 7, photographs of facial muscle tests have been reduced from full page to one column size, preserving all of the content in fewer pages. Cranial nerve charts were moved from Chapter 3 to Chapter 7. The cases of facial muscle involvement recorded on the cranial charts are new in this edition. Two new full-page illustrations of the Infrahyoid and Suprahyoid muscles have been added.

A textbook on muscle testing and function is not complete if it does not stress the importance of muscle testing in postural evaluation, and the function of muscles in maintaining the upright position of the body. In each of the first two editions, ten pages relating to posture were included but were scattered throughout the text. In this edition, there is a new Chapter 8 which deals specifically with muscle function as related to posture. Because there is a high incidence of low back pain that is postural in origin, and because many other painful conditions are associated with faulty posture, it becomes increasingly important that people understand the muscle problems that may give rise to, or be associated with, these painful conditions. With better evaluation techniques and a clearer understanding of muscle function, more adequate treatment procedures will evolve.

Most of the material for this chapter has been taken from *Posture and Pain* (Kendall, Kendall & Boynton). The muscle balance present in good posture and the imbalances that often accompany faulty posture are illustrated and described. The chapter is divided into sections dealing with various aspects of posture. Following each section is a chart which includes analyses of the muscle imbalances that frequently accompany the faulty alignment, and an outline of the exercises and other therapeutic measures used in the correction of the postural problems.

Revising *MUSCLES, Testing and Function* has been a rewarding experience and has offered the rare opportunity for me to work on a new edition more than thirty years after the publication of the first edition. My late husband, Henry O. Kendall, and I have had more than fifty years of experience each, with more than forty of them representing the overlapping years when we worked together evaluating and treating musculoskeletal conditions. The ability to do muscle testing and to understand muscle function is rooted in a knowledge of anatomy and physiology, and in the practical application of this knowledge through experience. The material presented in the first and second editions has "stood the test of time", and I believe that the expanded third edition will meet the challenges of the future.

Florence P. Kendall

Preface to the Second Edition

A scientific achievement of our time has been the tremendous reduction in the incidence of poliomyelitis. Probably no other neuromuscular condition afforded a comparable opportunity to study muscle function and to observe the effects of the loss of such function. Those of us who have had the opportunity to learn from experience in this field have the responsibility to leave for posterity a record of the information we have obtained. This book represents our effort to record some of this information.

This second edition of *MUSCLES, Testing and Function* contains almost twice as much material as the first edition without an increase in the size of the book.

All the original photographs have been preserved, and, with the newer printing techniques, reproductions are far superior to those in the first edition. The fact that no changes in the photographs were necessary stands as a tribute to the validity of the tests as previously presented.

Our desire to make the text most useful has been fulfilled to a great extent by the addition of drawings to accompany the photographs of all the muscle tests in Chapters IV and V. These drawings have been created by a team of medical illustrators (listed on the title page) to whom we are deeply indebted for their skill, their endless patience, and their response to the challenges presented to them. These illustrations are unique in several respects. Not only have we asked that the bones and muscles be illustrated on the one drawing, but that the same view be shown in the drawing as in the photograph. To meet these requirements, the authors and the artists used radiographs and posed live subjects and a skeleton in test positions to enable the artists to make sketches. The artists were asked to maintain the relationship of the bones and muscles to surface anatomy in the fervent hope that the muscles would "come alive" to the reader.

Chapter I has been rewritten and is presented in a concise and direct manner. A new chapter has been added, Chapter II on Joint Motions and Joint Positions. We believe this a valuable addition to the text and a necessary component of a book which deals with the action of muscles. The authors have taken the liberty of expanding the basic description of *anatomical position* to make more precise its use as the position of reference for joint movements.

The Spinal Nerve and Muscle Charts presented in Chapter III basically are the same as in Chapter II of the previous edition, but have been made more useful by some changes and additions. A numeral has been used instead of an X to represent the spinal segment, making it possible to identify immediately the segments supplying each muscle. Dots have been substituted for X's in the peripheral nerve section, thus removing the X's entirely from the face of the chart. Drawings which show the dermatomes and the areas of distribution of cutaneous nerves have been added to the Cranial and Spinal Nerve and Muscle Charts.

The information about spinal segment supply to muscles from Spalteholz[7] and Foerster,[8] as listed on pages 43 and 45 is as it appeared in the first edition. The most recent texts by Gray[4] and Haymaker and Woodhall[9] were reviewed, and data were updated in a few instances. Information from texts by Cunningham[6] and deJong[5] replaced that from Morris and Quain.

The complete retabulation of data from the above sources resulted in only slight changes, and does not affect the validity of the original charts. The illustrations for the Spinal Nerve and Motor Point Charts are basically the same as in the first edition but have been improved in the light of detailed information obtained from *Neuro-vascular Hila of Limb Muscles* by Brash.[17]

Included in Chapter III is a drawing of the brachial plexus shown in relation to body contour and bony structures. Each spinal nerve that enters into the brachial plexus is represented by a different color or shade. It is hoped that this illustration will be a useful visual aid for those trying to comprehend this complex network.

Numerous illustrations from the authors' book on *Posture and Pain,*[20] along with new material, have been added to Chapter VI on trunk muscles. An effort has been made to present, in terms that are easily understood, the answers to many controversial questions about the action of the abdominal and hip flexor muscles. This chapter should be of particular interest to teachers of physical education as well as persons in the medical and allied health fields who are concerned with the problems of body mechanics.

As therapeutic exercise plays a role in the treatment of disease or injury, proper exercise plays a role in preventive medicine. Those engaged in the use of exercise as a therapeutic or preventive measure need to be knowledgeable about muscles. It is the aim of this text to help the reader better comprehend muscle function.

Nomina Anatomica, third edition,[24] has been the guide for terminology, and a comparative list of some NA and BNA terms appear on page 278. A glossary of terms relating to foot deformities appears on page 27. Other terms have been defined within the text. Whereas the first edition incorporated in the preface an acknowledgement of the texts used as references, this edition includes a bibliography.

Exceptional skill in performing a procedure is an art. Knowledge, evolved and formulated in the search for truth, is a science. We hope that the second edition of this book will enhance muscle testing as both an art and a science.

HENRY O. KENDALL
FLORENCE P. KENDALL

Preface to the First Edition

Muscle testing procedures are an integral part of physical diagnosis, and form the basis for determining individual needs in relation to specific forms of treatment in muscular and neuromuscular disorders. A detailed account of the testing procedures, and of the functional significance of muscle weakness and contracture, is embodied in this text.

We believe this book will be particularly useful to physicians, surgeons, physical therapists, occupational therapists, and physical educators. There has been an insistent demand for a comprehensive book on muscle testing by people in these various fields.

We have had the opportunity to do detailed muscle testing on several thousand patients, both paralytic and non-paralytic. In addition muscle tests and postural examinations have been done on approximately one thousand so-called normal individuals of various age groups, for the purpose of research. The information obtained through experience in muscle testing has proved indispensable to us in further analyzing problems relating to muscle function. We feel it is a privilege and responsibility to make available to others the information thus acquired.

We do not aim to set forth new theories nor outline new systems of procedure. Our purpose is to interpret existing facts in the light of our experience and study. We have accepted only such procedures as stand the test of detailed analyses on the basis of known anatomical facts.

Dr. Robert W. Lovett and his co-workers laid the firm foundation for much of the muscle testing done today. We do not seek to have this text supplant the earlier work, but we do seek to enlarge the scope of muscle testing and help clarify many of the problems that have continued to exist.

This book presents muscle testing in a simple, direct manner which should enable the reader to clearly visualize the procedures involved in testing, grading and recording muscle strength. It is written in text book style and is extensively illustrated. The first chapter which deals with general procedures is presented in non-technical terms. The remaining chapters, which consist of specific test procedures, require the use of anatomical terms for accuracy and brevity.

Extensive investigation has been done in the preparation of the Diagnostic Charts for Nerve Lesions (pp. 49 and 51), and the Spinal Motor Nerve and Anatomical Motor Point Charts (pp. 48 and 50). The authors have made every effort to extract the pertinent information from authoritative works on anatomy and neurology. Texts which have been consulted include the following: Grant, Sabotta, Cunningham, Gray, Morris, Quain, Spalteholz, Foerster, and Haymaker and Woodhall. The spinal segment nerve supply as charted is a compilation from texts by the last six above-named authorities. The appendix, which lists muscles grouped according to joint action, and their origins and insertions, has been compiled mainly from *Gray's Anatomy.*

U. S. Public Health Bulletin no. 242 (April 1938), written by the authors, has been used rather extensively. Chapter IV of this Bulletin has been revised, and with additions, constitutes Chapter V of this text.

The similarity which abdominal muscle tests bear to those described in *Physical Therapy for Lower Extremity Amputees* (War Department Technical Manual, 8–293) is accounted for by the fact that the authors of this text assisted in the preparation of the manual.

This book, as many others, has come into existence through the inspiration, cooperation, and assistance of many people.

We thank the medical directors and staff, and the co-workers in the physical therapy department at the Children's Hospital School, for their encouragement in study and research, their interest and assistance in the project of writing this book, and for their constructive criticism in reviewing the subject-matter of this text. Williams & Wilkins, the publishers, have given us valuable aid and advice. For the splendid photography we are indebted to Charles C. Krausse, Jr. In the test pictures every effort has been made to show the muscle, the direction of movement, and to indicate the place and direction of the examiner's pressure. These requirements have been met only because of the photographer's patience, skill, and meticulous attention to detail. We appreciate, too, the assistance of all who acted as subjects for the photographs. The anatomical drawings hve been made by William E. Loechel whose ability and interest have made it possible to obtain the detailed accuracy in the illustrations.

To the patients and friends who created a fund which enabled the undertaking of this project, we wish to express our sincere appreciation. The extent of illustrative material bears testimony of their generosity, but the inspiration that the friends imparted cannot be bound between the covers of a book.

For the facilities provided for Physical Therapy at the Children's Hospital School, we are indebted to Louise Cunningham Bowles, who gave the Thomas Henry Bowles Memorial Building. It is to Mrs. Bowles that we dedicate this book.

We wish to thank all others who have assisted us, and, were it possible, we would acknowledge each individually.

HENRY O. KENDALL
FLORENCE P. KENDALL

Acknowledgments

Throughout the years, many people have contributed, in one way or another, to the three editions of this text. Those who helped with the first and second editions have been acknowledged, but for some, because of the invaluable and lasting quality of their contributions, I wish to acknowledge them again. They are the photographer and artists who worked on the earlier editions and whose names appear on the title page.

Two of the artists, Ranice Crosby and Marjorie Gregerman, have worked on the third edition, also. Their meticulous attention to detail and their innovative and creative ideas have added new dimensions to this book. The multicolored portrayal of the nerves and spinal segment innervation in the complicated plexus drawings is evidence of their expertise. Furthermore, their helpful, calm, reassuring manner made working with them a great joy.

For some of the new photographs, I wish to thank Irvin Miller, a physical therapist and former student who came to the aid of his former teacher. Picture taking has its frustrating moments and together we survived many such episodes. My thanks, also, to the physical therapy students at the University of Maryland, four grandchildren, and other people who served as subjects for many of the pictures.

To Roger Michael, M.D., Chief of Orthopedic Surgery at Union Memorial Hospital and Assistant Professor of Orthopedic Surgery, Johns Hopkins University School of Medicine, I wish to express my appreciation for taking time from his busy schedule to read Chapter 6 and offer helpful comments. To Shirley Sahrmann, for her professional advice and encouragement, I extend my sincere thanks.

It has been my good fortune to be able to work closely with the publishers over the years. There have been numerous changes in personnel but the spirit of cooperation and dedication still prevails at Williams & Wilkins. The lasting value of the book is largely due to the demand for excellence by the publishers and printers. Especially, I wish to thank Sara Finnegan and George Stamathis for their help.

Not the least among the many who have helped are members of my family. From those three little girls who sometimes sat on the floor under the table (where we spread out our work) in order to be near without disturbing us, or "helped" by scribbling on some of the pages, or made the "funny faces" for the facial muscle tests—to the same three, now grown, who listen patiently to the many rewrites, or help with editing or proofreading—I say thanks for putting up with your parents for all those years! Special thanks goes to one of them who now joins me as co-author of this third edition.

Contents

Foreword to the Third Edition vii

Foreword to the First Edition ix

Preface to the Third Edition xi

Preface to the Second Edition xiii

Preface to the First Edition xv

Acknowledgments xvii

chapter 1   **Fundamental Principles in Manual Muscle Testing** 1

chapter 2   **Joint Motions** 17

chapter 3   **Spinal Nerve and Muscle Charts and Plexuses** 31

chapter 4   **Upper Extremity and Scapular Muscles** 59

chapter 5   **Lower Extremity Muscles** 129

chapter 6   **Trunk Muscles** 185

chapter 7   **Facial, Eye, and Neck Muscles; Muscles of Deglutition; Respiratory Muscles** 235

chapter 8   **Muscle Function in Relation to Posture** 269

Terminology 316

Bibliography 317

Index 321

chapter 1

Fundamental Principles in Manual Muscle Testing

Introduction
Terms used in descriptions of muscle tests
Substitution
Grading

Key to muscle grading (chart)
Use of word "normal" in relation to muscle testing
Suggested order of muscle tests

Fundamental Principles in Manual Muscle Testing

Muscle testing is an integral part of physical examination. It provides information, not obtained by other procedures, that is useful in differential diagnosis, prognosis, and treatment of neuromuscular and musculoskeletal disorders.

Many *neuromuscular* conditions are characterized by muscle weakness. Some show definite patterns of muscle involvement; others show spotty weakness without any apparent pattern. In some cases weakness is symmetrical, in others, asymmetrical. The site or level of a peripheral lesion may be determined because the muscles distal to the site of the lesion will show weakness or paralysis. Careful testing and accurate recording of test results will reveal the characteristic findings and aid in diagnosis.

Musculoskeletal conditions frequently show patterns of muscle imbalance. Some patterns are associated with handedness; some with habitually poor posture. Imbalance that affects body alignment is an important factor in many painful postural conditions. Muscle imbalance may result also from occupational or recreational activities in which there is persistent use of certain muscles without adequate exercise of opposing muscles. Examination to determine muscle length and strength is essential before prescribing therapeutic exercises because most of these exercises are designed to stretch short muscles or strengthen weak ones.

Muscle testing demands rigorous attention to every detail that might affect its accuracy. Failure to take into account apparently insignificant factors may alter test results. Findings are useful only if they are accurate, and inaccurate test results only confuse although they appear to enlighten. Muscle testing is a procedure which depends on the knowledge, skill, and experience of the examiner who should not betray, through carelessness or lack of skill, the confidence that others rightfully place in this procedure.

Test performance and evaluation of muscle strength are the two fundamental components of manual muscle testing. To become proficient in applying this procedure one must possess a comprehensive and detailed knowledge of muscle function. It must include a knowledge of joint motion, the origin and insertion of muscles, their agonistic and antagonistic actions, as well as their role in fixation and substitution. In addition, it requires the ability to palpate the muscle or its tendon, to distinguish between normal and atrophied contour, and to recognize abnormalities of position or movement.

One who has a comprehensive knowledge of the actions of muscles can learn in a relatively short time how to obtain the actual test performance. Experience is necessary, however, to detect the substitution movements that occur whenever weakness exists, and practice is necessary to acquire skill in accurate grading of muscle strength.

Continuing efforts will be made to devise a mechanical substitute for manual muscle testing. Technically, it may not be too difficult to measure objectively the strength of a muscle group, but it will be difficult to develop a machine that can test for individual muscle strength and detect muscle substitution.

Many factors are involved in the picture of weakness and return of strength. Weakness may be due to nerve involvement, disuse atrophy, stretch weakness, pain, or fatigue. Return of muscle strength may be due to recovery following the disease process, return of nerve impulse after trauma and repair, hypertrophy of unaffected muscle fibers, muscular development resulting from exercises to overcome disuse atrophy, or return of strength after stretch and strain have been relieved.

The muscle testing described in this book is directed toward examination of individual muscles insofar as is practical. Anyone in the field of muscle testing recognizes the overlap of muscle actions, as well as the interdependence of muscles in movement. Such close relationship in muscle function does not rule out the possibility or the practicability of testing individual muscles. (See figure below.)

Emphasis has been placed by kinesiologists on naming the prime mover in various joint movements. One might rather be concerned with the fact that every muscle is a prime mover in some specific action. In the search for that action, one is led into the field of individual muscle testing.

No two muscles in the body have exactly the same function. When any one muscle is paralyzed, stability of the part is impaired or some exact movement is lost. Observing the effects of this loss helps establish the function of the muscle. When paralysis of a muscle results in the inability to perform the muscle test, the validity of the test is substantiated.

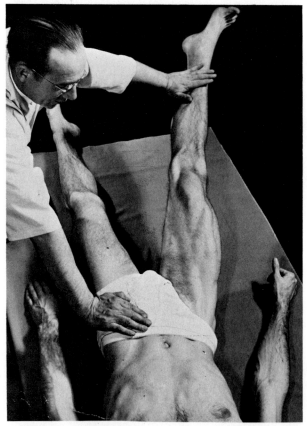

The above figure illustrates the test of the Iliopsoas with emphasis on the Psoas major muscle. The subject is an athlete in whom muscle action is particularly well demonstrated. As the photograph clearly shows, the Quadriceps (Rectus femoris in the hip action), Sartorius, and the adductors are all assisting in holding the hip-flexion position. Yet the line of pull of the muscle and the direction of the examiner's pressure place emphasis on the action of the right Psoas major, and the test disclosed a weakness of this muscle. (See p. 164)

Testing individual muscles rather than muscle groups is essential for diagnosis in neuromuscular conditions. In peripheral nerve and nerve root lesions, loss of function may follow the pattern of partial or complete peripheral nerve involvement, involvement of one or more cords of a plexus, or involvement of a spinal nerve root. Complete or partial paralysis of muscles in anterior poliomyelitis does not appear to follow any particular pattern when such weakness is spotty and extensive. There are instances, however, when careful examination indicates the level of spinal cord involvement. For example, paralysis of the Deltoid, Biceps, and Brachioradialis is indicative of involvement at the level of C5 and C6.

The word "isolation" has been used in connection with testing for separate or individual muscle action. Technically, it is well to limit the use of the word isolation to a comparatively few muscle tests. "Isolate" is defined (Funk & Wagnalls) as "to place in a detached or insulated position; to obtain in a free or uncombined state." The first usage does not apply in muscle testing; the second usage may apply since some tests obtain the action of muscles in an uncombined state. The action of such muscles as Flexor digitorum profundus and Flexor digitorum longus can be isolated in the flexion tests of the distal phalanges of the fingers and toes because no other muscles cross the joint to assist in the flexion action.

The action of various other muscles can be differentiated to the extent that the test is practical and of diagnostic value. Differentiation in testing depends on the relationship of the muscles or parts of muscles to each other. It may involve differentiating a one-joint muscle from a multi-joint muscle, a multi-joint muscle from another multi-joint muscle, a one-joint muscle from another one-joint muscle, or a part of a fan-shaped muscle from another part.

Differentiating the action of a *one-joint muscle* from that of a multi-joint muscle is done by placing the multi-joint muscle at a disadvantage. Muscles that cross two or more joints do not have the capacity to shorten sufficiently to complete the range of motion of all joints at the same time. When actively shortened as far as possible, or if passively placed in this shortened position, the muscle is incapable of assisting with further movement although further range of joint motion exists. An example frequently cited is that of testing the one-joint Gluteus maximus (buttocks muscle) with the two-joint Hamstrings in a shortened position by having the knee flexed during hip extension.

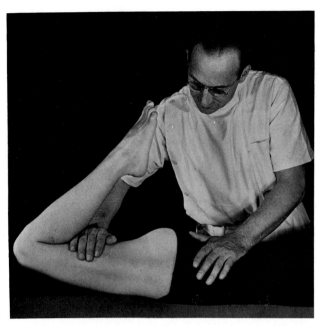

The Gluteus maximus is differentiated from the Hamstrings in the hip extension test by flexing the knee joint.

In muscle testing and exercise, the expression "putting a muscle on a slack" often is used in describing the condition of the muscle in the shortened position in which it is incapable of developing enough tension to exert an effective force. Within this context, the word slack has the same meaning as "actively insufficient" as described by O'Connell and Gardner:[1]

If a muscle which crosses two or more joints produces simultaneous movement at all of the joints that it crosses, it soon reaches a length at which it can no longer generate a useful amount of force. Under these conditions the muscle is said to be *actively insufficient.* An example of such insufficiency occurs when one tries to achieve full hip extension with maximal knee flexion. The two-joint hamstrings are incapable of shortening sufficiently to produce a complete range of motion of both joints simultaneously.

Within the meaning of the words slack and actively insufficient, as described here, it is emphasized that the definitions apply to muscles that cross two or more joints, but not to one-joint muscles *intact in the body.* A one-joint muscle is expected to complete the full range of motion of the joint over which it crosses, and to exert a strong effective force at completion of the range of motion, sufficient to grade 100% or normal. If both the one-joint and the multi-joint muscles were made ineffective by being placed in a shortened position, the action of the one-joint muscle could not be distinguished from that of the multi-joint muscle.

Haines[2] explains this characteristic of the multi-joint muscle on the basis of the length of the fibers. In addition to Hamstrings, he refers to the Biceps brachii and the Gastrocnemius as muscles of short action since the length of their fibers is too short to complete the range of movement (simultaneously) of all joints over which they act. Placing these muscles in their shortened position decreases their ability to develop tension until at maximum contraction, i.e., approximately 60% of rest length (Mountcastle[3]), the tension they exert approaches zero.

The fact that a one-joint muscle can be differentiated from a multi-joint muscle by placing the multi-joint muscle in a position of maximal shortening must be applied judiciously. Application of excessive pressure against the one-joint muscle being tested will require assistance from the multi-joint muscle, which, in its shortened position, will contract excessively producing the phenomenon of a muscle "cramp." Applying too much pressure when testing for the Supinator with the Biceps in a shortened position, for example, can cause a cramping of the Biceps that may leave the muscle sore for several days.

When a *multi-joint muscle* is being tested, the position of excessive shortening of that muscle must be avoided. For example, if the wrist is allowed to flex as the fingers flex, the strength of finger flexion is greatly diminished. If forced to hold against strong pressure in this fully flexed position, a feeling of excessive contraction (cramping) might result in the flexor muscles because of the excessive shortening, and one would be unable to obtain a true measure of the strength of the flexors being tested. At the same time there would be evidence of strain on the multi-joint finger extensors because they would be excessively elongated.

By the action of some muscles to *safeguard* others, nature has provided a mechanism to protect multi-joint muscles from overshortening when contracting, and to protect multi-joint antagonists from being overlengthened. To prevent overshortening of the contracting muscles (the agonists), they must be elongated over one or more joints while shortening over other joints. Likewise, to prevent overlengthening of the multi-joint antagonists, they must be shortened over one or more joints while being lengthened over other joints.

A familiar example of this action to safeguard other muscles is that of the wrist extensor action during finger flexion. By the action of the Extensor carpi radialis longus and brevis and the Extensor carpi ulnaris holding the wrist in extension, the finger flexors are elongated over the wrist joint while shortening over the finger joints. At the same time this position of the wrist prevents full stretching of the Extensor digitorum, the tendons of which pass over the wrist and all the finger joints. (This example has been used frequently to illustrate *synergistic* action. Since the word "synergy" means "combined action", and since this word is being used with a wide variety of meanings, this text does not attach a specific meaning to the word synergy.)

This action to safeguard other muscles is found in numerous places in the body. For example, in order to flex the elbow joint completely against strong resistance, the shoulder extensors act to elongate the Biceps over the shoulder joint. In the lower extremity, hip flexors act to elongate the Hamstrings over the hip joint when the knee is fully flexed against strong resistance.

The close relationship of muscles determines their action in substitution, assistance, and stabilization, during tests of individual muscles. The grouping of muscles according to joint action as seen in the charts on pp. 124 and 180, has been done to aid the examiner in understanding the allied actions of muscles.

The order of the tests as they appear in this text is not of particular significance except that muscles which are closely related in position or action tend to appear in sequence in order to distinguish test differences.

Although most muscle testing is applied to patients with varying degrees of weakness or paralysis, the authors have chosen to use normal subjects to show normal muscle function, and the contour and location of muscles in test-action.

TERMS USED IN THE DESCRIPTION OF MUSCLE TESTS

Descriptions of the muscle tests presented in Chapters 4, 5, and 6 are outlined according to the following: "Patient", "fixation", "test", and "pressure." Each of these topics is discussed in detail in order to point out its particular significance in relation to accurate muscle testing.

Patient. In the description of each muscle test, the word patient is followed by specifying the position in which the patient is placed in order to best accomplish the desired test. The position is important in relation to the test in two respects: (1) Insofar as practical, the position of the body should permit function against gravity for all muscles in which gravity is a factor in grading, and (2) the body should be placed in such a position that parts not being tested will remain as stable as possible. (This point is discussed further under fixation.)

In all muscle testing, the comfort of the patient and the intelligent handling of affected muscles are of more concern than rules or principles related to testing. Insisting on an antigravity position may result in absurd positioning of a patient. Side lying which offers the best test position for several muscles may be uncomfortable and result in strain of other muscles, particularly during the acute or subacute stage of paralysis. A good examiner should be able to obtain a satisfactory estimate of strength regardless of the position of the patient.

Fixation. There are numerous factors to be considered under the heading of fixation. In general, fixation refers to the firmness or stability of the body necessary to insure an accurate test of any muscle or muscle group. In the extremities the part proximal to the tested part must be stable. Stabilization (holding steady or holding down), support (holding up), and counterpressure (equal and opposite pressure) are all included under the word fixation which implies holding firm.

Fixation depends to a great extent on the firmness of the *examining table*. In a sense, the table offers much of the necessary support. Testing and grading of strength will not be accurate if the patient lies on a soft bed that "gives" as the examiner applies pressure.

Body weight may furnish the necessary fixation. Because the weight of the body is an important factor in offering stability, the horizontal position, whether supine, prone, or side lying, tends to offer the best fixation for most tests.

The *examiner* may stabilize the proximal part in tests of finger, wrist, toe, and foot muscles, but in other tests the body weight should help to stabilize the proximal part. In some instances the examiner may offer fixation in addition to the weight of the proximal part. He may need to hold a part firmly down on the table so that the pressure he applies on the distal part, plus the weight of that part, do not displace the weight of the proximal part. In rotation tests it is necessary for the examiner to apply counter pressure to ensure exact test performance. (See pp. 108, 111, 171, 173.)

In some tests, *muscles* furnish fixation for the accomplishment of the test. The muscles which stabilize the scapula during arm movements and the pelvis during leg movements are referred to as fixation muscles. They do not enter directly into the test movement but do stabilize the movable scapula to the fixed trunk, or the pelvis to the thorax, and make it possible for the tested muscle to have a firm origin from which to pull. In the same way, the anterior abdominal muscles must fix the thorax to the pelvis as the anterior neck flexors act to lift the head forward in flexion from a supine position. (See p. 231 regarding action of opposite hip flexors in stabilizing the pelvis during hip extension.)

Muscles which have an antagonistic action may give fixation by preventing excessive joint movement. This principle is illustrated by the fixation of the Lumbricales and Interossei at the metacarpophalangeal joint during finger extension. Normally, these muscles restrict hyperextension at this joint. In the presence of weak Lumbricales and Interossei, the pull of a strong Extensor digitorum results in hyperextension of these joints and passive flexion of the interphalangeal joints. If the examiner prevents hyperextension of the metacarpophalangeal joints, by fixation equivalent to that of the Lumbricales and Interossei, this hyperextension does not occur and the fingers may be extended normally. (See below.)

When the fixation muscles are too weak or too strong, the examiner, by assisting or restricting movement of the part, can simulate the normal stabilization. The examiner must be able to differentiate between the normal action of these muscles in fixation and the abnormal actions which occur when muscle imbalance is present.

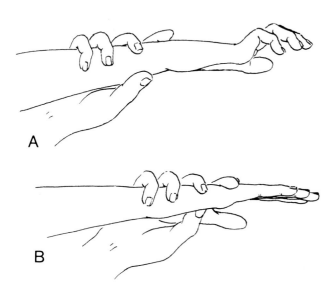

Hyperextension of the metacarpophalangeal joints due to weakness of the Lumbricales and Interossei prevents normal function of the Extensor digitorum in extending the interphalangeal joints, as seen in fig. A.

When the examiner offers fixation that normally is afforded by the Lumbricales and Interossei, a strong Extensor digitorum will extend the fingers, as in fig. B.

Test. The word test is followed by a description that may apply to the completed *test position* or to the *test movement*, whichever is used in the examination.

Test position is the position in which the part is placed by the examiner, and held (if possible) by the patient. Test movement is a movement of the part in a specified direction and through a specified arc of motion. The photographs of the muscle tests in Chapters 4 and 5 show the test position for each muscle.

The use of test position is time-saving and lends accuracy to muscle testing. When a test requires a combination of two or more joint positions or movements, it is difficult for a patient to assume the exact position through verbal description or through imitating a movement demonstrated to him. The examiner, however, can place the part precisely in the desired position. In addition to insuring accuracy of the test, the examiner at the same time is made aware of the existing range of motion. (See below.) If there is an attempt to substitute, when muscle weakness exists, the substitution movements become apparent immediately because there is a shift in the position of either the distal or proximal part, or both, to allow other muscles to hold a position resembling the test position. The use of test position also expedites the grading. As the effort is made to hold test position, the ability or inability to hold the position against gravity (50% or fair) is at once established. If the position is held, the examiner then applies pressure to grade above 50%; if it fails to hold, he tests for strength below 50%. (See "Key to muscle grading" p. 12.)

With few exceptions, a one-joint muscle is required to hold a position of completion of joint range. A muscle which crosses two or more joints is not expected or required to complete the range of motion of both or all joints simultaneously. To require it to do so is inadvisable both from the standpoint of overshortening the tested muscle and of overelongating the antagonists.

Muscle weakness must be distinguished from restriction of range of motion. Frequently a muscle does not complete the normal range of joint motion. It may be that the muscle is too weak to complete the movement; or it may be that the range of motion is restricted due to contracture of muscles, capsule, or ligamentous structures, or due to such factors as spasticity of opposing muscles. The examiner should passively carry the part through the range of motion to determine whether any restriction exists. If no restriction is present, failure to perform the muscle test may be interpreted as weakness. If the joint structures or opposing muscles restrict motion, the test will not be completely accurate but a good estimate of the muscle strength can be obtained. The inability to complete the range, due to joint restriction or shortness of opposing muscles, may be denoted by recording the grade of muscle strength in parentheses. For example, (50%) or (fair) indicates that the muscle can move the part through the existing range of motion or can hold the part at completion of the limited arc of motion against gravity.

The degree of actual muscle weakness is difficult to judge in cases of relaxed, unstable joints. Functionally the muscle may be weak and must be so graded, but it is important from the standpoint of treatment to detect the strong contraction of a muscle which appears extremely weak because of joint instability. Instances are not uncommon in which the Deltoid muscle shows a "fullness" of contraction throughout the muscle belly, and yet cannot begin to lift the weight of the arm. Such a muscle should be protected from strain by application of an adequate support with the express purpose of allowing the joint structures to shorten to their normal position. Muscle examinations should take into account such superimposed factors as relaxed joints. The patient may be deprived of adequate follow-up treatment by the failure to distinguish between real and apparent weakness.

Pressure. The word pressure has been used throughout this text except in the very few instances in which the term "resistance" is applicable. In the sense that resistance means a "force tending to hinder motion" one does not apply resistance in muscle testing. The subject is allowed to complete the arc of motion before any opposing force is applied by the examiner. Table friction and gravity do offer resistance to test movements (see pp. 11–13) but the examiner does not offer manual resistance.

Pressure is used to denote the outside force applied by the examiner to determine the strength of the muscle holding in test position. In the description of the muscle tests, pressure is specified as "against", and "in the direction of." Against refers to the position of the examiner's hand in relation to the patient; and in the direction of describes the direction of the application of force, directly opposite the line of pull of the muscle or its tendon. The place where pressure is applied and the direction of pressure are as important to the accuracy of the testing as is the test position itself.

In some of the muscle-test illustrations, the examiner's hand has been held extended for the purpose of indicating, photographically, that the direction of pressure is perpendicular to the palmar surface of the hand. It is not intended that such a hand position be imitated in the actual muscle testing, but the pressure should be applied in the direction as indicated.

The place at which pressure is applied bears relationship to the insertion of the muscle, to the joints involved, and to the principle of employing leverage in testing. In general, pressure is applied near the distal end of the part on which the muscle is inserted. The exceptions to this general rule occur in relation to muscles in which a longer leverage is required. For discussion of leverage see Gluteus medius, p. 169, hip adductors, p. 177, Serratus anterior, p. 119, and middle Trapezius, p. 114. The principle of leverage must be utilized in muscle testing. In many instances test results might be an indication of the strength of the examiner, rather than of the patient, if the examiner did not have the advantage of leverage.

Just as the *direction* of the pressure is an important part of accurate test performance, the *amount* of pressure is the determining factor in grading above 50% or fair. The amount of pressure varies according to the size of the patient, the part being tested, and the leverage. The ability of a muscle to hold against a slight or minimum amount of pressure indicates a grade of 60% (fair +) to 70% (good −); a medium amount, a grade of 80% (good) to 90% (good +); and a maximum amount, a grade of 100% (normal). In some instances an estimate of 100% may be considered as the ability of the muscle to hold the part against sufficient pressure to displace the body weight proximal to the tested part. In the sitting position a normal Quadriceps will hold the knee extended against so much pressure that the buttocks will be raised up from the table. While strong pressure is applied to determine the 100% grade, it is not necessary, and it may be injurious, to force the muscle to yield.

The 100% grade is not intended to indicate the maximum strength but, rather, what might be termed a "full" strength of the muscle. To become competent in judging this full strength, an examiner should acquaint himself with the strength of normal individuals of various age groups.

Pressure must be applied *gradually* in order to determine the degree of strength in muscles above 50%. The patient must be allowed to "set" his muscles against the examiner's pressure. The examiner cannot gauge the degree of strength unless pressure is applied gradually because slight pressure applied suddenly can "break" the pull of a muscle of any strength ranging from 60% to 100%.

There is, necessarily, a subjective evaluation based on the amount of pressure applied. Differences in strength are so apparent, however, that an observer who understands grading can estimate the strength with a high degree of accuracy while watching the examiner apply pressure.

Weakness, shortness, contracture, tautness, and tightness. Included with the descriptions of the muscles in this text is a discussion of the loss of movement or the position of deformity which results from muscle weakness or muscle shortness. Weakness is used as an overall term covering a range of strength from zero to 50% in nonweight-bearing muscles, but may be inclusive of 60% grades in weight-bearing muscles.

A weakness will result in loss of movement in the sense that the muscle cannot contract sufficiently to move the part through partial or complete range of motion. A contracture or shortness will result in loss of motion in the sense that the muscle cannot be elongated through its full range of motion. For the purpose of description in this text the words "contracture" and "shortness" have been used to distinguish between an almost complete loss of range of motion as denoted by contracture, and a partial loss as denoted by shortness.

When a muscle is weakened and its antagonist is not, a state of muscle imbalance exists. The stronger of the two opponents tends to shorten, and the weaker one tends to elongate. Either weakness or shortness will result in faulty alignment, weakness *permitting* a position of deformity, and shortness *creating* a position of deformity. In some instances, a fixed deformity does not occur as a result of weakness unless contractures develop in the stronger opponents. In the wrist, for example, a fixed deformity will not develop as a result of wrist extensor weakness unless the opposing flexors contract to hold the position of wrist flexion.

In many parts of the body, however, the deformities will develop as a result of the weakness even though the opposing muscles do not become contracted, because gravity and body weight exist as the opposing force. A kyphosis of the upper back will result from weakness of the upper back muscles whether or not the anterior trunk muscles become contracted. A position of pronation of the foot will become fixed if the inverters are weak because the body weight in standing will distort the bony alignment. The alignment will be distorted further if opposing peroneal muscles become contracted.

Muscles may exhibit tautness or tightness as well as shortness. While these words have similar meanings, they should not be used synonymously with respect to muscles.

Taut means stretched out fully, not slack. When a one-joint muscle is lengthened to the limit of range permitted by the joint, the muscle becomes taut. A muscle which crosses two joints becomes taut before it reaches the limits of range of both joints over which it passes. In so doing it prevents the simultaneous completion of movement at both joints. For example, normal Hamstring length permits full hip joint flexion if the knee is flexed, or full knee joint extension if the hip is only partially flexed, but does not permit full flexion of the hip with the knee extended. This characteristic of two-joint muscles is a useful and protective mechanism.

Within the definition of the word taut as described here, it has the same meaning as "*passive insufficiency*." Quoting from O'Connell and Gardner:[1]

Passive insufficiency of a muscle is indicated whenever a full range of motion at any joint or joints that the muscle crosses is limited by that muscle's length rather than by the arrangements of ligaments or structure of the joint itself. Thus passive insufficiency of the hamstrings is indicated when one is unable to touch the fingertips to the floor while maintaining full extension of the knees, a situation popularly known as tight hamstrings.

In shortened muscles, as in the above example of the Hamstrings in forward bending, tautness appears before the muscle has reached its normal length. In stretched muscles, the parts to which the muscle is attached move to the outer limit of the range of motion before the muscle becomes taut.

The word *tight* often is used interchangeably with the words short and taut which are not equivalent to each other in meaning. To the touch (palpation), Hamstrings that are *short* and drawn taut will feel tight. Hamstrings that are *stretched* and drawn taut will also feel tight on palpation. From the standpoint of prescribing treatment, it is very important to recognize the difference between stretched muscles and shortened muscles. In addition, some muscles are short and remain in what appears to be a state of semicontraction. On palpation, they feel firm or even rigid without being drawn taut. For example, posterior neck and upper Trapezius muscles often are tight in people with bad posture of the head and shoulders.

SUBSTITUTION

When a muscle or muscle group attempts to compensate for the lack of function of a weak or paralyzed muscle the result is a substitution movement. Muscles which normally act together in movements may act in substitution. These include fixation muscles, agonists, and antagonists.

Substitution by *fixation muscles* occurs specifically in relation to movements of the shoulder joint, hip joint, and flexion of the neck. Muscles which move the scapula may produce a secondary movement of the arm; muscles which move the pelvis may produce a secondary movement of the thigh. These substitution movements appear somewhat similar to, but are not actually movements of the shoulder or hip joint. (See below, figures A & B.) In cases of neck flexor weakness, as the abdominal muscles begin to flex the upper trunk and raise the shoulders up from the table, the head may be raised from the table, but the movement will not be one of neck flexion.

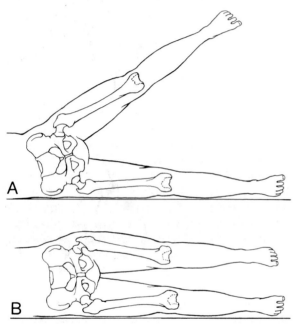

True abduction of the hip joint is accomplished by the hip abductors with the normal fixation by the lateral abdominal muscles as shown in fig. A. Apparent abduction may be accomplished, when hip abductors are weak, by the substitution action of the lateral abdominals. The pelvis is hiked up laterally and the leg is lifted from the table but, as shown in fig. B, there is no true hip joint abduction.

Antagonists may produce movements similar to test movements. If finger flexors are weak, action of the wrist extensors may produce passive finger flexion by the tension placed on flexor tendons.

Substitution by other *agonists* results in either: (1) A movement of the part in the direction of the stronger agonist; or (2) a shift of the body in such a way so as to favor the pull of that agonist. For example, during the Gluteus medius test in side lying, the thigh will tend to flex if the Tensor fasciae latae is attempting to substitute for the Gluteus

medius; or the trunk may rotate back so that the Tensor fasciae latae can hold a position that appears to be the desired test position.

When there is restriction of joint motion due to contracted muscles, a movement to relieve tension of a tight muscle may appear similar to a movement of substitution. (See discussion of substitution under "Note" in Hamstring test, p. 156.)

For accurate muscle examinations, no substitutions should be permitted. The position or movement described as the test should be done without shifting the body or turning the part to allow other muscles to substitute for the weak or paralyzed muscle.

An experienced examiner, who is aware of the ease with which normal muscles perform the tests, will readily detect substitutions. When test position is employed instead of test movement, even an inexperienced examiner can detect the sudden shift of the body which results from the effort to compensate for the muscle weakness.

GRADING

Grades are an expression of the examiner's evaluation of the strength or weakness of the muscles. To a considerable extent, the evaluation is subjective but the use of gravity resistance provides an aid for objectively measuring the strength. Robert W. Lovett, M.D. introduced a method of testing using gravity resistance, and a description of muscle grading based on the Lovett system[4] published in 1932, listed the following:

"*Gone*—no contraction felt.
Trace—muscle can be felt to tighten, but cannot produce movement.
Poor—produces movement with gravity eliminated, but cannot function against gravity.
Fair—can raise part against gravity.
Good—can raise part against outside resistance as well as against gravity.
Normal—can overcome a greater amount of resistance than a good muscle.

While symbols may vary, the movement and weight factors set forth by Lovett form the basis of most present-day muscle grading, including that presented in this text. The authors of this text had used the letter symbols of the Lovett system previous to the use of percentages and made a direct transfer to the percentage symbols.

In an effort to help standardize the systems of grading, a chart (p. 12) has been arranged which shows the percentage grades and the equivalent symbols that correspond with specified test performances.

The use of words or letters gives the feeling of being less precise and an easier way of grading but the use of figures has some definite advantages. In research work in which studies are made of the amount of change in muscle strength, numerical grades can be used directly in computations. Percentages are a widely understood numerical evaluation and need little interpretation. Such words as good and normal may be used in the interpretation of grades if not used in the scale of grading.

Detailed grading of muscle strength is more important in relation to prognosis than to diagnosis. Diagnosis of the extent of involvement may be made by such simple grading as zero, weak, normal. (See also p. 38.) On the other hand, more precise grading helps establish the rate and degree of return of muscle strength, and is useful in determining a prognosis. A muscle might appear "weak" for months, while the record shows that it has progressed from 10% to 60% during that period.

The gravity factor in grading. Gravity is a form of resistance basic to manual muscle testing, and is used in tests of trunk, neck, and extremity muscles. Interestingly, however, it is a factor in only about 60% of the extremity muscle tests because it is not required in tests of finger and toe muscles, and of forearm rotators which together constitute approximately forty percent of the extremity tests described.

It need not be considered in tests of finger and toe muscles because the weight of the part is so small in comparison to the strength of the muscle that the effect of gravity is unimportant. For the supinators and pronators of the forearm there is not appreciable weight lifted against gravity in comparison with the strength of the muscle.

In tests of the Rhomboids, middle and lower Trapezius, Serratus anterior, and Latissimus dorsi, there is a partial use of the gravity factor.

Tests of facial muscles involve neither gravity nor arc of motion factors. Grading these muscles is illustrated on p. 253.

In some systems of grading, the phrase "complete range of motion against gravity" has been used to designate the performance of a fair or 50% muscle. Actually, this is an ambiguous phrase. Frequently the complete range of motion in a test is not against gravity. In the Hamstring test in the prone position, the movement from horizontal up to vertical may be considered against gravity, but completion of the range of motion for the Hamstrings would consist of flexion beyond the range of motion against gravity. The completion of movement would no longer be Hamstring action but would involve gravity flex-

ion with or without a gradual relaxation of the Quadriceps. In the test of Quadriceps in the sitting position, complete range of motion would entail a movement starting from a position of full knee flexion. Actually the part of the range of motion which is used in a test movement is the antigravity part of the full arc of motion. Thus, a more appropriate phrase to describe this meaning is "completion of the anti-gravity arc of motion."

Introduced into testing and grading is the principle that the muscle strength required to hold the *test position* may be considered the equivalent of the muscle strength required to complete the *test movement*. In some muscle tests the bone on which the muscle is inserted moves from a position of suspension in the vertical plane (gravity-eliminated) toward the horizontal plane (antigravity). The Quadriceps, Deltoid, and hip rotators tested in the sitting position, and the Triceps and shoulder rotators tested in the prone position compose this group. The leverage exerted by the weight of the part increases as the part moves toward completion of arc, and the muscle strength required to hold the test position against gravity usually is sufficient to perform the test movement against gravity.

In a few tests the bone on which the muscle is inserted moves from a horizontal position toward a vertical position and less strength is required to hold the test position than to perform the test movement. This occurs during tests of Hamstrings tested by knee flexion in the prone position, and of the elbow flexors, Triceps, and Pectoralis major tested in the supine position.

In the remaining muscle tests the bone on which the muscle is inserted maintains a relatively horizontal position, and one finds few exceptions to the general rule that the test movement can be performed if the test position can be held.

Basically, antigravity movements are movements upward in the vertical plane, and gravity-assisted movements are movements downward in the vertical plane. Movements in the horizontal plane, quite generally, have been referred to as gravity-eliminated movements. Since the effect of gravity is not eliminated but merely lessened during movements in the horizontal plane, it is appropriate that this term be changed.

Gravity-lessened test position. The term gravity-lessened is substituted for gravity-eliminated in this text. The ability to perform full arc of motion in a gravity-lessened position is not close to the ability to perform the test against gravity for some muscles, notably those of the hip joint. Hip abductors, for example, may complete the movement of

Key to Muscle Grading

Test Performance	Kendalls	Lovett[4]			Nat. Found. for Inf. Par., & a Study[5]	Aids to Invest. of Per. N. Inj.[6]	Neurolog.
	Percent	Word & letter		Abbrev. of percent	Percent	Numerals	Rating
The ability to hold the test position against gravity and maximum pressure, or the ability to move the part into test position and hold against gravity and maximum pressure.	100	Normal	N	10	100	5	++++
	95	Normal−	N−	—		5−	
Same as above except holding against moderate pressure.	90	Good+	G+	9		4+	
	80	Good	G	8	75	4	+++
Same as above except holding against minimum pressure.	70	Good−	G−	7		4−	
	60	Fair+	F+	6		3+	
The ability to hold the test position against gravity, or the ability to move the part into test position and hold against gravity.	50	Fair	F	5	50	3	++
The gradual release from test position against gravity; or the ability to move the part toward test position against gravity almost to completion, or to completion with slight assistance; or the ability to complete the arc of motion with gravity lessened.	40	Fair−	F−	4		3−	
The ability to move the part through partial arc of motion with gravity lessened: Moderate arc, 30% or poor+; small arc, 20% or poor.	30	Poor+	P+	3		2+	
To avoid moving a patient into gravity-lessened position, these grades may be estimated on the basis of the amount of assistance given during anti-gravity test movements: A 30% or poor+ muscle requires moderate assistance, a 20% or poor muscle requires more assistance.	20	Poor	P	2	25	2	+
In muscles that can be seen or palpated, a feeble contraction may be felt in the muscle, or the tendon may become prominent during the muscle contraction, but there is no visible movement of the part.	10	Poor−	P−			2−	
	5	Trace	T	1		1	
No contraction felt in the muscle.	0	Zero	0	0	0	0	0

Restriction of range of motion may be denoted by putting the grade in parentheses.

abduction in a supine position and hold against pressure that may be considered an 80% or good grade in this position while the muscles will grade only 50% or fair in the antigravity, side-lying position. Some notation must be made on the chart to indicate that the grades of 50%, 60%, or 70% in the gravity-lessened position are not the equivalent of the same percentage grade in the antigravity position. Notations such as (BL) for back lying, (FL) for face lying, (SL) for side lying, or (GL) for gravity-lessened may be used beside the grade, and the grade may be recorded in red if all grades less than 50% antigravity are so recorded.

When testing hip extensors or hip flexors in side lying, a horizontal movement through all or part of the arc of motion furnishes a means of obtaining an objective grade of fair-minus or poor, but there are variable factors which make such grading inaccurate. The surface of the table may be smooth or rough, changing the amount of friction and resistance considerably. The strength of hip adductors (if the under leg is being tested) may make a material difference in results of the flexor and extensor tests. If the adductors are paralyzed, the full weight of the extremity will rest on the table and make flexion and extension difficult. If the adductor muscles are strong they will tend to raise the extremity so the full weight does not rest on the table, thereby reducing the friction, and the flexion and extension movements will be made easier.

Turning a patient from the antigravity test position into a gravity-lessened position for the purpose of obtaining an objective grade of poor or fair-minus has been an accepted procedure in most muscle testing. When used, these grades require that partial or complete arc of motion be performed. The partial arc of movement may be at the beginning of the arc, or may be present about the center of the arc of motion. The angle at which movement is initiated is not a matter of concern in grading except that the examiner should be sure that the movement starts from a relaxed position. If the part is carried to the beginning of the arc of motion and slight tension put on the muscle, there may be a "springing back" to position which can be confused with active movement.

The fair-minus or 40% grade represents a gradual release from test position against gravity and is a relatively objective grade. The trace and poor-minus grades can be determined in any given position. Consequently, the most that can be gained by moving the patient from an antigravity position (presuming the test was first tried in that position) to a gravity-lessened position is to ascertain whether the grade is poor (20%) or poor-plus (30%).

In practice, frequent change of the patient's position or repetition of the test in various positions is fatiguing to the patient and time-consuming for the examiner. Patients should not be subjected to unnecessary procedures in examination if the result obtained is relatively unimportant. For these reasons, this text recommends that an assistive movement in an antigravity position be used, when applicable, for grading poor and poor-plus (20% and 30%). This modification seems practical and sensible. Estimating the amount of assistance given weak muscles is comparable to estimating the pressure applied in grading the stronger muscles. The grades are determined in the antigravity position by measuring subjectively the assistance required in the test movement. The patient relaxes the part so that the examiner can feel the weight of the part. As the patient then attempts the test movement against gravity, the examiner subjectively measures the decrease in weight of the part. One might describe the grading by saying that a 30% muscle requires moderate assistance by the examiner; a 20% muscle requires more assistance by the examiner. The examiner should guide the movement and give assistance, but should not precede the patient in the effort to move the part. If the examiner's assistance is applied too quickly, he will not be able to gauge the patient's muscle strength.

USE OF THE WORD "NORMAL" IN RELATION TO MUSCLE TESTING

The word normal has a variety of meanings. It may mean average, typical, natural, or standard. As used in various methods of grading, it has been defined as that degree of strength which will perform a movement against gravity and hold against full or maximum resistance. By virtue of this definition the word normal has been used as a standard.

If one adheres to the usage in this sense, then a grade of poor will be recorded for a small child who cannot lift his head in flexion from a supine position. Knowing that it is natural for small children to exhibit weakness of the anterior neck muscles, an examiner might say this child's neck is normal— using normal in the sense that it is natural. Upon doing a leg-lowering test for abdominal strength in a large group of adolescent children, and finding that a grade of fair-plus or good-minus (60% to 70%) is average strength for the group, one might say this grade of strength is normal for this age. Thus we have three different uses of the word normal applied rather freely in muscle testing—as standard, as natural, and as average.

Since normal is defined as a standard when used in the scale of grading, grades of strength should relate to that standard and appropriate terms other than normal should be used in interpretations.

One of the advantages of the use of percentage grades is that it leaves the term normal free for use in interpretation of grades. In the following discussion it will be used in this manner.

Since most 100% grades are based on adult standards, it is necessary to acknowledge when a grade less than 100% is normal for children of a given age. This is particularly true with respect to the strength of the anterior neck and anterior abdominal muscles. The size of the head and trunk in relation to the lower extremities, as well as the long span and normal protrusion of the abdominal wall affect the relative strength of these muscles. Anterior neck muscles may grade about 30% in a three-year-old child, about 50% in a five-year-old, and gradually increase up to the 100% standard of performance by as early as ten or twelve years of age. Many adults will exhibit no more than 60% strength, but this need not be interpreted as neurogenic because usually it is found to be associated with faulty posture of the head and upper back.

In the trunk-raising test, some children as early as five years of age may do the 60% test, and some as early as seven or eight may do the 100% test. There is a good deal of variation in this accomplishment, but grades of 60% to 80% may be considered normal for ages between six and ten. Most children from the ages of ten or eleven will be able to do the 100% trunk raising test which is considered normal for adults of both sexes.

The leg-lowering test for abdominal strength is not applicable to very young children. The weight of the legs is small in relation to the trunk, and the back does not arch as legs are raised or lowered. Furthermore, at the age of six or seven when the test would have some significance, it is not easy for a child to differentiate muscle action and try to hold the back flat while lowering the legs. From about age eight or ten, it is possible to use the test for many children. As adolescence approaches and the legs grow long in relation to the trunk, the picture reverses from that of early childhood and the leverage exerted by the legs as they are lowered is greater in relation to the trunk. At this age, grades of 60% or 70% on the leg-lowering tests are normal for many children, especially those who have grown tall rather quickly, and should be interpreted as "normal for age." After about fourteen to sixteen, a desirable normal is considered as 100% for males, while 80% may be considered normal for females.

The prime example of a 100% standard that is an infant accomplishment instead of an adult one is that of toe flexor strength. In general, children have more strength in their toe flexors than adults. It is not uncommon to find that women, who have worn high heels and rather narrow-toed shoes, have weakness of toe flexors in which the grade is no more than 40%. With the standard being the ability to flex the toes and hold against strong resistance or pressure, the adult must be graded against that standard, but this weakness of toe flexors should not be interpreted as normal for age. One becomes so accustomed to toe flexor weakness among adults that it might be assumed that a degree of weakness is normal in the sense that normal is average. Marked weakness of toe flexors almost invariably is associated with some degree of disability of the foot and the word normal should not apply to such weakness, unless one is ready to accept the disability as normal.

The toe flexor weakness represents a loss of strength from childhood and should be regarded as an unnatural, *acquired* weakness. Other muscles exhibit this type of weakness which is often associated with strain from faulty postural positions, from occupational causes, or associated with handedness. Acquired weakness usually does not drop below the grade of 50%, but 50% and 60% grades of strength might be interpreted as neurogenic if one were not aware that such degrees of weakness could result from stretch and strain of the muscles.

The following muscles tend to show evidence of acquired postural weakness:

Toe flexors (brevis and lumbricales)
Middle and lower Trapezius
Upper back extensors
Anterior abdominal muscles (as tested by leg-lowering test)

Anterior neck muscles

Left lateral trunk muscles
Right hip abductors
Right hip lateral rotators
Left hip adductors
Right Peroneus longus and brevis   in right-handed
Left Tibialis posterior   individuals
Left Flexor hallucis longus
Left Flexor digitorum longus

Right lateral trunk muscles   in left-handed
Left hip abductors   individuals, but
Left hip lateral rotators   the pattern is
Right hip adductors   not as common
Left Peroneus longus and brevis   as that occurring
Right Tibialis posterior   in right-handed
Right Flexor hallucis longus   individuals
Right Flexor digitorum longus

The anterior neck and anterior abdominal muscles seem to exhibit both natural and acquired weakness. This is a significant factor in understanding the prevalence and persistence of abdominal and anterior neck weakness among many adults. If postural strain is superimposed on the natural weakness, before normal increase in strength has occurred, the muscle will tend to remain weak.

An examiner must build a basis for comparison of test results through experience in muscle testing. Such experience is necessary on both paralytic and normal individuals. The experience of many individuals engaged in muscle testing has been limited to examination of patients with disease or injury. The result is that their idea of normal strength tends to be a measure of what appears to be good functional recovery following weakness.

The authors recommend that an examiner make an effort to test a series of individuals of various ages, both male and female. He should test persons with faulty posture as well as those with good posture. If it is not possible to examine a large number of normal individuals, every effort should be made to examine the trunk and unaffected extremities in cases involving only one or two extremities.

Testing and grading procedures are modified in the examination of infants and children to the age of five or six. The ability to grade a child's muscle strength up to 50% usually is not difficult, but the grading above that depends on the cooperation of the child in holding against resistance or pressure. Young children seldom cooperate to any extent in strong test movements. Very often tests must be recorded as "apparently normal" which indicates that strength may be normal, although one cannot be sure.

SUGGESTED ORDER OF MUSCLE TESTS

1. Supine
Toe extensors
Toe flexors
Tibialis anterior
Tibialis posterior
Peroneals
Tensor fasciae latae
Sartorius
Iliopsoas
Abdominals
Neck flexors
Finger flexors
Finger extensors
Thumb muscles
Wrist extensors
Wrist flexors
Supinators
Pronators
Biceps
Brachioradialis
Triceps (supine test)
Pectoralis major, upper part
Pectoralis major, lower part
Pectoralis minor
Medial rotators of shoulder (supine test)
Teres minor and Infraspinatus
Lateral rotators of shoulder (supine test)
Serratus anterior
Anterior Deltoid (supine test)

2. Side Lying
Gluteus medius
Gluteus minimus
Hip adductors
Lateral abdominals

3. Prone
Gastrocnemius and Plantaris
Soleus
Hamstrings (medial and lateral)
Gluteus maximus
Neck extensors
Back extensors
Quadratus lumborum
Latissimus dorsi
Lower Trapezius
Middle Trapezius
Rhomboids
Posterior Deltoid (prone test)
Triceps (prone test)
Teres major
Medial rotators of shoulder (prone test)
Lateral rotators of shoulder (prone test)

4. Sitting
Quadriceps
Medial rotators of hip
Lateral rotators of hip
Hip flexors (group test)
Deltoid (anterior, middle, and posterior)
Coracobrachialis
Upper Trapezius
Serratus anterior (preferred test)

5. Standing
Serratus anterior
Ankle plantar flexors

chapter 2

Joint Motions

Anatomical Position, Lines, Planes, and Axes
    Anatomical position
    Line of gravity
    Plumb line
    Planes
    Axes
Definitions of Joint Movements
    Flexion and extension
    Abduction and adduction
    Lateral flexion
    Rotation
    Hyperextension
Movements of Scapula and Joints of Upper Extremity
    Shoulder girdle
    Scapula
    Shoulder joint
    Elbow joint

Radioulnar joint
Wrist joint
Carpometacarpal joints of fingers
Metacarpophalangeal joints of fingers
Interphalangeal joints of fingers
Carpometacarpal joint of thumb
Metacarpophalangeal and interphalangeal joints of thumb
Movements of the Pelvis and of Joints of Lower Extremity
    Pelvis
    Hip joint
    Knee joint
    Ankle joint
    Subtalar joint and transverse tarsal joints
    Metatarsophalangeal joints
    Interphalangeal joints of toes
Movements of Vertebral Column

Joint Motions

Anatomical position. The anatomical position of the body is an erect posture, face forward, arms at sides, palms of hands forward with fingers and thumbs in extension. This is the position of reference for definitions and descriptions of body planes and axes. It is designated as the *zero position* for defining and measuring joint motions for most of the joints of the body. Further descriptions of the positions of the various body segments appear below. (In Fig. A, arms and hands are in normal, not anatomical position.)

Center of gravity. Every mass or body is composed of a multitude of small particles which are pulled toward the earth in accordance with the law of gravitation. This attraction of gravity upon the particles of the body produces a system of practically parallel forces and the resultant of these forces acting vertically downward is the weight of the body. It is possible to locate a point at which a single force, equal in magnitude to the weight of the body and acting vertically upward, may be applied so that the body will remain in equilibrium in any position. This point is called the center of gravity of the body and may be defined as the point at which the entire weight of the body may be considered to be concentrated. In an ideally aligned posture in a so-called average adult human being the center of gravity is considered to be slightly anterior to the first or second sacral segment.

Line of gravity. The line of gravity is a vertical line through the center of gravity.

Plumb line. A plumb line is a cord with a plumb-bob attached to one end. It may be used to represent the projection of the gravity line to the external surface of the body, and used as an aid in analyzing alignment in static posture. In examining such postures the plumb line must be suspended in line with a fixed point. The only fixed point in a standing posture is at the base where the feet are in contact with the floor.

In a lateral view of an ideally aligned posture, starting at the base, the plumb line will coincide with the following points or skeletal parts:

Slightly anterior to the lateral malleolus
Slightly anterior to the axis of the knee joint
Slightly posterior to the axis of the hip joint
Bodies of the lumbar vertebrae
Shoulder joint
Bodies of most of the cervical vertebrae
External auditory meatus
Slightly posterior to apex of the coronal suture

In posterior view, starting with a fixed point midway between the heels, the plumb line will be equidistant from the medial aspects of the heels, legs, and thighs; be equidistant from the scapulae; and coincide with the midline of the trunk and head.

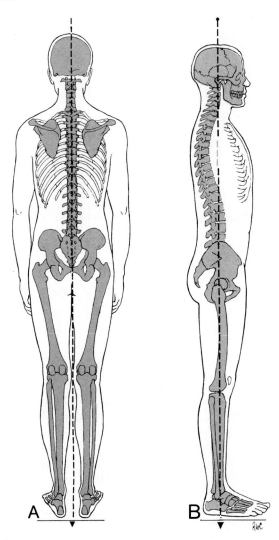

A   B

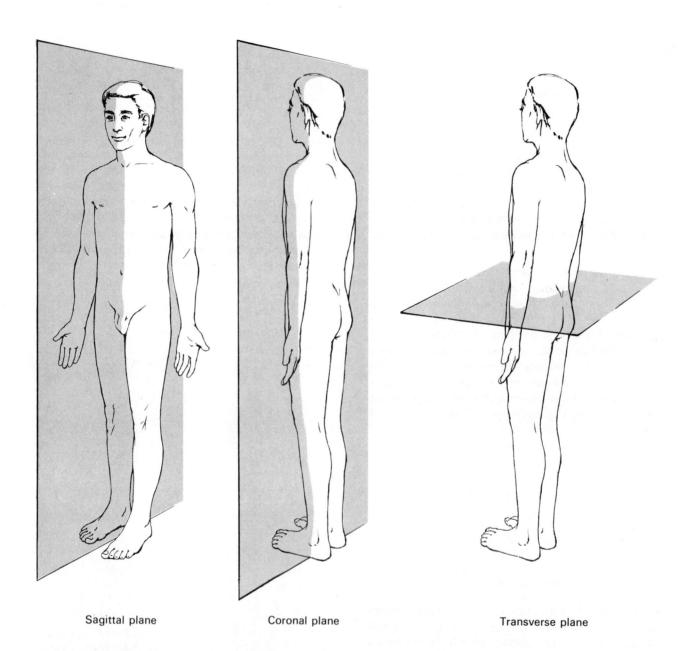

Sagittal plane   Coronal plane   Transverse plane

Planes. The three basic planes of reference are derived from the dimensions in space, and are at right angles to each other.

A *sagittal plane* is vertical and extends from front to back, deriving its name from the direction of the sagittal suture of the skull. It may also be called an anterior-posterior plane. The median sagittal plane, *midsagittal*, divides the body into right and left halves.

A *coronal plane* is vertical and extends from side to side, deriving its name from the direction of the coronal suture of the skull. It is also called the frontal or lateral plane, and divides the body into an anterior and a posterior portion.

A *transverse plane* is horizontal and divides the body into upper (cranial) and lower (caudal) portions.

The point at which the three midplanes of the body intersect is the center of gravity.

Axes. Axes are lines, real or imaginary, about which movement takes place. Related to the planes of reference, there are three basic types of axes at right angles to each other.

A *sagittal axis* lies in the sagittal plane and extends horizontally from front to back. The movements of abduction and adduction take place about this axis in a coronal plane.

A *coronal axis* lies in the coronal plane and extends horizontally from side to side. The movements of flexion and extension take place about this axis in a sagittal plane.

A *longitudinal axis* is vertical extending in a cranial-caudal direction. The movements of medial and lateral rotation take place about this axis in a transverse plane.

The exceptions to these general definitions occur with respect to movements of the scapula, clavicle, and thumb (see pp. 22, 24, and 25).

DEFINITIONS OF JOINT MOVEMENTS

Joints are of three types: Fibrous or immovable, cartilaginous or slightly movable, and synovial or freely movable. The following defintions of joint motions refer to synovial joints.

The plane of movement and the axis of motion are at right angles to each other. Flexion and extension occur in the sagittal plane about an axis that lies in the coronal plane. Abduction, adduction, and lateral flexion occur in the coronal plane about an axis that lies in the sagittal plane. Medial and lateral rotation, and horizontal abduction and adduction occur in the transverse plane about a longitudinal axis.

Flexion and extension. Flexion and extension are movements in the sagittal plane. Flexion is movement in the anterior direction for the head, neck, trunk, upper extremity and hip. However, flexion of the knee, ankle, foot, and toes refers to movement in the posterior direction. Extension is movement in a direction opposite flexion.

The developmental pattern of the upper extremity differs from that of the lower extremity. At an early stage the limbs of the embryo are directed ventrally, the flexor surfaces medially, and the great toes and thumbs cranially. With further development, the limbs rotate 90° at their girdle articulation so that the thumbs turn laterally and the flexor surfaces of the upper extremities ventrally, while the great toes turn medially and the flexor surfaces of the lower extremities dorsally. As a result of this 90° rotation of the limbs in opposite directions, movement which approximates the hand and the anterior surface of the forearm is termed flexion since it is performed by flexor muscles, and movement which approximates the foot and anterior surface of the leg is termed extension since it is performed by extensor muscles.

Abduction and adduction. Abduction and adduction are movements in the coronal plane. Abduction is movement away from, and adduction is movement toward, the midsagittal plane of the body for all parts of the extremities except the thumb, finger, and toes. For the fingers, abduction and adduction are movements away from and toward the axial line which extends through the third digit. For the toes the axial line extends through the second digit. For the thumb, see specific definitions p. 24.

Lateral flexion. Lateral flexion is a term used to denote lateral movements of the head, neck, and trunk in a coronal plane. It is usually combined with rotation.

Rotation. Rotation refers to movement which takes place about a longitudinal axis for all areas of the body except the scapula and the clavicle. (See p. 22.) In the extremities rotation occurs about the anatomical axis, except in the case of the femur which rotates about a mechanical axis. (See p. 178.) In the extremities the anterior surface of the extremity is used as a reference area. Rotation of the anterior surface toward the midsagittal plane of the body is medial rotation, away from the midsagittal plane is lateral rotation.

Since the head, neck, thorax, and pelvis rotate about longitudinal axes in the midsagittal area, rotation cannot be named in reference to the midsagittal plane. Rotation of the head is described as rotation of the face toward the right or left. Rotation of the thorax and pelvis are described, generally, as being clockwise or counterclockwise. With the transverse plane as a reference, and 12 o'clock at midpoint anteriorly, clockwise rotation occurs when the left side of the thorax or pelvis is more forward than the right; counterclockwise occurs when the right side is more forward. It may be said, also, that if the thorax or pelvis face toward the right, the rotation is clockwise, and vice versa.

Tilt. Tilt is a term used in describing certain movements of the scapula and pelvis. (See pp. 22 and 25.)

Gliding. Gliding movements occur when joint surfaces are flat or only slightly curved and one articulating surface slides on the other.

Circumduction. A movement successively combining flexion, abduction, extension, and adduction in which the part being moved describes a cone. The proximal end of the extremity forms the apex of the cone, serving as a pivot, while the distal end circumscribes a circle. Such movements are possible only in ball-and-socket, condyloid, and saddle types of joints.

Hyperextension. Hyperextension is the term used in describing excessive or unnatural movement in the direction of extension, as, for example, hyperextension of the knees. It is used also in reference to the increased lumbar curvature in a lordosis with anterior pelvic tilt, and to the increased cervical curvature in a forward head position. In such instances, the range of motion through which the lumbar or cervical vertebrae move is not excessive, but the position is of a greater degree of extension than is desirable from a postural standpoint. (See pp. 283 and 300.)

MOVEMENTS OF SCAPULA AND JOINTS OF UPPER EXTREMITY

Shoulder girdle. The shoulder girdle is composed of the clavicles and the scapulae. The clavicle articulates laterally with the acromion process of the scapula and medially with the sternum, the latter joint providing the only bony connection with the axial skeleton.

The sternoclavicular joint permits motion in anterior and posterior directions about a longitudinal axis, in cranial and caudal directions about a sagittal axis, and in rotation about a coronal axis. These movements are slightly enhanced and transmitted, by the acromioclavicular joint, to the scapula. Additional motions of the shoulder girdle that will be described are those of the scapula.

Scapula. The scapula articulates with the humerus at the glenohumeral joint and with the clavicle at the acromioclavicular joint.

In anatomical position, with the upper back in good alignment, the scapulae lie against the thorax approximately between the levels of the second and seventh ribs; the medial borders are essentially parallel and about three or four inches apart.

Muscles which attach the scapula to the thorax and to the vertebral column provide support and motion for it. They are obliquely oriented so that their directions of pull can produce rotatory as well as linear motions of the bone. As a result, the movements ascribed to the scapula do not occur individually as pure movements. Since the contour of the thorax is rounded, some degree of rotation or tilt accompanies abduction and adduction and, to a lesser extent, elevation and depression.

While there are no pure linear movements, seven basic movements of the scapula are described:

Adduction is basically a gliding movement in which the scapula moves toward the vertebral column.
Abduction is basically a gliding movement in which the scapula moves away from the vertebral column, and following the contour of the thorax assumes a posterolateral position in full abduction.
Lateral or upward rotation is movement about a sagittal axis in which the inferior angle moves laterally and the glenoid cavity moves cranially.
Medial or downward rotation is movement about a sagittal axis in which the inferior angle moves medially and the glenoid cavity moves caudally.
Anterior tilt is movement about a coronal axis in which the coracoid process moves in an anterior and caudal direction while the inferior angle moves in a posterior and cranial direction. The coracoid process may be said to be depressed anteriorly. This movement is associated with elevation.
Elevation is basically a gliding movement in which the scapula moves cranially as in "shrugging" the shoulder.
Depression is basically a gliding movement in which the scapula moves caudally, and is the reverse of elevation and of anterior tilt.

Shoulder joint. The shoulder joint, also called the glenohumeral joint, is a spheroid or ball-and-socket joint formed by the articulation of the head of the humerus and the glenoid cavity of the scapula. In addition to six basic joint movements, it is necessary to define circumduction and two movements in the horizontal plane.

Flexion and extension are movements about a coronal axis. *Flexion* is movement in the anterior direction and may begin from a position of 45° extension (arm extended backward). It describes an arc forward through the zero anatomical position and on to the 180° overhead position. However, the 180° overhead position is attained only by the combined movement of the shoulder joint and the shoulder girdle. The glenohumeral joint can be flexed only to approximately 120°. The remaining 60° is attained as a result of the abduction and lateral rotation of the scapula which allows the glenoid cavity to face more anteriorly and the humerus to flex to a fully vertical position. The scapular motion is at first variable but after 60° of flexion there is a relatively constant relationship between the movement of the humerus and the scapula. Inman et al.[7] found that between the 30° and 170° range of flexion the glenohumeral joint provided 10° and scapular rotation 5° for every 15° of motion.

Extension is movement in the posterior direction and technically refers to the arc of motion from 180° flexion to 45° extension. If the elbow joint is flexed, the range of shoulder joint extension will be increased because the tension on the Biceps will be released.

Abduction and adduction are movements about a sagittal axis. *Abduction* is movement in a lateral direction through a range of 180° to a vertical overhead position. This end position is the same as that attained in flexion, and coordinates shoulder girdle and glenohumeral joint movements. *Adduction* is movement toward the midsagittal plane in a medial direction and technically refers to the arc of motion from full elevation overhead through the zero anatomical position to a position obliquely upward and across the front of the body.

Horizontal abduction and adduction are movements in a transverse plane about a longitudinal axis. *Horizontal abduction* is movement in a lateral and posterior direction; *horizontal adduction* is movement in an anterior and medial direction. The end position of complete horizontal adduction is the same as that for adduction obliquely upward across the body. In one instance, the arm moves horizontally to that position; in the other instance, it moves obliquely upward to that position.

The range of horizontal abduction, being determined to a great extent by the length of the Pectoralis major muscle, is extremely variable. With the humerus in 90° flexion as the zero position for measurement, the normal range should be about 90° in horizontal abduction and about 40° in horizontal adduction, most readily judged by the ability to place the palm of the hand on top of the opposite shoulder.

Medial and lateral rotation are movements about a longitudinal axis through the humerus. *Medial rotation* is movement in which the anterior surface of the humerus turns toward the midsagittal plane. *Lateral rotation* is movement in which the anterior surface of the humerus turns away from the midsagittal plane.

The extent of medial or lateral rotation varies with the degree of elevation in abduction or flexion. For purposes of joint measurement the zero position is one in which the shoulder is at 90° abduction, the elbow is bent at right angles and the forearm is at right angles to the coronal plane. From this position, lateral rotation of the shoulder describes an arc of 90° to a position in which the forearm is parallel with the head. Medial rotation describes an arc of approximately 70° if shoulder girdle movement is not permitted. If the scapula is allowed to tilt an-teriorly, the forearm may describe an arc of 90° to a position in which it is parallel with the side of the body.

As the arm is abducted or flexed from the anatomical position, lateral rotation continues to be free but medial rotation is limited. As the arm is adducted or extended the range of medial rotation remains free and that of lateral rotation decreases. In treatment to restore motion in a restricted shoulder joint, one must be concerned with obtaining lateral rotation as a prerequisite to full flexion or full abduction.

Circumduction combines consecutively the movements of flexion, abduction, extension, and adduction as the upper limb circumscribes a cone with its apex at the glenohumoral joint. This succession of movements can be performed in either direction and is used to increase the overall range of motion of the shoulder joint as in Codman's or shoulder wheel exercise.

Elbow joint. The elbow is a ginglymus or hinge joint formed by the articulation of the humerus with the ulna and the radius.

Flexion and extension occur about a coronal axis and are the two movements permitted by this joint. *Flexion* is movement in the anterior direction, from the position of a straight elbow, 0°, to a fully bent position, approximately 145°. *Extension* is movement in a posterior direction from the fully bent position to the position of a straight elbow.

Radioulnar joints. The radioulnar joints are trochoid or pivot joints, formed by the articulations of the radius and ulna, proximally and distally. The axis of motion extends from the head of the radius proximally, to the head of the ulna distally, and allows rotation of the radius about the axis.

Supination and pronation are rotation movements of the forearm. In *pronation*, the distal end of the radius moves from a lateral position, as in the anatomical position, to a medial position; in *supination*, it moves from a medial to a lateral position. The palm of the hand faces anteriorly in supination and posteriorly in pronation.

In measuring joint motion, as well as in treating it to restore range of motion, it is necessary to inhibit shoulder rotation movements which otherwise are easily substituted for, and made to appear as, forearm movements. To ensure forearm movements only, place the arms directly at the sides of the body with elbows bent at right angles, forearms extended forward. Turn palms directly upward for full supination and directly downward for full pronation.

The neutral or zero position is midway between supination and pronation, that is, from anatomical position with elbow extended the thumb is directed forward; with the elbow bent at right angle, the thumb is directed upward. The normal range of motion is 90° in either direction from zero.

Wrist joint. The wrist is a condyloid joint formed by the radius and the distal surface of the articular disc articulating with the scaphoid, lunate, and triquetrum.

Flexion and extension are movements about a coronal axis. From the anatomical position, *flexion* is movement in an anterior direction approximating the palmar surface of the hand toward the anterior surface of the forearm. *Extension* is movement in a posterior direction approximating the dorsum of the hand toward the posterior surface of the forearm. Starting with the wrist straight (as in anatomical position) as zero position, the range of flexion is approximately 80° and that of extension approximately 70°. The fingers will tend to extend when measuring wrist flexion, and to flex when measuring wrist extension.

Abduction (radial deviation) and adduction (ulnar deviation) are movements about a sagittal axis. With the hand in anatomical position, moving it toward the ulnar side is also moving it medially toward the midline of the body and hence, is *adduction*. Moving the hand toward the radial side is *abduction*. With the anatomical position as zero, the range of adduction is approximately 35° and that of abduction approximately 20°.

Circumduction combines the successive movements of flexion, abduction, extension, and adduction of the radiocarpal joint and the midcarpal joint. The movements of these joints are closely related and permit the hand to describe a cone. The movement is not as free as that of the glenohumeral joint since abduction is more limited than adduction because the radial styloid process extends farther caudally than the ulnar styloid process.

Carpometacarpal joints of fingers. The carpometacarpal joints of the fingers are formed by the articulation of the distal row of carpal bones with the second, third, fourth, and fifth metacarpal bones and permit gliding movements. The joint between the hamate bone and the fifth metacarpal is somewhat saddle-shaped and allows, in addition, flexion, extension, and slight rotation.

Metacarpophalangeal joints of fingers. The metacarpophalangeal joints of the fingers are condyloid joints formed by the articulations of the distal ends of the metacarpals with the adjacent ends of the proximal phalanges.

Flexion and extension occur about a coronal axis with *flexion* in an anterior direction and *extension* in a posterior direction. With the extended position as zero, the metacarpophalangeal joints flex to approximately 90°. In most people some extension beyond zero is possible but for practical purposes the straight extension of this joint, when interphalangeal joints are also extended, is considered normal extension.

Abduction and adduction occur about a sagittal axis. The line of reference for abduction and adduction of the fingers is the axial line through the third digit. *Abduction* is movement in the plane of the palm away from the axial line, spreading fingers wide apart. The third digit may move in abduction both ulnarly and radially from the axial line. *Adduction* is movement in the plane of the palm toward the axial line, that is, closing the extended fingers together sideways.

Circumduction is the combination of flexion, abduction, extension, and adduction movements performed consecutively, in either direction, at the metacarpophalangeal joints of the fingers. Extension in these condyloid joints is somewhat limited so that the base of the cone described by the fingertip is relatively small.

Interphalangeal joints of fingers. The interphalangeal joints of the fingers are ginglymus or hinge joints formed by articulations of the adjacent surfaces of the phalanges.

Flexion and extension occur about a coronal axis and describe an arc from 0° extension to approximately 100° flexion for the proximal interphalangeal joints and 80° for the distal interphalangeal joints.

Carpometacarpal joint of thumb. The carpometacarpal joint of the thumb is a reciprocal reception or saddle joint, formed by the articulation of the trapezium with the first metacarpal. The zero position of *extension* is one in which the thumb has moved in a radial direction and is in the plane of the palm. *Flexion* is movement in an ulnar direction with a range of approximately 40° to 50° from zero extension. The thumb can be fully flexed only if accompanied by some degree of abduction and medial rotation.

Adduction and abduction are movements perpendicular to the plane of the palm, *adduction* being toward and *abduction* away from the palm. With the position of adduction as zero, the range of abduction is approximately 80°.

The range of rotation at the carpometacarpal joint is slight and does not occur independently. The *slight rotation*, however, that results from a combination of basic movements is of significance.

In the thumb and little finger, *opposition* is a combination of abduction and flexion with medial rotation of the carpometacarpal joints, and flexion of the metacarpophalangeal joint. To ensure opposition of the thumb and little finger the palmar surfaces (rather than the tips) of the distal phalanges must be brought in contact with each other. Touching the tips of the thumb and little finger to each other can be done without any true opposition.

The movements of opposition are accomplished by the combined actions of the respective opponens and metacarpophalangeal flexors: In the thumb, the Opponens pollicis, Abductor pollicis brevis, and Flexor pollicis brevis; in the little finger, the Opponens digiti minimi, the Flexor digiti minimi, the fourth Lumbricalis, and the fourth Palmar interosseus, assisted by the Abductor digiti minimi.

Circumduction is a movement which includes flexion, abduction, extension, and adduction, performed in sequence, by this saddle joint. The first metacarpal bone describes a cone and the tip of the thumb circumscribes a circle.

Metacarpophalangeal and interphalangeal joints of thumb. The metacarpophalangeal joint of the thumb is a condyloid joint formed by the articulation of the distal end of the first metacarpal with the adjacent end of the proximal phalanx. The interphalangeal joint of the thumb is a ginglymus or hinge joint formed by the articulation of the proximal and distal phalanges.

Flexion and extension are movements in an ulnar and radial direction, respectively. The zero position of extension is reached when the thumb moves in the plane of the palm to maximum radial deviation. From the position of zero extension, the metacarpophalangeal joint permits approximately 60° flexion, and the interphalangeal joint, approximately 80° flexion. The metacarpophalangeal joint permits, also, slight abduction, adduction, and rotation.

MOVEMENTS OF PELVIS AND JOINTS OF LOWER EXTREMITY

Pelvis. The *neutral position of the pelvis* is one in which the anterior-superior spines are in the same transverse plane, and in which they and the symphysis pubis are in the same vertical plane. An *anterior pelvic tilt* is a position of the pelvis in which the vertical plane through the anterior-superior spines is anterior to a vertical plane through the symphysis pubis. A *posterior pelvic tilt* is a position of the pelvis in which the vertical plane through the anterior-superior spines is posterior to a vertical plane through the symphysis pubis. In a standing position an anterior pelvic tilt is associated with hyperextension of the lumbar spine and flexion of the hip joints, while posterior pelvic tilt is associated with flexion of the lumbar spine and extension of the hip joints. (See pp. 194 and 280–285.)

In a *lateral pelvic tilt* the pelvis is not level from side to side, but one anterior-superior spine is higher than the other. In standing, a lateral tilt is associated with lateral flexion of the lumbar spine and adduction and abduction of the hip joints. For example, in a lateral tilt of the pelvis in which the right side is higher than the left, the lumbar spine is laterally flexed toward the *right* resulting in a curve convex to the *left*. The right hip joint is in adduction, and the left is in abduction. (See p. 291.)

Hip joint. The hip joint is a spheroid or ball-and-socket joint formed by the articulation of the acetabulum of the pelvis with the head of the femur.

Ordinarily, descriptions of joint movement refer to movement of the distal part upon a fixed proximal part. In the upright weight-bearing position, movement of the proximal part upon the more fixed distal part becomes equal, if not primary, in importance. For this reason, movements of the pelvis on the femur are mentioned as well as movements of the femur on the pelvis.

Flexion and extension are movements about a coronal axis. *Flexion* is movement in an anterior direction. The movement may be one of moving the thigh toward the fixed pelvis as in supine alternate-leg raising; or the movement may be one of bringing the pelvis toward the fixed thighs as in coming up from a supine to a sitting position, bending forward from a standing position, or tilting the pelvis anteriorly in standing. *Extension* is movement in a posterior direction. The movement may be one of bringing the thigh posteriorly as in leg-raising backward, or one of bringing the trunk posteriorly as in returning from a standing forward-bent position, or as in tilting the pelvis posteriorly in standing.

The range of hip joint flexion from zero is about 125°, the range of extension is about 10°, making a total range of about 135°. The knee joint should be flexed when measuring hip joint flexion to avoid restriction of motion by the Hamstring muscles, and extended when measuring hip joint extension to avoid restriction of motion by the Rectus femoris.

Abduction and adduction are movements about a sagittal axis. *Abduction* is movement away from the midsagittal plane in a lateral direction. In a supine position the movement may be one of moving the thigh laterally on a fixed trunk or moving the trunk so that the pelvis tilts laterally (downward) toward a fixed thigh. *Adduction* is movement of the thigh toward the midsagittal plane is a medial direction. In a supine position, the movement may be one of moving the thigh medially on a fixed trunk or moving the trunk so that the pelvis tilts laterally (upward) away from a fixed thigh. (For abduction and adduction of the hip joints accompanying lateral pelvic tilt, see pp. 291 and 292.)

From zero, the range of abduction is approximately 45° and of adduction, 10°, making the total range about 55°.

Lateral and medial rotation are movements about a longitudinal axis. *Medial rotation* is movement in which the anterior surface of the thigh turns toward the midsagittal plane. *Lateral rotation* is movement in which the anterior surface of the thigh moves away from the midsagittal plane. Rotation may result, also, from movement of the trunk on the femur. For example, when standing on the right leg, counterclockwise rotation of the pelvis will result in lateral rotation of the hip joint.

Knee joint. The knee joint is a modified ginglymus or hinge joint formed by the articulation of the condyles of the femur with the condyles of the tibia and by the patella articulating with the patellar surface of the femur.

Flexion and extension are movements about a coronal axis. *Flexion* is movement in a posterior direction, approximating the posterior surfaces of the leg and thigh. *Extension* is movement in an anterior direction to a position of straight alignment of the thigh and leg (0°). From the position of zero extension, the range of flexion is approximately 140°. The hip joint should be flexed when measuring full knee joint flexion to avoid restriction of motion by the Rectus femoris, but should not be fully flexed when measuring knee joint extension in order to avoid restriction by the Hamstring muscles.

Hyperextension is an abnormal or unnatural movement beyond the zero position of extension. For the sake of stability in standing, the knee normally is expected to be in a position of a very few degrees of extension beyond zero. If extended beyond these few degrees, the knee is said to be hyperextended.

Lateral and medial rotation are movements about a longitudinal axis. Rotation of the anterior surface of the leg toward the midsagittal plane is *medial rotation*, away from the midsagittal plane is *lateral rotation*.

The extended knee (in zero position) is essentially locked, preventing any rotation. Rotation occurs with flexion, combining movement between the tibia and the menisci as well as movement between the tibia and the femur.

With the thigh fixed, the movement that accompanies *flexion* is medial rotation of the tibia on the femur; with the leg fixed, the movement is lateral rotation of the femur on the tibia.

With the thigh fixed, the movement that accompanies *extension* is lateral rotation of the tibia on the femur; with the leg fixed, the movement is medial rotation of the femur on the tibia.

Ankle joint. The ankle joint is a ginglymus or hinge joint formed by the articulation of the tibia and fibula with the talus. The axis about which motion takes place extends obliquely from the posterolateral aspect of the fibular malleolus to the anteromedial aspect of the tibial malleolus.

Flexion and extension are the two movements that occur about the oblique axis. *Flexion* is movement of the foot in which the plantar surface moves in a caudal and posterior direction. *Extension* is movement of the foot in which the dorsal surface moves in an anterior and cranial direction.

Confusion has arisen about the terminology of these two ankle joint movements. An apparent discrepancy occurs because decreasing an angle frequently is associated with flexion while increasing it is associated with extension. Bringing the foot upward to "bend the ankle" seems to connote flexion, while pointing the foot downward to "straighten the ankle" connotes extension. To avoid confusion, there has been rather widespread acceptance of the terms *dorsiflexion* for extension and *plantar flexion* for flexion. This text will adhere to the use of these generally accepted terms. The knee should be flexed when measuring dorsiflexion. With the knee flexed, the ankle joint can be dorsiflexed about 20°. With the knee extended, the Gastrocnemius may limit the range of motion to about 15°. The range of motion in plantar flexion is approximately 45°.

Subtalar joint and transverse tarsal joints. The subtalar joint is a modified plane or gliding joint formed by the articulation of the talus and calcaneus. The talus also articulates with the navicular, and the talonavicular joint is involved in the movements ascribed to the subtalar joint.

Supination and pronation are movements permitted by the subtalar and talocalcaneonavicular joints. *Supination* is rotation of the foot in which the sole of the foot moves in a medial direction; *pronation* is rotation in which the sole of the foot moves in a lateral direction.

The transverse tarsal joints are formed by the articulations of the talus with the navicular, and the calcaneus with the cuboid.

Adduction and abduction of the forefoot are movements permitted by the transverse tarsal joints, *adduction* being movement of the forefoot in a medial direction and *abduction* in a lateral direction.

Inversion is a combination of supination and forefoot adduction. It is more free in plantar flexion than in dorsiflexion.

Eversion is a combination of pronation and forefoot abduction. It is more free in dorsiflexion than in plantar flexion.

Metatarsophalangeal joints. The metatarsophalangeal joints are condyloid, formed by the articulation of the distal ends of the metatarsals with the adjacent ends of the proximal phalanges.

Flexion and extension are movements about a coronal axis. *Flexion* is movement in a plantar direction, *extension* is movement in a cranial direction. The range of motion in adults is variable, but 30° flexion and 40° extension may be considered an average range for good function of the toes.

Adduction and abduction are movements about a sagittal axis. The line of reference for adduction and abduction of the toes is the axial line projected distally in line with the second metatarsal and extending through the second digit. *Adduction* is movement toward the axial line, and *abduction* is movement away from it, as in spreading the toes apart. Because abduction of the toes is restricted by the wearing of shoes, this movement is markedly limited in most adults and little attention is paid to the ability to abduct.

Interphalangeal joints of toes. The interphalangeal joints are ginglymus or hinge joints formed by the articulations of adjacent surfaces of phalanges.

Flexion and extension are movements about a coronal axis with *flexion* being movement in a plantar direction and *extension* movement in a cranial direction.

MOVEMENTS OF VERTEBRAL COLUMN

Vertebral articulations include the bilateral synovial joints of the vertebral arches where the inferior facets of one vertebra articulate with the superior facets of the adjacent vertebra, and the fibrous joints between successive vertebral bodies united by intervertebral fibrocartilaginous discs. Movement between two adjacent vertebrae is slight and is determined by the slope of the articular facets and the flexibility of the intervertebral discs. The range of motion of the column as a whole, however, is considerable and the movements permitted are flexion, extension, lateral flexion, and rotation.

The articulations of the first two vertebrae of the column are exceptions to the general classification. The atlantooccipital articulation, between the condyles of the occipital bone and the superior facets of the atlas, is classified as a condyloid joint. The movements permitted are flexion and extension with very slight lateral motion. The atlantoaxial articulation is composed of three joints, the two lateral fitting the general description of the joints of the vertebral column. The third, a median joint, formed by the articulation of the dens of the axis with the fovea dentis of the atlas is classified as a trochoid joint and permits rotation.

The normal curves of the spine, anterior in the cervical region, posterior in the thoracic region, and anterior in the lumbar region, are named according to the *convexity* of the curve. Likewise, a lateral curvature is named according to the direction of the convexity, e.g., a left lumbar curve is one which is convex toward the left. To the extent that the spine curves convexly forward or backward during movement, the same rule applies about naming the direction of movement according to the direction of the convexity of the curve.

Flexion of the spine, which occurs in a sagittal plane, is movement in which the head and trunk bend forward as the spine moves in the direction of curving convexly backward. (See p. 28, fig. A.). From a supine position, normal flexion will allow curling the trunk enough to lift the scapulae from the supporting surface. The seventh cervical vertebra will be lifted upward about eight or ten inches.

Flexion varies according to the region of the spine. In the *cervical* region, flexion of the spine is movement in the direction of *decreasing the normal forward curve.* Movement continues to the point of straightening or flattening this region of the spine, but normally does not progress to the point that the spine curves convexly backward. In the *thoracic* region, flexion of the spine is movement in the direction of *increasing the normal backward curve.* In normal flexion, the spine curves convexly backward producing a continuous, gently rounded contour throughout the thoracic area. In the *lumbar* region, flexion of the spine is movement in the direction of *decreasing the normal forward curve.* It progresses to the point of straightening or flattening the low back, but normally the lumbar spine does not curve convexly backward.

Extension of the spine, which occurs in a sagittal plane, is movement in which the head and trunk bend backward while the spine moves in the direction of curving convexly forward. (See fig. B below.) From a prone position, normal extension will allow the head and chest to be raised enough to lift the xiphoid process of the sternum about two to four inches from the table. (See p. 230.)

Extension varies according to the regions of the spine. In the *cervical* region, extension is movement in the direction of *increasing the normal forward curve.* It occurs by tilting the head back, bringing the occiput toward the seventh cervical vertebra. It may occur, also, in sitting or standing, by slumping into a round upper back, forward-head position—a position that also results in approximating the occiput toward the seventh cervical vertebra. In the *thoracic* region, extension is movement of the spine in the direction of *decreasing the normal backward curve.* Movement may progress to, but normally not beyond, the point of straightening or flattening the thoracic spine. In the *lumbar* region, extension is movement in the direction of *increasing the normal forward curve.* It occurs by bending the body backward or by tilting the pelvis forward.

According to Stedman's Medical Dictionary, to flex means to bend, to extend means to straighten. With respect to the thoracic spine, these meanings apply. The normal *posterior curve is a position of slight flexion.* As the spine moves from flexion to the *straight position, it is in extension.* However, there appears to be an ambiguity when describing the positions and movements of the cervical and lumbar spines. In each, the normal anterior *curve*

is a position of slight extension. As the spine moves from extension to the *straight position, it is in flexion.* Because it is confusing to describe a straight spine as being in flexion, it is better to use the term flat, rather than flexed, when referring to the cervical and lumbar regions.

Hyperextension of the spine is movement beyond the normal range of motion in extension, or may refer to a position greater than the normal anterior curve. Hyperextension may vary from slight to extreme. (See p. 281 and p. 300).

Lateral flexion and rotation are described separately although they occur in combination and are not considered pure movements.

Lateral flexion of the spine, which occurs in a coronal plane, is movement in which the head and trunk bend toward one side while the spine curves convexly toward the opposite side. A curve convex toward the left is the equivalent of lateral flexion toward the right. From a standing position with feet about three inches apart, body erect, and arms at sides, normal lateral flexion (bending directly sideways) will allow for the fingertips to reach approximately to the level of the knee. (See fig. C below.)

Lateral flexion varies according to the regions of the spine. It is most free in the cervical and lumbar regions, being restricted in the thoracic region by the rib cage.

Rotation is movement in a transverse plane. It is most free in the thoracic region and slight in the lumbar region. Rotation in the cervical region permits about 90° range of motion of the head and is referred to as rotation of the face toward the right or left. Rotation of the thorax on the pelvis is described as *clockwise* (forward on left side) or *counterclockwise* (forward on right side). (Fig. D.)

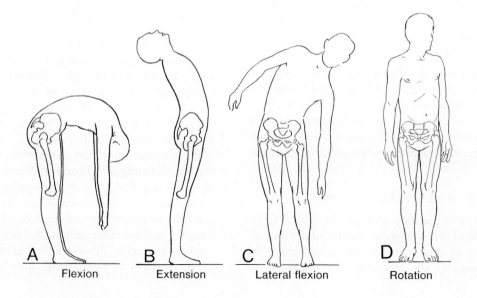

A	B	C	D
Flexion	Extension	Lateral flexion	Rotation

JOINT MEASUREMENT CHART

Name. . . . . .Cl. #. . . . .

Diagnosis. . . . . .Age. . . . .

Onset. . . . . .Doctor. . . . .

UPPER EXTREMITY

					Date	Motion	Avg. Range		Date					
					Exam. by		Anat.	Geom.	Exam. by					
						Extension	45°							
						Flexion	180°							
						Range	225°							
						Abduction	180°							
					Left Shoulder	Adduction	0°		Right Shoulder					
						Range	180°							
						Lat. Rotation	90°							
						Med. Rotation	70°							
						Range	160°							
						Extension	* 0°	180°						
					Left Elbow	Flexion	145°	35°	Right Elbow					
						Range	145°	145°						
						Supination	90°							
					Left Forearm	Pronation	90°		Right Forearm					
						Range	180°							
						Extension	70°							
						Flexion	80°							
						Range	150°							
					Left Wrist	Ulnar Deviat.	35°		Right Wrist					
						Radial Deviat.	20°							
						Range	55°							

* Use either anatomical or geometric basis for measurement. Cross off one not used. 180° is the plane of reference for the geometric basis of measurement. The zero position is the plane of reference for all the others. When a part moves in the direction of zero but fails to reach the zero position, the degrees designating the joint motion obtained are recorded with a minus sign and subtracted in computing the range of motion.

Notes: _____

_____

_____

_____

_____

_____

_____

_____

_____

_____

_____

JOINT MEASUREMENT CHART

Name...Cl. #.................................

Diagnosis...Age...................................

Onset...Doctor................................

LOWER EXTREMITY

				Date	Motion	Avg. Range		Date				
				Exam. by		Anat.	Geom.	Exam. by				
				Left Hip	Extension	* 10°	190°	Right Hip				
					Flexion	125°	55°					
					Range	135°	135°					
					Abduction	45°						
					Adduction	10°						
					Range	55°						
					Lat. Rotation	45°						
					Med. Rotation	45°						
					Range	90°						
				Left Knee	Extension	* 0°	180°	Right Knee				
					Flexion	140°	40°					
					Range	140°	140°					
				Left Ankle	Plantar Flex.	* 45°	135°	Right Ankle				
					Dorsal Flex.	20°	70°					
					Range	65°	65°					
				Left Foot	Inversion	35°		Right Foot				
					Eversion	20°						
					Range	55°						

*Use either anatomical or geometric basis for measurement. Cross off one not used. 180° is the plane of reference for the geometric basis of measurement. The zero position is the plane of reference for all the others. When a part moves in the direction of zero but fails to reach the zero position, the degrees designating the joint motion obtained are recorded with a minus sign and subtracted in computing the range of motion.

Notes: _____

_____

_____

_____

_____

_____

_____

_____

_____

_____

_____

_____

_____

_____

chapter 3

Spinal Nerve and Muscle Charts and Plexuses

Spinal nerve and muscle charts
Spinal nerve and motor point charts
Use of charts in differential diagnosis
   Six cases charted

Spinal segment distribution to nerves
and muscles
Nerve plexuses

Spinal Nerve and Muscle Charts

The recording of test results is an important part of muscle examinations. Records are important from the standpoint of diagnosis, treatment, and prognosis. An examination performed without recording the details can be of value at the moment, but one has an obligation to the patient, to the institution if one is involved, and to oneself to record the findings.

Charts used for recording should permit complete tabulation of test results, and the arrangement of the information should facilitate interpretation.

Six types of charts are illustrated in this text and each has its specific use and value. Except for the Spinal Nerve and Muscle Charts described in this chapter, the charts appear in the respective chapters to which they apply.

SPINAL NERVE AND MUSCLE CHARTS

There are two charts in this category, one for the neck, diaphragm, and upper extremity (p. 35), the other for the trunk and lower extremity (p. 37). Each chart has an accompanying anatomical drawing of the spinal nerves and motor points (pp. 34, 36). These charts have been designed especially for use as an aid in differential diagnosis of lesions of the spinal nerves. The motor involvement as determined by manual muscle tests can aid in determining whether there is a lesion of the nerve at root, plexus, or peripheral level. The chart may be useful, also, in determining the level of a spinal cord lesion. (See plexus drawings pp. 54–57).

In the upper and lower extremity charts the names of the muscles appear in the left hand column and are grouped, as indicated by heavy black lines, according to their nerve innervations listed to the left of the muscle names. The space between the column of muscle names and the nerves is used for recording the grade of muscle strength.

The Sternocleidomastoid and the Trapezius muscles are listed on the Spinal Nerve and Muscle Chart, p. 35, and the Cranial Nerve and Muscle Chart, p. 250. While these muscles receive their motor innervation mainly from the spinal portion of the 11th cranial nerve (accessory), additional spinal nerve branches are distributed to them: C2, C3 to the Sternocleidomastoid; C3, C4 to the Trapezius. It has not been determined, in man, whether these cervical spinal nerve branches consist of motor and sensory or only sensory fibers. Clinical findings in cases of pure accessory nerve lesions have led neurologists to assume that these spinal nerve fibers are chiefly concerned in the innervation of the caudal part of the Trapezius, its cranial and middle parts as well as the Sternocleidomastoid being supplied predominantly by the accessory nerve (Brodel, 1981).[8] Some authors report that these cervical nerves supply chiefly the upper part of the Trapezius. In still other individuals it would appear that these nerve fibers do not contribute any motor fibers to the Trapezius, the motor innervation of the entire muscle being dependent on the spinal portion of the accessory nerve. Apparently, considerable individual variations exist in the innervation of the Trapezius (Peele, 1977).[9]

Peripheral nerve section. Peripheral nerves and their segmental origins are listed across the top of the center of the chart and follow the order of proximal-distal branching insofar as possible. For the peripheral nerves which arise from cords of the brachial plexus, the appropriate cord is indicated. The key at the top of the charts explains the abbreviations used.

Below this section, in the body of the chart, the dots indicate the peripheral nerve supply to each muscle. (See p. 58 for sources of material for this section.)

Spinal segment section. In the section headed "Spinal Segment", a number denotes the spinal segment origin of nerve fibers innervating each of the muscles listed in the left-hand column. (See pp. 48–51 for sources of material for this section.)

Sensory section. On the the right side of the charts are diagrams showing the dermatomes and the distribution of cutaneous nerves for the upper extremity on the one chart, and for the trunk and lower extremity on the other. The dermatome illustrations are redrawn from Keegan and Garrett[10] on the extremity charts, and from Gray[11] on the cranial chart. The cutaneous nerve illustrations are redrawn from Gray.[11] (For cranial chart, see p. 250)

It is possible to use the illustrations for charting areas of sensory involvement by shading or using colored pencil to outline the areas of the involvement for any given patient. Only drawings of the right extremity are used on the extremity charts but labeling can indicate, when necessary, that the recorded information pertains to the left side.

Spinal Nerve and Motor Point Chart

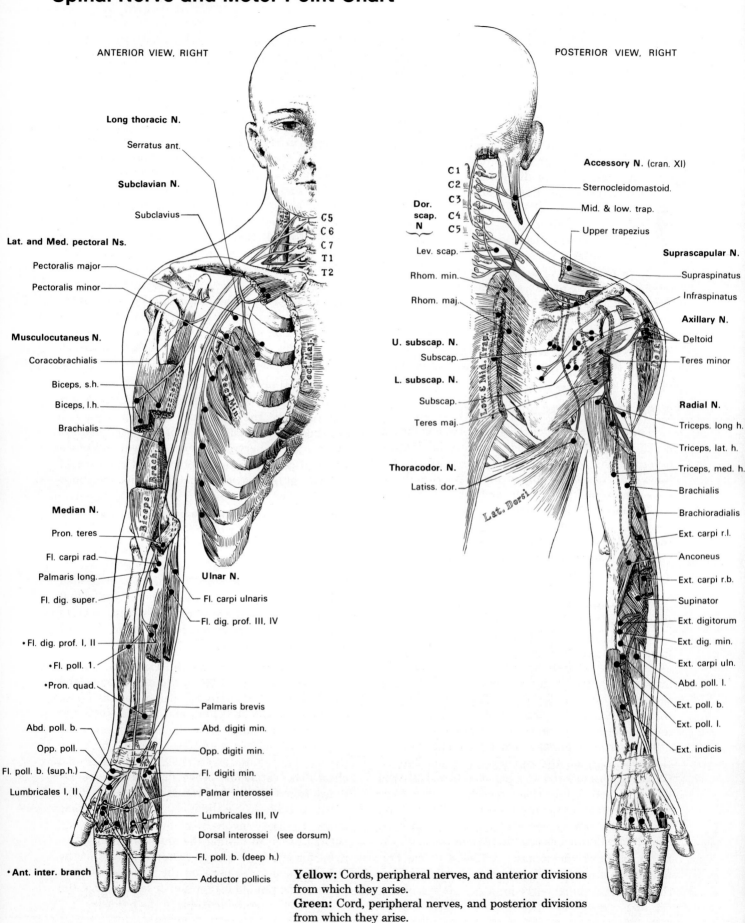

ANTERIOR VIEW, RIGHT

Long thoracic N.
Serratus ant.

Subclavian N.
Subclavius

Lat. and Med. pectoral Ns.
Pectoralis major
Pectoralis minor

Musculocutaneus N.
Coracobrachialis
Biceps, s.h.
Biceps, l.h.
Brachialis

Median N.
Pron. teres
Fl. carpi rad.
Palmaris long.
Fl. dig. super.
•Fl. dig. prof. I, II
•Fl. poll. 1.
•Pron. quad.

Abd. poll. b.
Opp. poll.
Fl. poll. b. (sup.h.)
Lumbricales I, II

•Ant. inter. branch

C5
C6
C7
T1
T2

Ulnar N.
Fl. carpi ulnaris
Fl. dig. prof. III, IV

Palmaris brevis
Abd. digiti min.
Opp. digiti min.
Fl. digiti min.
Palmar interossei
Lumbricales III, IV
Dorsal interossei (see dorsum)
Fl. poll. b. (deep h.)
Adductor pollicis

POSTERIOR VIEW, RIGHT

C1
C2
C3
Dor.
scap.   C4
N   C5

Lev. scap.
Rhom. min.
Rhom. maj.

U. subscap. N.
Subscap.

L. subscap. N.
Subscap.
Teres maj.

Thoracodor. N.
Latiss. dor.

Accessory N. (cran. XI)
Sternocleidomastoid.
Mid. & low. trap.
Upper trapezius

Suprascapular N.
Supraspinatus
Infraspinatus

Axillary N.
Deltoid
Teres minor

Radial N.
Triceps. long h.
Triceps, lat. h.
Triceps, med. h.
Brachialis
Brachioradialis
Ext. carpi r.l.
Anconeus
Ext. carpi r.b.
Supinator
Ext. digitorum
Ext. dig. min.
Ext. carpi uln.
Abd. poll. l.
Ext. poll. b.
Ext. poll. l.
Ext. indicis

Yellow: Cords, peripheral nerves, and anterior divisions from which they arise.
Green: Cord, peripheral nerves, and posterior divisions from which they arise.

SPINAL NERVE AND MUSCLE CHART
NECK, DIAPHRAGM AND UPPER EXTREMITY

Name _____ Date _____

KEY
D. = Dorsal Prim. Ramus
V. = Vent. Prim. Ramus
P.T. = Pre-Trunk
S.T. = Superior Trunk
P. = Posterior Cord
L. = Lateral Cord
M. = Medial Cord

Region	Muscle	Peripheral Nerve	C1	C2	C3	C4	C5	C6	C7	C8	T1
Cervical Nerves	HEAD & NECK EXTENSORS	Cervical (D.)	1	2	3	4	5	6	7	8	1
Cervical Nerves	INFRAHYOID MUSCLES	Cervical (V.)	1	2	3						
Cervical Nerves	RECTUS CAP. ANT. & LAT.	Cervical (V.)	1	2							
Cervical Nerves	LONGUS CAPITUS	Cervical (V.)	1	2	3	(4)					
Cervical Nerves	LONGUS COLLI	Cervical (V.)		2	3	4	5	6	(7)		
Cervical Nerves	LEVATOR SCAPULAE	Cervical (V.), Dor. Scap			3	4	5				
Cervical Nerves	SCALENI (A. M. P.)	Cervical (V.)			3	4	5	6	7	8	
Cervical Nerves	STERNOCLEIDOMASTOID	Cervical (V.)	(1)	2	3						
Cervical Nerves	TRAPEZIUS (U. M. L.)	Cervical (V.)		2	3	4					
Cervical Nerves	DIAPHRAGM	Phrenic			3	4	5				
Brachial Plexus — Pre-Tr.	SERRATUS ANTERIOR	Long Thor.					5	6	7	8	
Brachial Plexus — Pre-Tr.	RHOMBOIDS MAJ & MIN.	Dor. Scap				4	5				
Brachial Plexus — Trunk	SUBCLAVIUS	N. to Subcl.					5	6			
Brachial Plexus — Trunk	SUPRASPINATUS	Suprascap				4	5	6			
Brachial Plexus — Trunk	INFRASPINATUS	Suprascap				(4)	5	6			
Brachial Plexus — P. Cord	SUBSCAPULARIS	U. Subscap, L. Subscap					5	6	7		
Brachial Plexus — P. Cord	LATISSIMUS DORSI	Thoracodor						6	7	8	
Brachial Plexus — P. Cord	TERES MAJOR	L. Subscap					5	6	7		
Brachial Plexus — M & L.L.	PECTORALIS MAJ. (UPPER)	Lat. Pect					5	6	7		
Brachial Plexus — M & L.L.	PECTORALIS MAJ. (LOWER)	Lat. Pect, Med. Pect						6	7	8	1
Brachial Plexus — M & L.L.	PECTORALIS MINOR	Med. Pect						(6)	7	8	1
Axil.	TERES MINOR	Axillary					5	6			
Axil.	DELTOID	Axillary					5	6			
Musculo-cutan.	CORACOBRACHIALIS	Musculocu.						6	7		
Musculo-cutan.	BICEPS	Musculocu.					5	6			
Musculo-cutan.	BRACHIALIS	Musculocu.					5	6			
Radial	TRICEPS	Radial						6	7	8	1
Radial	ANCONEUS	Radial							7	8	
Radial	BRACHIALIS (SMALL PART)	Radial					5	6			
Radial	BRACHIORADIALIS	Radial					5	6			
Radial	EXT. CARPI RAD. L. & B.	Radial					5	6	7	8	
Radial	SUPINATOR	Radial					5	6	(7)		
Radial	EXT. DIGITORUM	Radial						6	7	8	
Radial	EXT. DIGITI MINIMI	Radial						6	7	8	
Radial	EXT. CARPI ULNARIS	Radial						6	7	8	
Radial	ABD. POLLICIS LONGUS	Radial						6	7	8	
Radial	EXT. POLLICIS BREVIS	Radial						6	7	8	
Radial	EXT. POLLICIS LONGUS	Radial						6	7	8	
Radial	EXT. INDICIS	Radial						6	7	8	
Median	PRONATOR TERES	Median						6	7		
Median	FLEX CARPI RADIALIS	Median						6	7	8	
Median	PALMARIS LONGUS	Median						(6)	7	8	1
Median	FLEX. DIGIT. SUPERFICIALIS	Median							7	8	1
Median — A. Inter.	FLEX. DIGIT. PROF. I & II	Median							7	8	1
Median — A. Inter.	FLEX. POLLICIS LONGUS	Median						(6)	7	8	1
Median — A. Inter.	PRONATOR QUADRATUS	Median							7	8	1
Median	ABD. POLLICIS BREVIS	Median						6	7	8	1
Median	OPPONENS POLLICIS	Median						6	7	8	1
Median	FLEX. POLL. BREV. (SUP. H.)	Median						6	7	8	1
Median	LUMBRICALES I & II	Median						(6)	7	8	1
Ulnar	FLEX. CARPI ULNARIS	Ulnar							7	8	1
Ulnar	FLEX. DIGIT. PROF. III & IV	Ulnar							7	8	1
Ulnar	PALMARIS BREVIS	Ulnar							(7)	8	1
Ulnar	ABD. DIGITI MINIMI	Ulnar							(7)	8	1
Ulnar	OPPONENS DIGITI MINIMI	Ulnar							(7)	8	1
Ulnar	FLEX. DIGITI MINIMI	Ulnar							(7)	8	1
Ulnar	PALMAR INTEROSSEI	Ulnar								8	1
Ulnar	DORSAL INTEROSSEI	Ulnar								8	1
Ulnar	LUMBRICALES III & IV	Ulnar							(7)	8	1
Ulnar	ADDUCTORS POLLICIS	Ulnar								8	1
Ulnar	FLEX. POLL. BREV. (DEEP H.)	Ulnar								8	1

SENSORY

Anterior view labels: C2, C3, C4, 5, T1, Supra-clavicular, C5, T1, Axillary, Dorsal antebrach. cutan., Inter-costobrach. and med brach, Lat. antebrach. cutan., Medial antebrach. cutan., C6, C7, C8, Superfic. br. of radial, Ulnar, Median

Posterior view labels: C2, 3, 4, 5, 6, 7, T1, C8, Inter-costobrach. and post. brach. cutan., Supra-clavicular, Axillary, Dorsal antebrach. cutan., Lat. antebrach. cutan., Medial antebrach. cutan., Ulnar, C6, C8, C7, Radial, Median

Dermatomes redrawn from Keegan and Garrett. Anat. Rec. 102, 409-437, 1948
Cutaneous Distribution of peripheral nerves redrawn from *Gray's Anatomy of the Human Body*, 28th ed

Spinal Nerve and Motor Point Chart

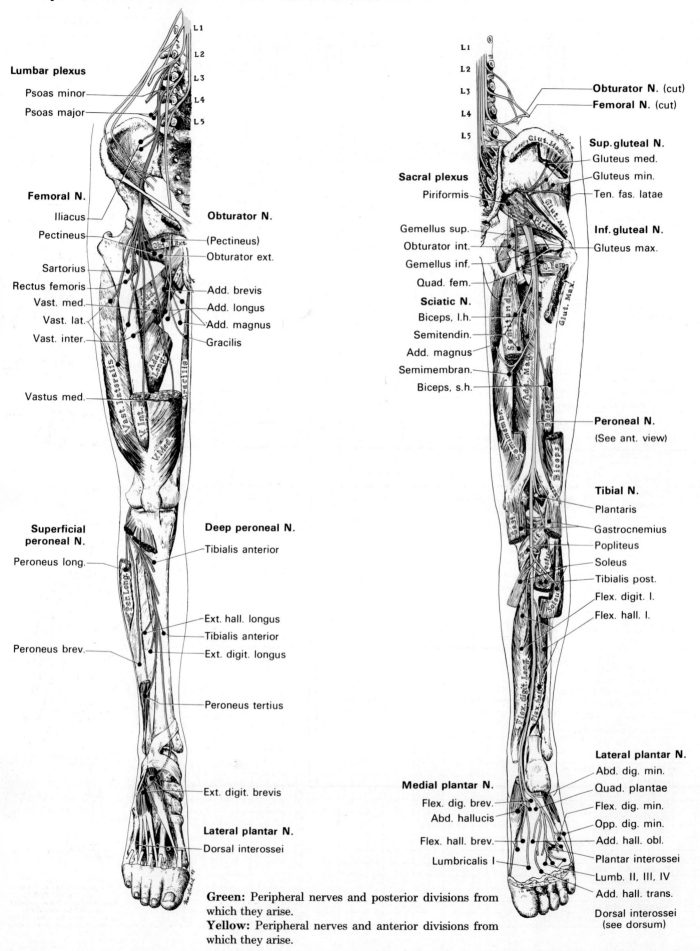

L1
L2
L3
L4
L5

Lumbar plexus
Psoas minor
Psoas major

Femoral N.
Iliacus
Pectineus
Sartorius
Rectus femoris
Vast. med.
Vast. lat.
Vast. inter.

Vastus med.

Obturator N.
(Pectineus)
Obturator ext.
Add. brevis
Add. longus
Add. magnus
Gracilis

Superficial peroneal N.
Peroneus long.

Peroneus brev.

Deep peroneal N.
Tibialis anterior

Ext. hall. longus
Tibialis anterior
Ext. digit. longus

Peroneus tertius

Ext. digit. brevis

Lateral plantar N.
Dorsal interossei

L1
L2
L3
L4
L5

Obturator N. (cut)
Femoral N. (cut)

Sup. gluteal N.
Gluteus med.
Gluteus min.
Ten. fas. latae

Sacral plexus
Piriformis
Gemellus sup.
Obturator int.
Gemellus inf.
Quad. fem.

Inf. gluteal N.
Gluteus max.

Sciatic N.
Biceps, l.h.
Semitendin.
Add. magnus
Semimembran.
Biceps, s.h.

Peroneal N.
(See ant. view)

Tibial N.
Plantaris
Gastrocnemius
Popliteus
Soleus
Tibialis post.
Flex. digit. l.
Flex. hall. l.

Medial plantar N.
Flex. dig. brev.
Abd. hallucis
Flex. hall. brev.
Lumbricalis I

Lateral plantar N.
Abd. dig. min.
Quad. plantae
Flex. dig. min.
Opp. dig. min.
Add. hall. obl.
Plantar interossei
Lumb. II, III, IV
Add. hall. trans.

Dorsal interossei
(see dorsum)

Green: Peripheral nerves and posterior divisions from which they arise.
Yellow: Peripheral nerves and anterior divisions from which they arise.

SPINAL NERVE AND MUSCLE CHART
TRUNK AND LOWER EXTREMITY

Name _____   Date _____

KEY
→
D = Dorsal Primary Ramus
V = Ventral Primary Ramus
A = Anterior Division
P = Posterior Division

Group	Muscle	D (T1-12,L1-5,S1-3)	V (T1,2,3,4)	V (T5,6)	V (T7,8)	V (T9,10,11,12)	Iliohypogastric (T12,L1)	Ilioinguinal (T(12),L1)	Lumb. Plex. (T(12),L1,2,3,4)	Femoral (L(1),2,3,4)	Obturator (L(1),2,3,4)	Sup. Glut. (L4,5,S1)	Inf. Glut. (L5,S1,2)	Sac. Plex. (L4,5,S1,2,3)	Sciatic (L4,5,S1,2,3)	C. Peroneal (L4,5,S1,2)	Tibial (L4,5,S1,2,3)	L1	L2	L3	L4	L5	S1	S2	S3
	ERECTOR SPINAE	●																1	2	3	4	5	1	2	3
Thoracic Nerves	SERRATUS POST SUP		●																						
Thoracic Nerves	TRANS. THORACIS		●	●	●																				
Thoracic Nerves	INT. INTERCOSTALS		●	●	●	●																			
Thoracic Nerves	EXT. INTERCOSTALS		●	●	●	●																			
Thoracic Nerves	SUBCOSTALES		●	●	●	●																			
Thoracic Nerves	LEVATOR COSTARUM		●	●	●	●																			
Thoracic Nerves	OBLIQUUS EXT. ABD.			(●)	●	●																			
Thoracic Nerves	RECTUS ABDOMINIS			●	●	●																			
Thoracic Nerves	OBLIQUUS INT. ABD.				●	●	●	(●)										1							
Thoracic Nerves	TRANSVERSUS ABD.				●	●	●	(●)										1							
Thoracic Nerves	SERRATUS POST. INF.					●																			
Lumb. Plexus	QUAD. LUMBORUM								●									1	2	3					
Lumb. Plexus	PSOAS MINOR								●									1	2						
Lumb. Plexus	PSOAS MAJOR								●									1	2	3	4				
Femoral	ILIACUS									●								(1)	2	3	4				
Femoral	PECTINEUS									●	(●)								2	3	4				
Femoral	SARTORIUS									●									2	3	(4)				
Femoral	QUADRICEPS									●									2	3	4				
Obturator (Ant.)	ADDUCTOR BREVIS										●								2	3	4				
Obturator (Ant.)	ADDUCTOR LONGUS										●								2	3	4				
Obturator (Ant.)	GRACILIS										●								2	3	4				
Obturator (Post.)	OBTURATOR EXT.										●									3	4				
Obturator (Post.)	ADDUCTOR MAGNUS										●			●					2	3	4	5	1		
Gluteal (Sup.)	GLUTEUS MEDIUS											●									4	5	1		
Gluteal (Sup.)	GLUTEUS MINIMUS											●									4	5	1		
Gluteal (Sup.)	TENSOR FAS. LAT.											●									4	5	1		
Gluteal (In.)	GLUTEUS MAXIMUS												●									5	1	2	
Sacral Plexus	PIRIFORMIS													●								(5)	1	2	
Sacral Plexus	GEMELLUS SUP.													●								5	1	2	
Sacral Plexus	OBTURATOR INT.													●								5	1	2	
Sacral Plexus	GEMELLUS INF.													●							4	5	1	(2)	
Sacral Plexus	QUADRATUS FEM.													●							4	5	1	(2)	
Sciatic (Tibial)	BICEPS (LONG H.)														●							5	1	2	3
Sciatic (Tibial)	SEMITENDINOSUS														●						4	5	1	2	
Sciatic (Tibial)	SEMIMEMBRANOSUS														●						4	5	1	2	
Sciatic (P.)	BICEPS (SHORT H.)														●							5	1	2	
Common Peroneal (Deep)	TIBIALIS ANTERIOR															●					4	5	1		
Common Peroneal (Deep)	EXT. HALL. LONG.															●					4	5	1		
Common Peroneal (Deep)	EXT. DIGIT. LONG.															●					4	5	1		
Common Peroneal (Deep)	PERONEUS TERTIUS															●					4	5	1		
Common Peroneal (Deep)	EXT. DIGIT. BREVIS															●					4	5	1		
Common Peroneal (Sup)	PERONEUS LONGUS															●					4	5	1		
Common Peroneal (Sup)	PERONEUS BREVIS															●					4	5	1		
Tibial (Tibial)	PLANTARIS																●				4	5	1	(2)	
Tibial (Tibial)	GASTROCNEMIUS																●						1	2	
Tibial (Tibial)	POPLITEUS																●				4	5	1		
Tibial (Tibial)	SOLEUS																●					5	1	2	
Tibial (Tibial)	TIBIALIS POSTERIOR																●				(4)	5	1		
Tibial (Tibial)	FLEX. DIGIT. LONG.																●					5	1	(2)	
Tibial (Tibial)	FLEX. HALL. LONG.																●					5	1	2	
Tibial (Med. Pl.)	FLEX. DIGIT. BREVIS																●				4	5	1		
Tibial (Med. Pl.)	ABDUCTOR HALL.																●				4	5	1		
Tibial (Med. Pl.)	FLEX. HALL. BREVIS																●				4	5	1		
Tibial (Med. Pl.)	LUMBRICALIS I																●				4	5	1		
Tibial (Lat. Plant.)	ABD. DIGITI MIN.																●						1	2	
Tibial (Lat. Plant.)	QUAD. PLANTAE																●						1	2	
Tibial (Lat. Plant.)	FLEX. DIGITI MIN.																●						1	2	
Tibial (Lat. Plant.)	OPP. DIGITI MIN.																●						1	2	
Tibial (Lat. Plant.)	ADDUCTORS HALL.																●						1	2	
Tibial (Lat. Plant.)	PLANT. INTEROSSEI																●						1	2	
Tibial (Lat. Plant.)	DORSAL INTEROSSEI																●						1	2	
Tibial (Lat. Plant.)	LUMB. II, III, IV																●				(4)	(5)	1	2	

SENSORY

Dermatomes redrawn from
Keegan and Garrett, Anat. Rec. 102. 409-437. 1948
Cutaneous Distribution of peripheral nerves
redrawn from Gray's Anatomy of the Human Body. 28th ed

SPINAL NERVE AND MOTOR POINT CHARTS

Drawings which show the course of the nerves from spinal cord to motor points appear on pp. 34 and 36. These illustrations facilitate interpretation of the muscle test findings as recorded on the Spinal Nerve and Muscle Chart, and aid in determining the site or level of a lesion. Every effort has been made to preserve anatomical accuracy in showing the course of the nerves in relation to the muscles. Cross-section illustrations[12] were used to check the position of the nerves at various levels. For the most part, the motor point location, which is the site where a nerve enters the muscle it innervates, was based on the work of Brash.[13]

Anterior divisions of the plexuses and their branches are shown in yellow; posterior divisions and their branches, in green. The motor points are designated as black dots.

USE OF CHARTS IN DIFFERENTIAL DIAGNOSIS

Muscle strength grades are recorded in the column to the left of the muscle names. The grades may be in percentages or in other numeral or letter symbols.

For use in diagnosis of spinal nerve lesions, the range of grades need not be as well defined as for use in prognosis. The use of Zero, Trace, Poor, Fair, Good, and Normal or their equivalents, 0%, 5%, 20%, 50%, 80%, 100%, respectively, is sufficient to determine which muscles are involved without finer distinctions of grading.

After the grades have been recorded, the nerve involvement is plotted, when applicable, by circling the dot under peripheral supply and the number(s) under spinal segment distribution that corresponds with each involved muscle.

The involvement of peripheral nerves and/or parts of the plexus is ascertained from the encircled dots by following the vertical lines upward to the top of the chart, or the horizontal lines to the left-hand margin. (See p. 35.) When there is evidence of involvement at spinal segment level, the level of lesion may be indicated by a heavy black line drawn vertically to separate the involved from the uninvolved spinal segments. (See p. 44.)

Generally, muscles grading zero through fair may be considered as involved from a neurological standpoint, except in the case of the fibers of lower and middle Trapezius which often show postural stretch-weakness that grades as low as fair or 50%. If only the Trapezius shows this degree of weakness it should not be assumed, necessarily, that it is indicative of spinal accessory nerve involvement.

As a rule, muscles graded good or 80% and above may be considered as not being involved from a neurological standpoint. This degree of weakness may be the result of such factors as disuse atrophy, stretch-weakness, or lack of fixation by other muscles. It should be borne in mind, however, that a grade of good might indicate a deficit of a spinal segment which minimally innervates the muscle.

The use of the Spinal Nerve and Muscle Charts is illustrated by the case studies presented below.

Case 1.[14]* (Chart on p 40.) A thirty-year-old male fell from a moving automobile and was unconscious for about twenty minutes. He was treated in the emergency room of a local hospital for minor abrasions and then released. During the next three weeks he was seen and treated by several physicians because of paralysis and edema of his right arm and pains in his chest and neck.

Twenty-two days later, he was admitted to the University of Maryland Hospital. A neuromuscular evaluation, including a manual muscle test and an electromyographic study, was made at that time and showed extensive involvement of the right upper extremity.

The decision was made to defer surgical exploration and to treat the patient conservatively by the application of an airplane splint and follow-up therapy in the outpatient clinic. Unfortunately, the patient did not report to the outpatient clinic until five months later. Subsequently, a detailed manual muscle test as well as electrodiagnostic and further electromyographic studies were made.

Manual Muscle Test. The Chart on p. 40 indicates, at a glance, that the muscles supplied by the ulnar nerve were Normal, those by the median were either Normal or Good, and those by the radial, musculocutaneous, and axillary were either Poor or Zero. At the brachial plexus level the involvement was more complicated, as noted by the grades ranging all the way from Normal to Zero. However, concurrent charting of the involved peripheral nerves and spinal segments furnished additional information and provided the basis for determination of the sites of the lesions as follows:

1. *A lesion of the posterior cord of the brachial plexus.* The muscles supplied by the upper and lower subscapular, thoracodorsal, axillary, and radial nerves, which arise from the posterior cord, show complete paralysis or major weakness.

Involvement of the Subscapularis muscle places the site of lesion proximal to the point where the upper subscapular nerve arises ("c" in figure on page 41.)

* Reprinted with permission of *Physical Therapy* 48:733–739, July 1968.

2. *A lesion of either the upper trunk (formed by C5 and C6 roots of the plexus) or the anterior division of the upper trunk before it joins with the anterior division of the middle trunk (C7) to form the lateral cord.* Confirmation of this statement requires an explanation of how it is ascertained that the lesion is in this area, and that it is no more proximal than "a" nor more distal than "b."

The complete paralysis of the Biceps and Brachialis (from C5 and C6) raises the question of the level of involvement of these muscles—musculocutaneous nerve (C5, C6, C7), lateral cord (C5, C6, C7), trunk, or spinal nerve root?

The fact that the Coracobrachialis showed some strength rules out complete involvement at the musculocutaneous level. A complete lesion at the level of the lateral cord (C5, C6, C7) is refuted by several findings that indicate that the C7 component is not involved.

The Flexor digitorum superficialis (sublimis), the Flexor digitorum profundus I and II, and the Lumbricales I and II, which have C7, C8, and T1 supply through the median nerve, graded Normal. Other muscles supplied by the median nerve, which have C6, C7, C8, and T1 supply, graded Good and undoubtedly would have exibited more weakness had C7 been involved.

The sternal part of the Pectoralis major and Pectoralis minor, which are supplied chiefly by the medial pectoral (C8 and T1) and to some extent by the lateral pectoral (C5, C6, and C7), graded Good and Good plus. Had C7 been involved the weakness would undoubtedly have been greater.

The presence of some strength in the Coracobrachialis is thus explained on the basis of the C7 components being intact and further confirms that such is the case. The stretch-weakness, superimposed on this muscle by the shoulder joint subluxation and the weakness of the Deltoid and Biceps, could account for the Coracobrachialis grading no more than Poor.

Thus, with C7 not involved, the most distal point of lesion may be considered as "b".

The possibility of C5 and C6 being involved more proximal than "a" at the level of the roots of the plexus is ruled out because the Rhomboids and Serratus anterior muscles graded Normal. Whether the lesion is proximal or distal to the point where the suprascapular nerve arises depends on whether the involvement of the Supraspinatus and Infraspinatus muscles is on a neurogenic or stretch-weakness basis.

The Supraspinatus and Infraspinatus (C4, C5, C6) graded Fair, and if this partial weakness resulted from a neurological deficit the lesion must be proximal to the point where the suprascapular nerve arises. Most logically, the presence of Fair strength would then be interpreted as a result of regeneration in the seven months since onset.

On the other hand, there is the possibility that the weakness in these muscles is of a secondary stretch-weakness type, and that it was not neurogenic. The patient had not worn the airplane splint that was applied twenty-three days after injury, and there was subluxation of the joint and stretching of the capsule. The weakness was not as pronounced as in the other muscles supplied by C5 and C6, a fullness of contraction could be felt on palpation, and these muscles had been subjected to undue stretch. If the weakness resulted from stretch, the initial site of lesion would have been distal to the point where the suprascapular nerve arises.

3. *No involvement of the medial cord of the plexus.* The muscles supplied by the ulnar nerve, which is the terminal branch of the medial cord, graded Normal. The sternal part of the Pectoralis major and the Pectoralis minor (C5-T1), and some muscles receiving median nerve supply (C6-T1) graded Good. It is logical to assume that the slight weakness is attributable to the C5 and C6 deficit, and not to any involvement of the medial cord.

Sensory and Reflex Tests. Sensation to pinprick was absent over the area of sensory distribution of the axillary, musculocutaneous, and radial nerves. There were no deep tendon reflexes of the Biceps or Triceps muscles.

SPINAL NERVE AND MUSCLE CHART
NECK, DIAPHRAGM AND UPPER EXTREMITY

Name: **Case #1** Date:

SENSORY

Dermatomes redrawn from Keegan and Garrett Anat Rec 102 409 437 1948. Cutaneous Distribution of peripheral nerves redrawn from *Gray's Anatomy of the Human Body.* 28th ed

KEY

D.	Dorsal Prim. Ramus
V.	Vent. Prim. Ramus
P.T.	Pre-Trunk
S.T.	Superior Trunk
P.	Posterior Cord
L.	Lateral Cord
M.	Medial Cord

Peripheral Nerves (column headers)

Type	Nerve	Cervical roots
D.	Cervical	1-8
V.	Cervical	1-8
V.	Cervical	1-4
V.	Phrenic	3,4,5
P.T.	Long Thor.	5,6,7,(8)
P.T.	Dor. Scap.	4,5
S.T.	N. to Subcl.	5,6
S.T.	Suprascap.	4,5,6
P.	U. Subscap.	(4)5,6,(7)
P.	Thoracodor.	(5)6,7,8
P.	L. Subscap.	5,6,(7)
L.	Lat. Pect.	5,6,7
M.	Med. Pect.	(6)7,8 1
P.	Axillary	5,6
L.	Musculocu.	(4)5,6,7
P.	Radial	5,6,7,8 1
LM.	Median	5,6,7,8 1
M.	Ulnar	7,8 1

Muscle — Spinal Segment (Right)

Group	Muscle	Grade	C1	C2	C3	C4	C5	C6	C7	C8	T1
Cervical Nerves	HEAD & NECK EXTENSORS	⌐	1	2	3	4	5	6	7	8&	1
	INFRAHYOID MUSCLES	—	1	2	3						
	RECTUS CAP ANT & LAT	—	1	2							
	LONGUS CAPITUS	—	1	2	3	(4)					
	LONGUS COLLI	—		2	3	4	5	6	(7)		
	LEVATOR SCAPULAE	N			3	4	5				
	SCALENI (A M P)	—			3	4	5	6	7	8	
	STERNOCLEIDOMASTOID	N	(1)	2	3						
	TRAPEZIUS (U M L)	N		2	3	4					
	DIAPHRAGM	—			3	4	5				
Brachial Plexus (Pre-Tr)	SERRATUS ANTERIOR	N					5	6	7	8	
	RHOMBOIDS MAJ & MIN	N				4	5				
(Trunk)	SUBCLAVIUS	—					5	6			
	SUPRASPINATUS	F				(4)	5	6			
	INFRASPINATUS	F+				(4)	5	6			
(P.Cord)	SUBSCAPULARIS	O?						6	7	8	
	LATISSIMUS DORSI	P?						6	7	8	
	TERES MAJOR	O?					5	6	7		
(M & LL)	PECTORALIS MAJ (UPPER)	O					5	6	7		
	PECTORALIS MAJ (LOWER)	G						6	7	8	1
	PECTORALIS MINOR	G+						(6)	7	8	1
Axil	TERES MINOR	P					5	6			
	DELTOID	Tr					5	6			
Musculo-cutan	CORACOBRACHIALIS	P						6	7		
	BICEPS	O					5	6			
	BRACHIALIS	O					5	6			
Radial	TRICEPS	P						6	7	8	1
	ANCONEUS	—							7	8	
	BRACHIALIS (SMALL PART)	—					5	6			
	BRACHIORADIALIS	O					5	6			
	EXT CARPI RAD L & B	O					5	6	7		
	SUPINATOR	O					5	6	(7)		
	EXT DIGITORUM	O						6	7	8	
	EXT DIGITI MINIMI	O						6	7	8	
	EXT CARPI ULNARIS	O						6	7	8	
	ABD POLLICIS LONGUS	O						6	7	8	
	EXT POLLICIS BREVIS	O						6	7	8	
	EXT POLLICIS LONGUS	O						6	7	8	
	EXT INDICIS	O						6	7	8	
Median (A Inter)	PRONATOR TERES	G						6	7		
	FLEX CARPI RADIALIS	G						6	7	8	
	PALMARIS LONGUS	G						(6)	7	8	1
	FLEX DIGIT SUPERFICIALIS	N							7	8	1
	FLEX DIGIT PROF I & II	N							7	8	1
	FLEX POLLICIS LONGUS	N						(6)	7	8	1
	PRONATOR QUADRATUS	N							7	8	1
	ABD POLLICIS BREVIS	G						6	7	8	1
	OPPONENS POLLICIS	N						6	7	8	1
	FLEX POLL BREV (SUP H)	G						6	7	8	1
	LUMBRICALES I & II	N						(6)	7	8	1
Ulnar	FLEX CARPI ULNARIS	N							7	8	1
	FLEX DIGIT PROF III & IV	N							7	8	1
	PALMARIS BREVIS	—							(7)	8	1
	ABD DIGITI MINIMI	N							(7)	8	1
	OPPONENS DIGITI MINIMI	N							(7)	8	1
	FLEX DIGITI MINIMI	N							(7)	8	1
	PALMAR INTEROSSEI	N								8	1
	DORSAL INTEROSSEI	N								8	1
	LUMBRICALES III & IV	N							(7)	8	1
	ADDUCTORS POLLICIS	N								8	1
	FLEX POLL BREV (DEEP H)	N								8	1

Post. cord

Sensory diagram labels: C2, C3, C4, 5, T1, Supra clavicular, C5, T1, Axillary, Dorsal antebrach cutan, Inter costobrach and med brach, Lat antebrach cutan, Medial antebrach cutan, C6, C7, C8, Superfic br of radial, Ulnar, Median, Post. cord, Inter-costobrach and post brach cutan, Radial.

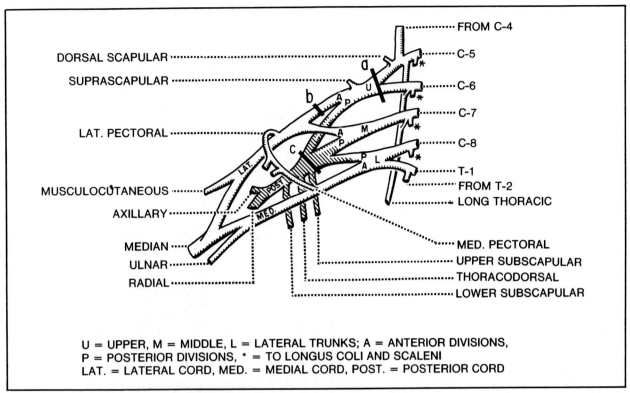

Brachial plexus with possible sites of lesions, a, b, c.

For the cases that follow, only a brief interpretation of the manual muscle test is presented. No significance need be attached to the fact that letter grades were used in some instances and numerals in others. Either system can be used and grades can be translated as indicated on the "Key to Muscle Grading", p. 12

Case 2. A manual muscle test was done prior to surgery and findings indicated the following:

Slight involvement of the muscles supplied by the radial nerve below the level of the innervation to the Triceps.

Moderate involvement of the lateral cord below the level of the lateral pectoral nerve.

Probably complete involvement of the medial cord above the level of the medial pectoral nerve, interrupting C8 and T1 supply.

The fact that the Pectoralis minor, Flexor carpi ulnaris, and the Flexor digitorum profundus III & IV show some strength can mislead one to assume that C8 and T1 are intact. These muscles, along with some of the intrinsic muscles of the hand, receive C7 innervation also, and there may be slight evidence of power in these muscles from C7 without the medial cord being intact.

At the time of surgery, it was found that the medial cord had been interrupted by a bullet above the level of the medial pectoral nerve as had been indicated by the muscle testing.

Case 3. The radial, median, and ulnar nerves all are involved at approximately the same level of the forearm just below the elbow. (Refer to Motor Point Chart, p. 43.) This type of involvement may be the result of pressure from a tourniquet, bandaging, or a cast. While the etiology is not clear-cut in this particular instance, the history does indicate that bandaging may have been a factor.

Case 4. Muscle test findings indicate a probable C5 lesion. The test findings in this case compare very closely with those in a known C5 lesion.

Case 5. The patient, on whom muscle and sensory tests were done six weeks after onset, had fallen through a glass door, injuring the left leg. There was a laceration at the level indicated by the proximal margin of the area of anesthesia as shown on the lower right-hand drawing on the chart. Muscle test findings indicated the following:

SPINAL NERVE AND MUSCLE CHART
NECK, DIAPHRAGM AND UPPER EXTREMITY

Name: *Case #2*   *Left*   Date:

KEY
- D. = Dorsal Prim. Ramus
- V. = Vent. Prim. Ramus
- P.T. = Pre-Trunk
- S.T. = Superior Trunk
- P. = Posterior Cord
- L. = Lateral Cord
- M. = Medial Cord

Cervical Nerves

Muscle	Peripheral Nerve (segments)	Spinal Segment
HEAD & NECK EXTENSORS	Cervical D. 1-8	C1 2 3 4 5 6 7 8 T1
INFRAHYOID MUSCLES	Cervical V. 1-8	C1 2 3
RECTUS CAP. ANT. & LAT.	Cervical V. 1-4	C1 2
LONGUS CAPITUS	Cervical V. 1-4	C1 2 3 (4)
LONGUS COLLI	Cervical V. 1-8	C2 3 4 5 6 (7)
LEVATOR SCAPULAE	Cervical V. 1-4 / Dor. Scap.	C3 4 5
SCALENI (A. M. P.)	Cervical V. 1-8	C3 4 5 6 7 8
STERNOCLEIDOMASTOID	Cervical V. 1-4	(C1) 2 3
TRAPEZIUS (U. M. L.)	Cervical V. 1-4	C2 3 4
DIAPHRAGM	Phrenic 3,4,5	C3 4 5

Brachial Plexus

Grade	Muscle	Peripheral Nerve (segments)	Spinal Segment
N	SERRATUS ANTERIOR	Long. Thor. 5,6,7,(8)	C5 6 7 8
N	RHOMBOIDS MAJ. & MIN.	Dor. Scap. 4,5	C4 5
—	SUBCLAVIUS	N. to Subcl. 5,6	C5 6
N	SUPRASPINATUS	Suprascap. 4,5,6	C4 5 6
N	INFRASPINATUS	Suprascap. 4,5,6	(C4) 5 6
N	SUBSCAPULARIS	U. Subscap. (4),5,6,(7)	C5 6 7
N	LATISSIMUS DORSI	Thoracodor. (5),6,7,8	C6 7 8
N	TERES MAJOR	L. Subscap. 5,6,(7)	C5 6 7
N	PECTORALIS MAJ. (UPPER)	Lat. Pect. 5,6,7	C5 6 7
G	PECTORALIS MAJ. (LOWER)	Med. Pect. (6),7,8,1	C6 7 8 T1
P	PECTORALIS MINOR	Med. Pect. (6),7,8,1	(C6) 7 8 T1

Axillary

Grade	Muscle	Peripheral Nerve (segments)	Spinal Segment
N	TERES MINOR	Axillary 5,6	C5 6
N	DELTOID	Axillary 5,6	C5 6

Musculo-cutan.

Grade	Muscle	Peripheral Nerve (segments)	Spinal Segment
G⁻	CORACOBRACHIALIS	Musculocu. (4),5,6,7	C6 7
F+	BICEPS	Musculocu. (4),5,6,7	C5 6
	BRACHIALIS	Musculocu. (4),5,6,7	C5 6

Radial

Grade	Muscle	Peripheral Nerve (segments)	Spinal Segment
N	TRICEPS	Radial 5,6,7,8,1	C6 7 8 T1
N	ANCONEUS	Radial	C7 8
—	BRACHIALIS (SMALL PART)	Radial	C5 6
G⁻	BRACHIORADIALIS	Radial	C5 6
N	EXT. CARPI RAD. L. & B.	Radial	C5 6 7 8
N	SUPINATOR	Radial	C5 6 (7)
G⁻	EXT. DIGITORUM	Radial	C6 7 8
—	EXT. DIGITI MINIMI	Radial	C6 7 8
F+	EXT. CARPI ULNARIS	Radial	C6 7 8
G⁻	ABD. POLLICIS LONGUS	Radial	C6 7 8
G⁻	EXT. POLLICIS BREVIS	Radial	C6 7 8
G⁻	EXT. POLLICIS LONGUS	Radial	C6 7 8
—	EXT. INDICIS	Radial	C6 7 8

Median (A. Inter.)

Grade	Muscle	Peripheral Nerve (segments)	Spinal Segment
P	PRONATOR TERES	Median 5,6,7,8,1	C6 7
F+	FLEX. CARPI RADIALIS	Median	C6 7 8
—	PALMARIS LONGUS	Median	(C6) 7 8 T1
G⁻	FLEX. DIGIT. SUPERFICIALIS	Median	C7 8 T1
F+	FLEX. DIGIT. PROF. I & II	Median	C7 8 T1
F+	FLEX. POLLICIS LONGUS	Median	(C6) 7 8 T1
P	PRONATOR QUADRATUS	Median	C7 8 T1
F+	ABD. POLLICIS BREVIS	Median	C6 7 8 T1
F⁻	OPPONENS POLLICIS	Median	C6 7 8 T1
P	FLEX. POLL. BREV. (SUP. H.)	Median	C6 7 8 T1
O	LUMBRICALES I & II	Median	(C6) 7 8 T1

Ulnar

Grade	Muscle	Peripheral Nerve (segments)	Spinal Segment
F+	FLEX. CARPI ULNARIS	Ulnar 7,8,1	C7 8 T1
P	FLEX. DIGIT. PROF. III & IV	Ulnar	C7 8 T1
—	PALMARIS BREVIS	Ulnar	(C7) 8 T1
O	ABD. DIGITI MINIMI	Ulnar	(C7) 8 T1
O	OPPONENS DIGITI MINIMI	Ulnar	(C7) 8 T1
O	FLEX. DIGITI MINIMI	Ulnar	(C7) 8 T1
O	PALMAR INTEROSSEI	Ulnar	C8 T1
O	DORSAL INTEROSSEI	Ulnar	C8 T1
O	LUMBRICALES III & IV	Ulnar	(C7) 8 T1
O	ADDUCTORS POLLICIS	Ulnar	C8 T1
O	FLEX. POLL. BREV. (DEEP H.)	Ulnar	C8 T1

SENSORY

Labels (upper figure): C2, C3, C4, 5, T1, Supra-clavicular, Axillary, Dorsal antebrach cutan, Inter-costobrach and med brach, C5, T1, C6, C7, C8, Lat antebrach cutan, Medial antebrach cutan, Superfic. br. of radial, Ulnar, Median; *} med. and Lat. cord*

Labels (lower figure): C2, 3, 4, 5, 6, 7, C8, T1, *Post. cord*, Supra-clavicular, Inter-costobrach. and post. brach. cutan., Axillary, Dorsal antebrach. cutan., Lat. antebrach. cutan., Medial antebrach. cutan., Radial, Ulnar, Median, C6, C7, C8; *med. and Lat. cord*; *medial cord*

Dermatomes redrawn from Keegan and Garrett. Anat. Rec. 102, 409-437, 1948
Cutaneous Distribution of peripheral nerves redrawn from *Gray's Anatomy of the Human Body*, 28th ed

SPINAL NERVE AND MUSCLE CHART
NECK, DIAPHRAGM AND UPPER EXTREMITY

Case # 3

Left — MUSCLE

Name ___ Date ___

SENSORY

C2, C3, C4, 5, T1, Supra-clavicular, C5, T1, Axillary, Dorsal antebrach cutan, Inter-costobrach and med brach, Lat antebrach cutan, Medial antebrach cutan, C6, C7, C8, Superfic br of radial, Ulnar, Median

C2, 3, 4, 5, 6, 7, T1, C8, Supra-clavicular, Inter-costobrach and post brach cutan, Axillary, Dorsal antebrach cutan, Lat antebrach cutan, Medial antebrach cutan, C6, C7, C8, Radial, Ulnar, Median

Dermatomes redrawn from Keegan and Garrett. Anat Rec. 102. 409-437. 1948
Cutaneous Distribution of peripheral nerves redrawn from *Gray's Anatomy of the Human Body*, 28th ed

KEY: D. = Dorsal Prim. Ramus; V. = Vent. Prim. Ramus; P.T. = Pre-Trunk; S.T. = Superior Trunk; P. = Posterior Cord; L. = Lateral Cord; M. = Medial Cord

Grade	MUSCLE	C1	C2	C3	C4	C5	C6	C7	C8	T1
	HEAD & NECK EXTENSORS	1	2	3	4	5	6	7	8	1
	INFRAHYOID MUSCLES	1	2	3						
	RECTUS CAP. ANT. & LAT.	1	2							
	LONGUS CAPITUS	1	2	3	(4)					
	LONGUS COLLI		2	3	4	5	6	(7)		
	LEVATOR SCAPULAE			3	4	5				
	SCALENI (A. M. P.)			3	4	5	6	7	8	
	STERNOCLEIDOMASTOID	(1)	2	3						
	TRAPEZIUS (U. M. L.)		2	3	4					
	DIAPHRAGM			3	4	5				
	SERRATUS ANTERIOR					5	6	7	8	
	RHOMBOIDS MAJ. & MIN.				4	5				
	SUBCLAVIUS					5	6			
	SUPRASPINATUS				4	5	6			
	INFRASPINATUS				(4)	5	6			
	SUBSCAPULARIS					5	6	7		
	LATISSIMUS DORSI						6	7	8	
	TERES MAJOR					5	6	7		
	PECTORALIS MAJ. (UPPER)					5	6	7		
	PECTORALIS MAJ. (LOWER)						6	7	8	1
	PECTORALIS MINOR						(6)	7	8	1
	TERES MINOR					5	6			
	DELTOID					5	6			
—	CORACOBRACHIALIS						6	7		
N	BICEPS					5	6			
N	BRACHIALIS					5	6			
N	TRICEPS						6	7	8	1
N	ANCONEUS							7	8	
—	BRACHIALIS (SMALL PART)					5	6			
N	BRACHIORADIALIS					5	6			
N	EXT. CARPI RAD. L & B.					5	6	7	8	
N	SUPINATOR					5	6	(7)		
F	EXT. DIGITORUM						6	7	8	
O	EXT. DIGITI MINIMI						6	7	8	
O	EXT. CARPI ULNARIS						6	7	8	
O	ABD. POLLICIS LONGUS						6	7	8	
O	EXT. POLLICIS BREVIS						6	7	8	
O	EXT. POLLICIS LONGUS						6	7	8	
O	EXT. INDICIS						6	7	8	
G	PRONATOR TERES						6	7		
N	FLEX CARPI RADIALIS						6	7	8	1
G	PALMARIS LONGUS						(6)	7	8	1
O	FLEX. DIGIT. SUPERFICIALIS							7	8	1
O	FLEX. DIGIT. PROF. I & II							7	8	1
O	FLEX. POLLICIS LONGUS						(6)	7	8	1
O	PRONATOR QUADRATUS							7	8	1
O	ABD. POLLICIS BREVIS						6	7	8	1
O	OPPONENS POLLICIS						6	7	8	1
O	FLEX. POLL. BREV. (SUP. H.)						6	7	8	1
O	LUMBRICALES I & II						(6)	7	8	1
G	FLEX. CARPI ULNARIS							7	8	1
O	FLEX. DIGIT. PROF. III & IV							7	8	1
—	PALMARIS BREVIS							(7)	8	1
O	ABD. DIGITI MINIMI							(7)	8	1
O	OPPONENS DIGITI MINIMI							(7)	8	1
O	FLEX. DIGITI MINIMI							(7)	8	1
O	PALMAR INTEROSSEI								8	1
O	DORSAL INTEROSSEI								8	1
O	LUMBRICALES III & IV							(7)	8	1
O	ADDUCTORS POLLICIS								8	1
O	FLEX. POLL. BREV. (DEEP H.)								8	1

Peripheral nerve columns (headers): Cervical (D.) 1-8 / T.1; Cervical (V.) 1-8; Cervical (V.) 1-4; Phrenic (V.) 3,4,5; Long Thor. (P.T.) 5,6,7,(8); Dor. Scap (P.T.) 4,5; N. to Subcl. (S.T.) 5,6; Suprascap (S.T.) 4,5,6; U. Subscap (P.) (4),5,6,(7); Thoracodor (P.) (5),6,7,8; L. Subscap (P.) 5,6,(7); Lat. Pect. (L.) 5,6,7; Med. Pect. (M.) (6),7,8; Axillary (P.) 5,6; Musculocu. (L.) (4),5,6,7; Radial (P.) 5,6,7,8; Median (L.M.) 5,6,7,8; Ulnar (M.) 7,8

SPINAL NERVE AND MUSCLE CHART
NECK, DIAPHRAGM AND UPPER EXTREMITY

Case # 4

Name   Date

Left MUSCLE

PERIPHERAL NERVES (column headers):

Abbrev	Nerve	Roots
D.1	Cervical T.1	—
V.	Cervical	1-8
V.	Cervical	1-8
V.	Cervical	1-4
V.	Phrenic	3,4,5
P.T.	Long Thor.	5,6,7,(8)
P.T.	Dor. Scap	4,5
S.T.	N. to Subcl.	5,6
S.T.	Suprascap	4,5,6
P.	U. Subscap.	(4),5,6,(7)
P.	Thoracodor.	(5),6,7,8
P.	L. Subscap.	5,6,7
L.	Lat. Pect.	5,6,7
M.	Med. Pect.	(6),7,8,1
P.	Axillary	5,6
L.	Musculocu.	(4),5,6,7
P.	Radial	5,6,7,8,1
L.M.	Median	5,6,7,8,1
M.	Ulnar	7,8,1

Muscle chart — Strength Grade, Muscle, Peripheral Nerve, and Spinal Segment (C1–T1):

Grade	Muscle	Peripheral Nerve	C1	C2	C3	C4	C5	C6	C7	C8	T1
•	HEAD & NECK EXTENSORS	Cervical (D.)	1	2	3	4	5	6	7	8	1
	INFRAHYOID MUSCLES	Cervical	1	2	3						
	RECTUS CAP. ANT. & LAT.	Cervical	1	2							
	LONGUS CAPITUS	Cervical	1	2	3	(4)					
	LONGUS COLLI	Cervical		2	3	4	5	6	(7)		
	LEVATOR SCAPULAE	Cervical / Dor. Scap			3	4	5				
	SCALENI (A. M. P.)	Cervical			3	4	5	6	7	8	
	STERNOCLEIDOMASTOID	Cervical	(1)	2	3						
N?	TRAPEZIUS (U. M. L.)	Cervical		2	3						
*	DIAPHRAGM	Phrenic			3	4	5				
G	SERRATUS ANTERIOR	Long Thor.					5	6	7	8	
P	RHOMBOIDS MAJ & MIN	Dor. Scap				4	5				
—	SUBCLAVIUS	N. to Subcl.					5	6			
?	SUPRASPINATUS	Suprascap				4	5	6			
Tr	INFRASPINATUS	Suprascap				(4)	5	6			
—	SUBSCAPULARIS	U./L. Subscap					5	6	7		
G	LATISSIMUS DORSI	Thoracodor.						6	7	8	
G	TERES MAJOR	L. Subscap					5	6	7		
G-	PECTORALIS MAJ. (UPPER)	Lat. Pect.					5	6	7		
G	PECTORALIS MAJ. (LOWER)	Lat./Med. Pect.						6	7	8	1
G	PECTORALIS MINOR	Med. Pect.						(6)	7	8	1
Tr	TERES MINOR	Axillary					5	6			
Tr	DELTOID	Axillary					5	6			
—	CORACOBRACHIALIS	Musculocu.						6	7		
P+	BICEPS	Musculocu.					5	6			
	BRACHIALIS	Musculocu.					5	6			
G+	TRICEPS	Radial						6	7	8	1
	ANCONEUS	Radial							7	8	
P	BRACHIALIS (SMALL PART)	Radial					5	6			
F-	BRACHIORADIALIS	Radial					5	6			
G	EXT. CARPI RAD. L & B.	Radial					5	6	7	8	
F+	SUPINATOR	Radial					5	6	(7)		
N	EXT. DIGITORUM	Radial						6	7	8	
N	EXT. DIGITI MINIMI	Radial						6	7	8	
N	EXT. CARPI ULNARIS	Radial						6	7	8	
N	ABD. POLLICIS LONGUS	Radial						6	7	8	
N	EXT. POLLICIS BREVIS	Radial						6	7	8	
N	EXT. POLLICIS LONGUS	Radial						6	7	8	
N	EXT. INDICIS	Radial						6	7	8	
N	PRONATOR TERES	Median						6	7		
N	FLEX CARPI RADIALIS	Median						6	7	8	
N	PALMARIS LONGUS	Median						(6)	7	8	1
N	FLEX. DIGIT. SUPERFICIALIS	Median							7	8	1
N	FLEX. DIGIT. PROF. I & II	Median (A. Inter.)							7	8	1
N	FLEX. POLLICIS LONGUS	Median (A. Inter.)						(6)	7	8	1
N	PRONATOR QUADRATUS	Median (A. Inter.)							7	8	1
N	ABD. POLLICIS BREVIS	Median						6	7	8	1
N	OPPONENS POLLICIS	Median						6	7	8	1
N	FLEX. POLL. BREV. (SUP. H.)	Median						6	7	8	1
N	LUMBRICALES I & II	Median						(6)	7	8	1
N	FLEX. CARPI ULNARIS	Ulnar							7	8	1
N	FLEX. DIGIT. PROF. III & IV	Ulnar							7	8	1
—	PALMARIS BREVIS	Ulnar							(7)	8	1
N	ABD. DIGITI MINIMI	Ulnar							(7)	8	1
N	OPPONENS DIGITI MINIMI	Ulnar							(7)	8	1
N	FLEX. DIGITI MINIMI	Ulnar							(7)	8	1
N	PALMAR INTEROSSEI	Ulnar								8	1
N	DORSAL INTEROSSEI	Ulnar								8	1
N	LUMBRICALES III & IV	Ulnar							(7)	8	1
N	ADDUCTORS POLLICIS	Ulnar								8	1
N	FLEX. POLL. BREV. (DEEP H.)	Ulnar								8	1

Left-margin group labels: Cervical Nerves; Brachial Plexus (Pre-Tr, Trunk, P. Cord, M. & L.L.); Axil.; Musculo-cutan.; Radial; Median (A. Inter.); Ulnar.

SENSORY (dermatome figures): labels include C2, C3, C4, 5, T1, Supra-clavicular, Axillary, Dorsal antebrach cutan, Inter-costobrach and med brach, Lat antebrach cutan, Medial antebrach cutan, C5, C6, C7, C8, Superfic br of radial, Ulnar, Median, T1.

Second figure labels: C2, 3, 4, 5, 6, T1, C8, Supra-clavicular, Inter-costobrach and post. brach. cutan., Axillary, Dorsal antebrach. cutan., Lat. antebrach. cutan., Medial antebrach. cutan., C6, C7, C8, Radial, Ulnar, Median.

* Note: Pt's breathing seemed sl. labored. Pt. states breathing has been difficult for about a wk. after onset

SPINAL NERVE AND MUSCLE CHART
TRUNK AND LOWER EXTREMITY

Name: Case #5 Date:

Left

KEY (Peripheral Nerves)
- D = Dorsal Primary Ramus
- V = Ventral Primary Ramus
- A = Anterior Division
- P = Posterior Division

Grade	Muscle	D (T1-12,L1-5,S1-3)	V T1,2,3,4	V T5,6	V T7,8	V T9,10,11,12	Iliohypogastric T12 L1	Ilioinguinal T(12) L1	Lumb. Plex. T(12),L1,2,3,4	Femoral L(1),2,3,4	Obturator L(1),2,3,4	Sup. Glut. L4,5,S1	Inf. Glut. L5,S1,2	Sac. Plex. L4,5,S1,2,3	Sciatic L4,5,S1,2,3	C. Peroneal L4,5,S1,2	Tibial L4,5,S1,2,3	L1	L2	L3	L4	L5	S1	S2	S3
	ERECTOR SPINAE	•																1	2	3	4	5	1	2	3
	SERRATUS POST SUP		•																						
	TRANS. THORACIS		•	•	•																				
	INT. INTERCOSTALS		•	•	•	•																			
	EXT. INTERCOSTALS		•	•	•	•																			
	SUBCOSTALES		•	•	•	•																			
	LEVATOR COSTARUM		•	•	•	•																			
	OBLIQUUS EXT. ABD.			(•)	•	•																			
	RECTUS ABDOMINIS			•	•	•																			
	OBLIQUUS INT. ABD.				•	•	•	(•)										1							
	TRANSVERSUS ABD.				•	•	•	(•)										1							
	SERRATUS POST. INF.					•																			
	QUAD. LUMBORUM								•									1	2	3					
	PSOAS MINOR								•									1	2						
	PSOAS MAJOR								•									1	2	3	4				
	ILIACUS									•								(1)	2	3	4				
	PECTINEUS									•	(•)								2	3	4				
	SARTORIUS									•									2	3	(4)				
	QUADRICEPS									•									2	3	4				
	ADDUCTOR BREVIS										•								2	3	4				
	ADDUCTOR LONGUS										•								2	3	4				
	GRACILIS										•								2	3	4				
	OBTURATOR EXT.										•									3	4				
	ADDUCTOR MAGNUS										•				•				2	3	4	5	1		
	GLUTEUS MEDIUS											•									4	5	1		
	GLUTEUS MINIMUS											•									4	5	1		
	TENSOR FAS. LAT.											•									4	5	1		
	GLUTEUS MAXIMUS												•									5	1	2	
	PIRIFORMIS													•								(5)	1	2	
	GEMELLUS SUP.													•								5	1	2	
	OBTURATOR INT.													•								5	1	2	
	GEMELLUS INF.													•							4	5	1	(2)	
	QUADRATUS FEM.													•							4	5	1	(2)	
	BICEPS (LONG H.)														•							5	1	2	3
	SEMITENDINOSUS														•						4	5	1	2	
	SEMIMEMBRANOSUS														•						4	5	1	2	
	BICEPS (SHORT H.)														•							5	1	2	
30	TIBIALIS ANTERIOR															⊙					4	5	1		
0	EXT. HALL. LONG.															⊙					4	5	1		
0	EXT. DIGIT. LONG.															⊙					4	5	1		
0	PERONEUS TERTIUS															⊙					4	5	1		
0	EXT. DIGIT. BREVIS															⊙					4	5	1		
20	PERONEUS LONGUS															⊙					4	5	1		
20	PERONEUS BREVIS															⊙					4	5	1		
—	PLANTARIS																•				4	5	1	(2)	
100	GASTROCNEMIUS																•						1	2	
—	POPLITEUS																•				4	5	1		
100	SOLEUS																•					5	1	2	
70	TIBIALIS POSTERIOR																⊙				(4)	5	1		
0	FLEX. DIGIT. LONG.																⊙					5	1	(2)	
0	FLEX. HALL. LONG.																⊙					5	1	2	
100	FLEX. DIGIT. BREVIS																•				4	5	1		
—	ABDUCTOR HALL.																•				4	5	1		
100	FLEX. HALL. BREVIS																•				4	5	1		
G	LUMBRICALIS I																•				4	5	1		
—	ABD. DIGITI MIN.																•						1	2	
—	QUAD. PLANTAE																•						1	2	
—	FLEX. DIGITI MIN.																•						1	2	
—	OPP. DIGITI MIN.																•						1	2	
—	ADDUCTORS HALL.																•						1	2	
—	PLANT. INTEROSSEI																•						1	2	
—	DORSAL INTEROSSEI																•						1	2	
G	LUMB. II, III, IV																•				(4)	(5)	1	2	

Muscle groups (left margin): Thoracic Nerves; Lumb. Plexus; Femoral; Obturator (Ant./Post.); Gluteal (Sup./In.); Sacral Plexus; Sciatic (Tibial / P.); Common Peroneal (Deep/Sup.); Tibial (Med. Pl. / Lat. Plant.)

SENSORY (dermatome figures): T2, 4, 6, 8, 10, T12, L1, 2, 3, 4; S1, S2; L5, S1; L4.

Lumbo-inguinal, Ilio-inguinal, Post. divs. of lumbar, Ilio-hypogastric, T12, Lat. fem. cut., Ant. fem. cut., Post. fem. cut., Com. peron., Super. peron., Saph., Deep peron., Sural, Lat. plantar, Med. plantar, Tibial.

(Left leg)

SPINAL NERVE AND MUSCLE CHART
TRUNK AND LOWER EXTREMITY

Name: *Case #6* Date:

Right MUSCLE

KEY
D = Dorsal Primary Ramus
V = Ventral Primary Ramus
A = Anterior Division
P = Posterior Division

PERIPHERAL NERVES (spinal segment labels):
- D. — T1-12, L1-5, S1-3
- V. — T1, 2, 3, 4
- V. — T5, 6
- V. — T7, 8
- V. — T9, 10, 11, 12
- V. — Iliohypogastric T12, L1
- V. — Ilioinguinal T(12), L1
- V. — Lumb. Plex. T(12), L1, 2, 3, 4
- P. — Femoral L(1), 2, 3, 4
- A. — Obturator L(1), 2, 3, 4
- P. — Sup. Glut. L4, 5, S1
- P. — Inf. Glut. L5, S1, 2
- V. — Sac. Plex. L4, 5, S1, 2, 3
- A.P. — Sciatic L4, 5, S1, 2, 3
- P. — C. Peroneal L4, 5, S1, 2
- A. — Tibial L4, 5, S1, 2, 3

Nerve Group	Grade	Muscle	Peripheral Nerve(s) (dots)	Spinal Segment (L1 L2 L3 L4 L5 S1 S2 S3)
		ERECTOR SPINAE	D	1 2 3 4 5 1 2 3
Thoracic Nerves		SERRATUS POST SUP	T1-4	
		TRANS. THORACIS	T1-4, T5-6, T7-8	
		INT. INTERCOSTALS	T1-4, T5-6, T7-8, T9-12	
		EXT. INTERCOSTALS	T1-4, T5-6, T7-8, T9-12	
		SUBCOSTALES	T1-4, T5-6, T7-8, T9-12	
		LEVATOR COSTARUM	T1-4, T5-6, T7-8, T9-12	
		OBLIQUUS EXT. ABD.	(T5-6), T7-8, T9-12	
		RECTUS ABDOMINIS	T5-6, T7-8, T9-12	
		OBLIQUUS INT. ABD.	T7-8, T9-12, Iliohypogastric, (Ilioinguinal)	1
		TRANSVERSUS ABD.	T7-8, T9-12, Iliohypogastric, (Ilioinguinal)	1
		SERRATUS POST. INF.	T9-12	
Lumb. Plexus	—	QUAD. LUMBORUM	Lumb. Plex.	1 2 3
	—	PSOAS MINOR	Lumb. Plex.	1 2
	100	PSOAS MAJOR	Lumb. Plex.	1 2 3 4
Femoral	100	ILIACUS	Femoral	(1) 2 3 4
	—	PECTINEUS	Femoral, (Obturator)	2 3 4
	100	SARTORIUS	Femoral	2 3 (4)
	100	QUADRICEPS	Femoral	2 3 4
Obturator Ant.	100	ADDUCTOR BREVIS	Obturator	2 3 4
		ADDUCTOR LONGUS	Obturator	2 3 4
		GRACILIS	Obturator	2 3 4
Obturator Post.		OBTURATOR EXT.	Obturator	3 4
		ADDUCTOR MAGNUS	Obturator, Sciatic	2 3 4 5 1
Gluteal Sup.	40	GLUTEUS MEDIUS	Sup. Glut.	4 5 1
	40	GLUTEUS MINIMUS	Sup. Glut.	4 5 1
	50	TENSOR FAS. LAT.	Sup. Glut.	4 5 1
Gluteal Inf.	60	GLUTEUS MAXIMUS	Inf. Glut.	5 1 2
Sacral Plexus		PIRIFORMIS	Sac. Plex.	(5) 1 2
		GEMELLUS SUP	Sac. Plex.	5 1 2
	70	OBTURATOR INT.	Sac. Plex.	5 1 2
		GEMELLUS INF.	Sac. Plex.	4 5 1 (2)
		QUADRATUS FEM.	Sac. Plex.	4 5 1 (2)
Sciatic Tibial		BICEPS (LONG H.)	Sciatic	5 1 2 3
	70	SEMITENDINOSUS	Sciatic	4 5 1 2
		SEMIMEMBRANOSUS	Sciatic	4 5 1 2
Sciatic P.		BICEPS (SHORT H.)	Sciatic	5 1 2
Common Peroneal Deep	40	TIBIALIS ANTERIOR	C. Peroneal	4 5 1
	80	EXT. HALL. LONG.	C. Peroneal	4 5 1
	80	EXT. DIGIT. LONG.	C. Peroneal	4 5 1
	80	PERONEUS TERTIUS	C. Peroneal	4 5 1
	80	EXT. DIGIT. BREVIS	C. Peroneal	4 5 1
Common Peroneal Sup.	70	PERONEUS LONGUS	C. Peroneal	4 5 1
	70	PERONEUS BREVIS	C. Peroneal	4 5 1
Tibial	—	PLANTARIS	Tibial	4 5 1 (2)
	100	GASTROCNEMIUS	Tibial	1 2
	—	POPLITEUS	Tibial	4 5 1
	100	SOLEUS	Tibial	5 1 2
	70	TIBIALIS POSTERIOR	Tibial	(4) 5 1
	60	FLEX. DIGIT. LONG.	Tibial	5 1 (2)
	70	FLEX. HALL. LONG.	Tibial	5 1 2
Tibial Med. Pl.	70	FLEX. DIGIT. BREVIS	Tibial	4 5 1
	—	ABDUCTOR HALL.	Tibial	4 5 1
	70	FLEX. HALL. BREVIS	Tibial	4 5 1
	80	LUMBRICALIS I	Tibial	4 5 1
Tibial Lat. Plant.	—	ABD. DIGITI MIN.	Tibial	1 2
	—	QUAD. PLANTAE	Tibial	1 2
	—	FLEX. DIGITI MIN.	Tibial	1 2
	—	OPP. DIGITI MIN.	Tibial	1 2
	—	ADDUCTORS HALL.	Tibial	1 2
	—	PLANT. INTEROSSEI	Tibial	1 2
	—	DORSAL INTEROSSEI	Tibial	1 2
	60	LUMB. II, III, IV	Tibial	(4) (5) 1 2

SENSORY

Dermatomes redrawn from Keegan and Garrett. Anat Rec 102 409-437 1948
Cutaneous Distribution of peripheral nerves redrawn from Gray's Anatomy of the Human Body, 28th ed

Involvement of the nerve branches to the Flexor digitorum longus and Flexor hallucis longus without involvement of the tibial nerve and its terminal branches.

Involvement of the superficial peroneal nerve; and of the deep peroneal nerve, probably below the level of a proximal branch to the tibialis anterior.

The weakness of the Posterior tibial muscle may have been due to trauma of the muscle rather than nerve involvement since it made complete recovery within three and half months after onset. By that time, the Flexor digitorum longus and Flexor hallucis longus had made good recovery, and made complete recovery by the end of six months. Progress was slow and muscle weakness, grading Fair Plus or 60%, remained in all muscles supplied by the deep and superficial peroneal nerves.

Case 6. Muscle test findings indicate a possible L5 lesion. Numerous muscles which receive innervation from L4 were normal in strength, leading to the assumption that L4 was not involved. The patient was able to stand on one foot at a time and rise on toes without any difficulty, hence the normal grade for the Gastrocnemius. With innervation to this muscle from S1 and S2, the grade of normal rules out the probability of a disc below L5.

Subsequent examination by a neurologist confirmed a probable disc lesion, and complete recovery followed bed rest and a period of convalescence.

SPINAL SEGMENT DISTRIBUTION TO NERVES AND MUSCLES

For anatomists and clinicians, the determination of the spinal segment distribution to peripheral nerves and muscles has proven to be a difficult task. The pathway of the spinal nerves is obscured by the intertwining of the nerve fibers as they pass through the nerve plexuses. Since it is almost impossible to trace the course of an individual nerve fiber through the maze of its plexus, information regarding spinal segment distribution has been derived mainly from clinical observation. The use of this empirical method has resulted in a variety of findings regarding the segmental origins of these nerves and the muscles they innervate. An awareness of possible variations is important in the diagnosis and the location of a nerve lesion. To focus attention on the range of variations that exists, the authors have tabulated information on spinal segment distribution to peripheral nerves and muscles from six well-known sources. (See pp. 48–51.)

The symbols used in tabulating the reference material were: A large X to denote a major distribution, a small x to denote a minor distribution, and a parenthetical (x) to denote a possible or infrequent distribution.

For the chart on "Spinal Segment Distribution to Nerves" (p. 58), T2 was included in the brachial plexus by all of the sources but separate columns for T2 were not added to the upper extremity chart because T2 contains only cutaneous sensory fibers. The information in the compilation columns on the two charts (p. 58) has been converted from X's to numbers in the right-hand column. This information regarding spinal segment distribution to nerves appears at the top of the upper and lower extremity "Spinal Nerve and Muscle Chart" under the heading "Peripheral Nerves."

In the authors' compilation of spinal segment supply to muscles as it appears in the last column on the right of the tabulation (see pp. 49 and 51), the x's represent an arithmetical summary. As a general rule, if five or six authorities agreed that a spinal segment was distributed to a given muscle, the nerve supply was indicated by a large X; if three or four agreed, by a small x; if only two agree, by a small x in parenthesis; if mentioned by only one source, it was disregarded. (See Triceps tabulation.)

Triceps

	C6	C7	C8	T1
Gray[11]		X	X	
deJong[18]	X	X	X	(x)
Cunningham[15]	X	X	X	
Spalteholz[17]	x	X	X	(x)
Foerster & Bumke[20]	(x)	X	X	x
Haymaker & Woodhall[19]		X	X	x
Totals	4	6	6	4
Authors' Compilation of Triceps innervation	x	X	X	x
	C6	**C7**	**C8**	T1

When one of the six sources did not specify the spinal segment, agreement by four or five sources was indicated by a large X. This occurred for the Popliteus and some intrinsic muscles of the foot.

MUSCLE	GRAY[11] C1	C2	C3	C4	C5	C6	C7	C8	T1	deJONG[18] C1	C2	C3	C4	C5	C6	C7	C8	T1	CUNNINGHAM[15] C1	C2	C3	C4	C5	C6	C7	C8	T1
HEAD & NECK EXTENSORS	X	X	X	X	X	X	X	X	X	X	X	X	X	X	X	X	X		X	X	X	X	X	X	X	X	X
INFRAHYOID MUSCLES	X	X	X																X	X	X						
RECTUS CAP. ANT. & LAT.	X	X								X	X								X	X							
LONGUS CAPITUS	X	X	X							X	X	X	X						X	X	X	X					
LONGUS COLLI		X	X	X	X	X					X	X	X	X	X					X	X	X	X	X	X	X	
LEVATOR SCAPULAE			X	X	(x)							X	X	X							X	X	X				
SCALENI (A.M.P.)			X	X	X	X	X	X					X	X	X	X	X				X	X	X	X	X	x	
STERNOCLEIDOMASTOID		X	X							(x)	X	X								X							
TRAPEZIUS (U.M.L.)			X	X							(x)	X	X								X	X					
DIAPHRAGM			x	X	x							x	X	x							(x)	X	X				
SERRATUS ANTERIOR					X	X	X							X	X	X	X						X	X	X		
RHOMBOIDS, MAJ. & MINOR					X								X	X							x	X	X				
SUBCLAVIUS					X	X							(x)	X	X								X	X			
SUPRASPINATUS					X								(x)	X	X								X	X			
INFRASPINATUS					X	X								X	X								X	X			
SUBSCAPULARIS					X	X								X	X								X	X			
LATISSIMUS DORSI						X	X	X							X	X	X							X	X	X	
TERES MAJOR					X	X								X	X	(x)							X	X			
PECTORALIS MAJ. (UPPER)					X	X	X	X	X					X	X	X	X	X					X	X			
PECTORALIS MAJ. (LOWER)																									X	X	X
PECTORALIS MINOR							X	X								X	X	X							X	X	X
TERES MINOR					X								(x)	X	X								X	X			
DELTOID					X	X								X	X								X	X			
CORACOBRACHIALIS						X	X								X	X								X	X		
BICEPS					X	X								X	X								X	X			
BRACHIALIS					X	X								X	X								X	X			
TRICEPS							X	X							X	X	X	(x)						X	X	X	
ANCONEUS							X	X								X	X							X	X		
BRACHIALIS (SMALL PART)					X	X								X	X								X	X			
BRACHIORADIALIS					X	X								X	X								X	X			
EXT. CARPI RAD. L. & B.						X	X							(x)	X	X	(x)							X	X	X	
SUPINATOR						X								X	X	X								X	X		
EXT. DIGITORUM						X	X	X							X	X	X							X	X	X	
EXT. DIGITI MINIMI						X	X	X							X	X	X							X	X	X	
EXT. CARPI ULNARIS						X	X	X							X	X	X							X	X	X	
ABD. POLLICIS LONGUS						X	X								X	X	X							X	X	X	
EXT. POLLICIS BREVIS						X	X								X	X	X							X	X	X	
EXT. POLLICIS LONGUS						X	X	X							X	X	X							X	X	X	
EXT. INDICIS						X	X	X							X	X	X							X	X	X	
PRONATOR TERES						X	X								X	X								X	X		
FLEX. CARPI RADIALIS						X	X								X	X	(x)							X	X	X	
PALMARIS LONGUS						X	X									(x)	X								X	X	X
FLEX. DIGIT. SUPERFICIALIS							X	X	X							X	X	X							X	X	X
FLEX. DIGIT. PROF. I & II								X	X							X	X	X							X	X	X
FLEX. POLLICIS LONGUS								X	X							X	X	X							X	X	X
PRONATOR QUADRATUS								X	X							X	X	X							X	X	X
ABD. POLLICIS BREVIS						X	X									X	X									X	X
OPPONENS POLLICIS						X	X									X	X									X	X
FLEX. POLL. BREV. (SUP. H.)						X	X									X	X									X	X
LUMBRICALES I & II						X	X									X	X	X								X	X
FLEX. CARPI ULNARIS							X	X								(x)	X	X								X	X
FLEX. DIGIT. PROF. III & IV							X	X								X	X	X								X	X
PALMARIS BREVIS							X	X								(x)	X	X								X	X
ABD. DIGITI MINIMI								X	X							(x)	X	X								X	X
OPPONENS DIGITI MINIMI								X	X							(x)	X	X								X	X
FLEX. DIGITI MINIMI								X	X							(x)	X	X								X	X
PALMAR INTEROSSEI								X	X								X	X								X	X
DORSAL INTEROSSEI								X	X								X	X								X	X
LUMBRICALES III & IV								X	X								X	X								X	X
ADDUCTORS POLLICIS								X	X								X	X								X	X
FLEX. POLL. BREV. (DEEP H.)								X	X								X	X								X	X

MUSCLES: NECK, DIAPHRAGM AND UPPER EXTREMITY.

SPALTEHOLZ[17]		FOERSTER & BUMKE[20]		HAYMAKER & WOODHALL[19] (modified after Bing)		COMPILATION by Kendalls	
SPINAL SEGMENT		SPINAL SEGMENT		SPINAL SEGMENT		SPINAL SEGMENT	
C1 C2 C3 C4 C5 C6 C7 C8 T1		C1 C2 C3 C4 C5 C6 C7 C8 T1		C1 C2 C3 C4 C5 C6 C7 C8 T1		C1 C2 C3 C4 C5 C6 C7 C8 T1	

MUSCLE	GRAY[11] THORACIC T1,2,3,4	T5,6	T7,8	T9,10,11	T12	GRAY LUMBAR L1	L2	L3	L4	L5	GRAY SACRAL S1	S2	S3	deJONG[18] THORACIC T1,2,3,4	T5,6	T7,8	T9,10,11	T12	deJONG LUMBAR L1	L2	L3	L4	L5	deJONG SACRAL S1	S2	S3	CUNNINGHAM[15] THORACIC T1,2,3,4	T5,6	T7,8	T9,10,11	T12	CUNN LUMBAR L1	L2	L3	L4	L5	CUNN SACRAL S1	S2	S3
ERECTOR SPINAE	X	X	X	X	X	X	X	X	X	X	X	X	X	X	X	X	X	X	X	X	X	X	X	X	X	X	X	X	X	X	X	X	X	X	X	X	X	X	X
SERRATUS POST. SUP.	X													X													X												
TRANS. THORACIS	X	X	X											X	X	X											X	X	X										
INT. INTERCOSTALS	X	X	X	X										X	X	X	X										X	X	X	X									
EXT. INTERCOSTALS	X	X	X	X										X	X	X	X										X	X	X	X									
SUBCOSTALES	X	X	X											(not listed)												X	X	X											
LEVATOR COSTARUM	X	X	X											X													X	X	X											
OBLIQUUS EXT. ABD.		X	X	X											X	X	X											X	X	X									
RECTUS ABDOMINIS			X	X	X											X	X	X										X	X	X									
OBLIQUUS INT. ABD.			X	X	X	X										X	X	X	X									X	X	X	X								
TRANSVERSUS ABD.			X	X	X	X										X	X	X	X									X	X	X	X								
SERRATUS POST. INF.				X	X												X												X										
QUAD. LUMBORUM					X	X								(not listed)																X	X	X	X						
PSOAS MINOR																		X	X												X	(X)							
PSOAS MAJOR							X	X										(X)	X	X										X	X	X	(X)						
ILIACUS							X	X											X	X											X	X	X						
PECTINEUS							X	X	X										X	X	X										X	X						
SARTORIUS							X	X											X	X											X	X						
QUADRICEPS							X	X	X										X	X	X										X	X	X						
ADDUCTOR BREVIS							X	X	X										X	X	X										X	X	X						
ADDUCTOR LONGUS							X	X	X										X	X	X										X	X	X						
GRACILIS							X	X	X										X	X	X										X	X	X						
OBTURATOR EXT							X	X	X										X	X	X										X	X	X						
ADDUCTOR MAGNUS						X	X	X											X	X	X	X									X	X	X	X					
GLUTEUS MEDIUS								X	X	X											X	X	X									X	X	X					
GLUTEUS MINIMUS								X	X	X											X	X	X									X	X	X					
TENSOR FAS. LAT.								X	X	X											X	X	X									X	X	X					
GLUTEUS MAXIMUS									X	X	X											X	X	X										X	X	X		
PIRIFORMIS									X	X	X											X	X	X										X	X	X		
GEMELLUS SUPERIOR								X	X	X											X	X	X										X	X	X		
OBTURATOR INTERNUS								X	X	X											X	X	X										X	X	X		
GEMELLUS INFERIOR								X	X	X											X	X	X										X	X	X		
QUADRATUS FEMORIS								X	X	X											X	X	X										X	X	X		
BICEPS (LONG HEAD)									X	X	X	X										X	X	X	X								X	X	X	X	X
SEMITENDINOSUS									X	X	X											X	X	X									X	X	X	X	
SEMIMEMBRANOSUS									X	X	X										X	X	X										X	X	X	X	
BICEPS (SHORT HEAD)									X	X	X											X	X	X									X	X	X	X	
TIBIALIS ANTERIOR									X	X	X											X	X	X									X	X	X		
EXT. HALL. LONG.									X	X	X											X	X	X									X	X	X		
EXT. DIGIT. LONG.									X	X	X											X	X	X									X	X	X		
PERONEUS TERTIUS									X	X	X											X	X	X									X	X	X		
EXT. DIGIT. BREVIS									X	X	X											X	X	X									X	X	X		
PERONEUS LONGUS									X	X	X											X	X	X									X	X	X		
PERONEUS BREVIS									X	X	X											X	X	X									X	X	X		
PLANTARIS									X	X	X											X	X	X									X	X	X		
GASTROCNEMIUS										X	X	X											X	X	X										X	X	X	
POPLITEUS									X	X	X											X	X	X									X	X	X		
SOLEUS										X	X												X	X										X	X	X	
TIBIALIS POSTERIOR									X	X											X	X	X										X	X	X		
FLEX. DIGIT. LONG.									X	X											X	X	X										X	X	X		
FLEX. HALL. LONG.									X	X	X											X	X	X										X	X	X	
FLEX. DIGIT. BREVIS									X	X	X											X	X	X									X	X	X		
ABDUCTOR HALLUCIS									X	X	X											X	X	X									X	X	X		
FLEX. HALLUCIS BREVIS									X	X	X											X	X	X									X	X	X		
LUMBRICALIS I									X	X												X	X										X	X	X		
ABD. DIGITI MINIMI										X	X												X	X										X	X	X	
QUAD. PLANTAE										X	X												X	X										X	X	X	
FLEX. DIGITI MINIMI										X	X												X	X										X	X	X	
OPP. DIGITI MINIMI										X	X				(not listed)																			X	X	X	
ADDUCTORS HALLUCIS										X	X												X	X	X										X	X	X	
PLANT. INTEROSSEI										X	X												X	X	X										X	X	X	
DORSAL INTEROSSEI										X	X												X	X	X										X	X	X	
LUMBRICALES II, III, IV										X	X												X	X	X	X									X	X	X	

50

MUSCLES: TRUNK AND LOWER EXTREMITIES

SPALTEHOLZ[17]												
SPINAL SEGMENT												
THORACIC					LUMBAR					SACRAL		
T1, 2, 3, 4	T5, 6	T7, 8	T9, 10, 11	T12	L1	L2	L3	L4	L5	S1	S2	S3
x	x	x	x	x	x	x	x	x	x	X	X	X
x												
\	X											
x	x	x	x									
x	x	x	x									
x	x	x	x									
x	x	x	x									
	x	x	x	x								
	\	x	x	x	x							
			x	x	x	(x)						
	x	x	x	(x)	x	x						
			x									
				x	x	x	x					
					x	x						
						x	x	x				
							x	x				
						x	x					
						x	x					
						x	x	x				
						x	x	x				
						x	x					
						x	x	x				
							x	x				
						x	x	x	x	x		
								x	X	x	(x)	
								x	X	x	(x)	
								x	X	x	(x)	
								(x)	x	X	x	
										x	x	x
									x	X	x	(x)
									x	X	x	(x)
								x	X	x		
								x	X	x		
									(x)	x	X	x
								x	x	X	x	
								(x)	x	x		
								x	x	x		
								x	x	x		
								x	x	x		
								x	x	x		
								x	x	x		
								x	x	x		
								x	x	x		
								x	x	x		
									x	x	x	
										x	x	
								x	x	x		
									x	x	x	
									x	x		
									x	x		
									x	x	x	
									x	x		
									x	x		
									x	x	x	
									x	x		
									x	x		
									x	x		
									x	x		
										x	x	
										x	x	
										x	x	
										x	x	
										x	x	

FOERSTER & BUMKE[20]												
SPINAL SEGMENT												
THORACIC					LUMBAR					SACRAL		
T1, 2, 3, 4	T5, 6	T7, 8	T9, 10, 11	T12	L1	L2	L3	L4	L5	S1	S2	S3
X	X	X	X	X	X	X	X	X	X	X	X	X
(not listed)												
(not listed)												
X	X	X	X	X								
X	X	X	X	X								
(not listed)												
(not listed)												
(not listed)												
	x	X	X									
(not listed)												
(not listed)												
(not listed)												
				X	X	X	X					
					(not listed)							
					X	X	x					
					X	X	x					
					x	X	x					
					X	X	x					
						x	X	X				
						X	X	x				
						X	X	x				
					x	X						
								X	x			
						X	X	x				
								x	X	x		
					(not listed)							
								X	x			
									x	X	x	
									x	X	x	
									x	X	x	
									x	X	x	
									x	X	x	
									x	X	x	
									x	X	X	
									x	X	X	
									x	X	X	
										x	X	
								X	x			
									X	X		
									X	X		
									(not listed)			
									(not listed)			
									X	X		
									X	X		
										x	X	
										x	X	
									(not listed)			
										x	X	
								X	x			
										(x)	X	x
										(x)	X	x
											(not listed)	
										(x)	X	x
										(x)	X	x
									(not listed)			

SCHADE[21] & HAYMAKER & WOODHALL[19]												
SPINAL SEGMENT												
THORACIC					LUMBAR					SACRAL		
T1, 2, 3, 4	T5, 6	T7, 8	T9, 10, 11	T12	L1	L2	L3	L4	L5	S1	S2	S3
X	X	X	X	X	X	X	X	X	X	X	X	X
X												
X	X											
X	X	X	X									
X	X	X	X									
X	X	X	X									
X	X	X	X									
	X	X	X	X								
	X	X	X	X								
		\	X	X								
		X	X	X	X							
			X	X								
				X	X	X	X	X				
					x	X	X	x				
					x	X	X	x				
					x	X	X	x				
						X	X	x				
						X	X	x				
						x	X	X				
						X	X	x				
						X	X	x				
						X	X	x				
							X	X				
						x	X	X	X	x		
								X	X	X		
								X	X	X		
								X	X	X		
									x	X	X	
									x	X	x	
								x	X	X	x	
									x	X	X	
								x	X	X	x	
								X	X	X	x	
									x	X	X	
								x	X	X	X	
								x	X	X	X	
									x	X	X	
								X	x			
									x	X	X	
								x	X	X	x	
								x	X	x		
									X	X	X	
									X	X	x	
									X	X	x	
								X	X	X		
										X	X	
								X	X	X		
									x	X	X	
									X	X	X	
									X	X	x	
									X	X	X	
									X	X		
									X	X		
									X	X		
								x	X	X	X	
										X	X	
										X	X	
										X	X	
									(not listed)			
										X	X	
										X	X	
										X	X	
								x	X	X	X	

COMPILATION by Kendalls												
SPINAL SEGMENT												
THORACIC					LUMBAR					SACRAL		
T1, 2, 3, 4	T5, 6	T7, 8	T9, 10, 11	T12	L1	L2	L3	L4	L5	S1	S2	S3
X	X	X	X	X	X	X	X	X	X	X	X	X
X												
X	X	X										
X	X	X	X	(x)								
X	X	X	X	(x)								
X	X	X	X									
X	X	X	X									
	(x)	X	X	x								
	x	X	X	x								
		x	X	X	X							
		X	X	X	X							
			X	x								
				X	X	X	X					
					X	X						
					x	X	X	x				
					(x)	X	X	x				
						X	X	x				
						X	X	(x)				
						X	X	X				
						X	X	X				
						X	X	x				
						X	X	X				
							X	X				
						x	X	X	x	x		
								X	X	X		
								X	X	X		
								X	X	X		
									X	X	X	
									(x)	X	X	
									X	X	X	
									X	X	X	
								X	X	X	(x)	
								X	X	X	(x)	
									x	X	X	x
								x	X	X	X	
								x	X	X	x	
									X	X	X	
								X	X	x		
								x	X	X		
								X	X	X		
								X	X	X		
								x	X	X		
								x	X	X		
								x	X	X		
								x	X	X	(x)	
										X	X	
								X	X	X		
									x	X	X	
								(x)	X	X		
									X	X	(x)	
									X	X	X	
								x	X	X		
								x	X	X		
								x	X	X		
								x	X	X		
										X	X	
										X	X	
										X	X	
										X	X	
										X	X	
										X	X	
										X	X	
								(x)	(x)	X	X	

While the tabulation of data focuses attention on the range of variations that exists among these sources, the arithmetical summary indicates the extent of their agreement. Only in the case of three thumb muscles (Opponens, Abductor brevis, and superficial head of the Flexor brevis) were the six authorities divided in their opinion, resulting in an apparent overstatement of the number of roots of origin. The method used in compiling the information resulted in all segments being listed with small x's, i.e., C6, 7, 8 and Tl, without major emphasis on any one segment.

In most instances, the arithmetical summary preserved the major emphasis on the spinal segments that provide innervation to the muscles. When the summary did not do so, exceptions were made. For example, all sources included C3, 4, 5 innervation to the Diaphragm, but all placed emphasis on C4, so only C4 was given a large X. All sources included the following spinal segment innervations: C5 for the Supinator, C8 for the Extensor carpi radialis longus and brevis, L4 for the Adductor longus, and L4 as a component of the sacral plexus. However, all represented these innervations by a small x indicating a minor distribution, so the compilation preserved the lesser emphasis. All sources included T(12) innervation to the lumbar plexus but all indicated it was a minimal supply so T(12) remained in parenthesis in the compilation.

Innervation was omitted in the compilation, in two instances, because there was a discrepancy between the spinal segment innervation to the muscle and that to the peripheral nerve supplying the muscle. C8 innervation, mentioned by two of the sources as supplying the Subscapularis, was omitted because there was no indication that the upper or lower subscapular nerve received C8 innervation. Likewise, C(4), included by two sources for the Teres minor, was omitted since there was no indication that the axillary nerve received C4 innervation. In two other instances, innervation was added in the compilation. C6 and C7 were added to the medial pectoral nerve. Above the communicating loop, the medial pectoral nerve is composed of C8 and T1 fibers. Below the loop, C7 and possible C6 fibers (branching from the lateral pectoral nerve) join the medial pectoral nerve. While the medial cord of the plexus is derived from C8 and T1, the ulnar nerve, as the terminal branch of this cord, is listed as having a C7 component in addition to C8 and T1. Numerous anatomists record this information and some[5, 11, 12] indicate that the C7 component is variable.

The authors have modified the compilation in regard to spinal segment distribution to the upper and lower portions of the Pectoralis major. In the muscle sections of the books used as references for the compilation, only one text[7] divided the Pectoralis major muscle into upper and lower portions and listed the spinal segment innervation to each portion. However, Gray,[11] in the description of the lateral and medial pectoral nerves, indicated that the lateral pectoral supplies the more cranial part of the muscle while the medial pectoral, joined by two or three branches from the lateral, supplies the more caudal part. In addition, several other references[13-15] differentiate the peripheral supply to the upper and lower parts. In certain lesions of the cervical region of the spinal cord, it has been noted, clinically, that the upper part of the Pectoralis major has had normal strength while the lower part has been paralyzed. This observation suggests that there is a difference in spinal segment innervation to the parts of the muscle. On the basis of the above information, the compilation distinguishes between the upper and lower parts of the Pectoralis major in regard to spinal segment distribution.

The results of the compilation, pp. 49 and 51, have been used in the spinal segment column on the nerve-muscle charts. The X's have been converted to numbers which indicate the specific spinal segment. The major emphasis designated by large X's has been obtained by use of numbers in bold type; the minor emphasis, by numbers that are not bold; and the possible or infrequent innervation, by numbers in parentheses.

NERVE PLEXUSES

The word "plexus" comes from the Latin word which means a braid. A nerve plexus results from the dividing, reuniting, and intertwining of nerves into a complex network. When describing the origins, components, and terminal branches of a plexus, the words "nerves", "roots", and "cord" are used with dual meanings. There are spinal nerves and peripheral nerves, roots of the spinal nerves and roots of the plexus, the spinal cord and cords of the plexus. To avoid confusion, appropriate modifying words are used in the descriptions below.

The *spinal cord* lies within the vertebral column extending from the first cervical vertebra to the level of the second lumbar vertebra. Each of the thirty-one pairs of *spinal nerves* arise from the spinal cord by two *spinal nerve roots*. The *ventral root* composed of motor fibers, and the *dorsal root* composed of sensory fibers, unite at the intervertebral foramen to form the spinal nerve. (See p. 54 *top*.) A *spinal segment* is the part of the spinal cord that gives rise to each pair of spinal nerves. Each spinal nerve contains motor and sensory fibers from a single spinal segment.

Shortly after the spinal nerve exits through the foramen, it divides into a *dorsal primary ramus* and a *ventral primary ramus*. The dorsal rami are directed posteriorly, and the sensory and motor fibers innervate the skin and extensor muscles of the neck and trunk. The ventral rami, except those in the thoracic region, contain the nerve fibers that become part of the plexuses. (Four plexuses are described below and illustrated on pp. 54–57.) Emerging from the plexuses at various levels or as terminal branches are the *peripheral nerves*. As a result of the interchange of fibers within the plexus, peripheral nerves contain fibers from at least two, and, in some instances, as many as five, spinal segments.

The *cervical plexus* is formed by the ventral primary rami of spinal nerves C1 through C4 with a small contribution from C5. The peripheral nerves arising from it innervate most of the anterior and lateral muscles of the neck, and supply sensory fibers to part of the head and much of the neck. (See p. 54, *bottom*.)

The *brachial plexus* arises just lateral to the Scalenus anterior muscle. The ventral rami of C5, 6, 7, and 8, and the greater part of T1, plus a communicating loop from C4 to C5 and one from T2 (sensory) to T1 form, successively, the roots, trunks, divisions, cords, and branches of the plexus. Although variations occur in the structural patterns of the components of the plexus, the most common arrangement is illustrated on p. 55, *top*. A. (On the Spinal Nerve and Muscle Chart, Upper Extremity, p. 35, the roots of the plexus are labeled Pretrunk.)

Ventral rami containing C5 and C6 fibers unite to form the *superior* (upper) *trunk*, C7 fibers form the *middle trunk*, and C8 and T1 fibers unite to form the *inferior* (lower) *trunk*. Next the trunks separate into *anterior* and *posterior divisions*. The anterior divisions from the superior and middle trunks, composed of C5, 6 and 7 fibers, unite to form the *lateral cord*; the anterior division from the inferior trunk, composed of C8 and T1 fibers, forms the *medial cord*; the posterior divisions from all three trunks, composed of fibers from C5 through C8 (but not T1), unite to form the *posterior cord*.

The cords then divide and reunite into *branches* which become *peripheral nerves*. The posterior cord branches into the axillary and radial nerves. The medial cord, after receiving a branch from the lateral cord, terminates as the ulnar nerve. One branch of the lateral cord becomes the musculocutaneous nerve; the other branch unites with one from the medial cord to form the median nerve. Other peripheral nerves exit directly from various components of the plexus and some directly from the ventral rami. (See left-hand column and top of Spinal Nerve and Muscle Chart, p. 35.)

The anterior divisions, the lateral and medial cords, and the peripheral nerves arising from them innervate anterior or flexor muscles of the upper extremity. The posterior division, the posterior cord, and the peripheral nerves arising from them innervate the posterior or extensor muscles of the upper extremity. (See p. 55, *bottom*.)

The *lumbar plexus* is formed by the ventral primary rami of L1, 2, 3, and a part of L4 with, frequently, a small contribution from T12. Within the substance of the Psoas major muscle, the rami branch into anterior and posterior divisions. (See p. 56, *top*.) Peripheral nerves from the anterior divisions innervate adductor muscles on the medial side of the thigh; those from the posterior divisions innervate hip flexors and knee extensors on the anterior aspect of the thigh. (See p. 56, *bottom*.)

The *sacral plexus* arises from the smaller part of the ventral primary ramus of L4 and from the entire ventral rami of L5, S1, 2, 3. The L4 and L5 ventral rami unite to form the lumbosacral trunk which enters the pelvic cavity. There it is joined by the ventral rami of S1, 2, and 3, forming the plexus which then branches into anterior and posterior divisions. (See p. 57, *top*.) The anterior divisions and the peripheral nerves arising from them innervate the posterior aspect of the thigh and leg, and the plantar surface of the foot. The posterior divisions and the peripheral nerves arising from them innervate the abductor muscles on the lateral side of the thigh, a hip extensor muscle posteriorly, and the extensor (dorsiflexor) muscles of the ankle and toes anteriorly. (See p. 57, *bottom*.)

Cervical Plexus

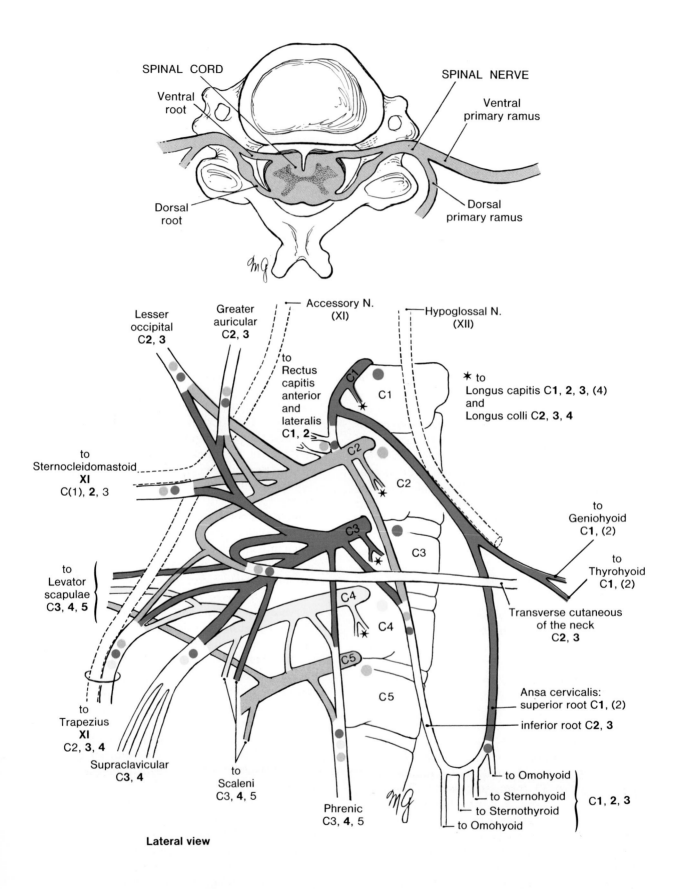

SPINAL CORD

Ventral root

Dorsal root

SPINAL NERVE

Ventral primary ramus

Dorsal primary ramus

Lesser occipital
C**2**, **3**

Greater auricular
C**2**, **3**

Accessory N. (XI)

Hypoglossal N. (XII)

to Rectus capitis anterior and lateralis C**1**, **2**

C1

C1

* to
Longus capitis C**1**, **2**, **3**, (4)
and
Longus colli C**2**, **3**, **4**

to Sternocleidomastoid
XI
C(1), **2**, 3

C2

C2

to Geniohyoid
C**1**, (2)

to Levator scapulae
C**3**, **4**, **5**

C3

C3

to Thyrohyoid
C**1**, (2)

C4

C4

Transverse cutaneous of the neck
C**2**, **3**

C5

C5

Ansa cervicalis:
superior root C**1**, (2)

inferior root C**2**, 3

to Trapezius
XI
C2, **3**, **4**

Supraclavicular
C**3**, **4**

to Scaleni
C3, **4**, 5

Phrenic
C3, **4**, 5

to Omohyoid
to Sternohyoid
to Sternothyroid
to Omohyoid

C**1**, **2**, **3**

Lateral view

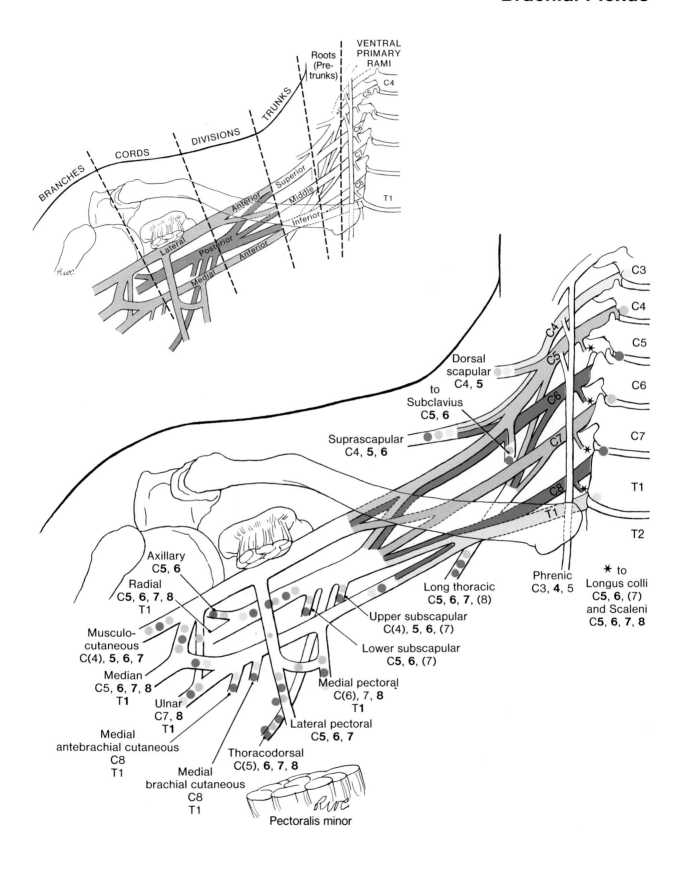

Roots (Pre-trunks)

VENTRAL PRIMARY RAMI

BRANCHES — CORDS — DIVISIONS — TRUNKS

C4
C5
C6
C7
C8
T1
T1

Superior
Middle
Inferior

Anterior
Posterior
Anterior

Lateral
Medial

C3
C4
C5
C6
C7
T1
T2

C4
C5
C6
C7
C8
T1

Dorsal scapular
C4, **5**

to Subclavius
C**5**, **6**

Suprascapular
C4, **5**, **6**

Axillary
C**5**, **6**

Radial
C**5**, **6**, **7**, **8**
T1

Musculo-cutaneous
C(4), **5**, **6**, **7**

Median
C5, **6**, **7**, **8**
T**1**

Ulnar
C7, **8**
T**1**

Medial antebrachial cutaneous
C8
T1

Medial brachial cutaneous
C8
T1

Thoracodorsal
C(5), **6**, **7**, **8**

Lateral pectoral
C**5**, **6**, **7**

Medial pectoral
C(6), **7**, **8**
T**1**

Lower subscapular
C**5**, **6**, (**7**)

Upper subscapular
C(4), **5**, **6**, (**7**)

Long thoracic
C**5**, **6**, **7**, (**8**)

Phrenic
C3, **4**, 5

✳ to Longus colli
C**5**, **6**, (**7**)
and Scaleni
C**5**, **6**, **7**, **8**

Pectoralis minor

Lumbar Plexus

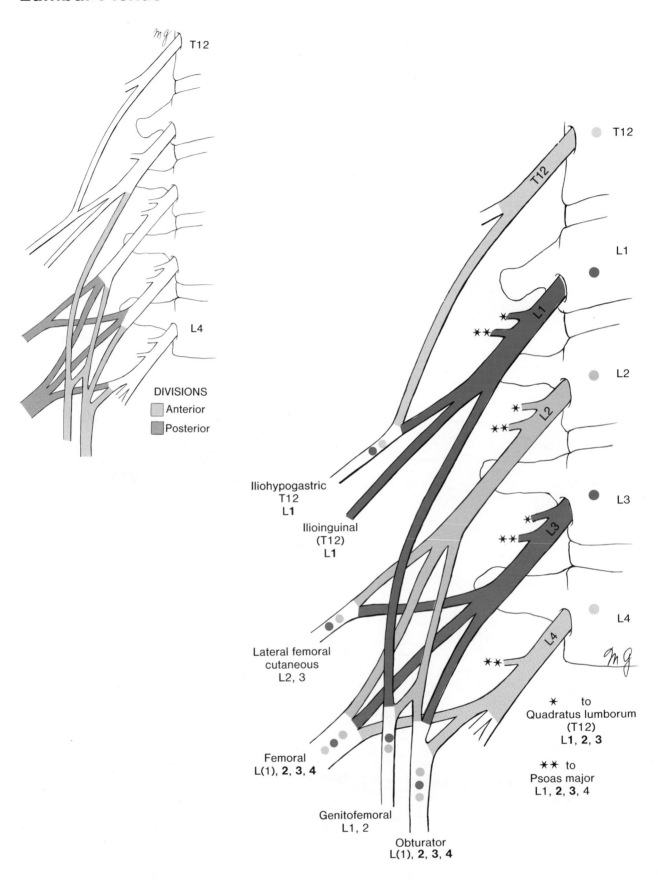

T12

L4

DIVISIONS
Anterior
Posterior

T12

L1

L2

L3

L4

Iliohypogastric
T12
L1

Ilioinguinal
(T12)
L1

Lateral femoral
cutaneous
L2, 3

Femoral
L(1), **2, 3, 4**

Genitofemoral
L1, 2

Obturator
L(1), **2, 3, 4**

★   to
Quadratus lumborum
(T12)
L1, 2, 3

★★ to
Psoas major
L1, **2, 3,** 4

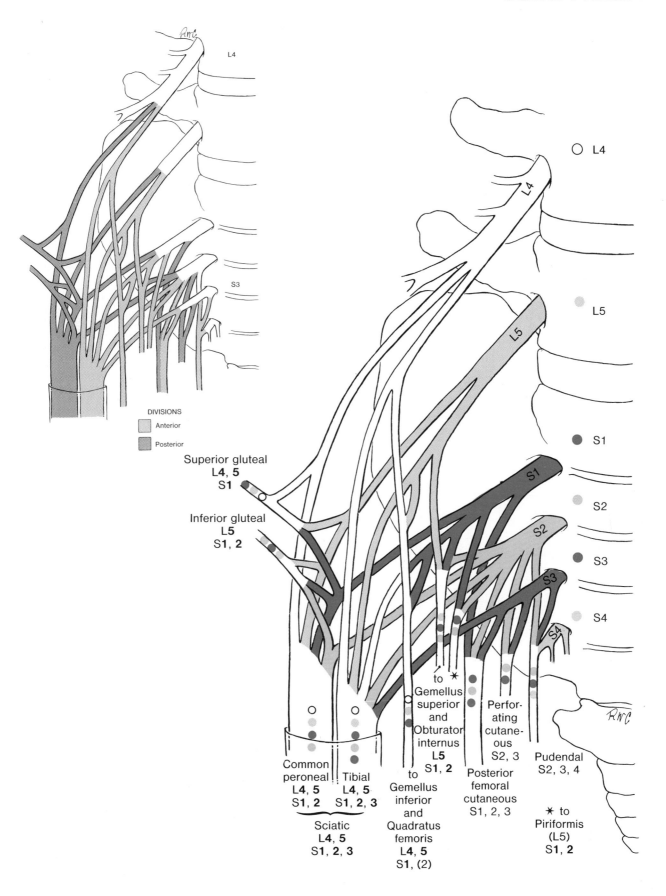

DIVISIONS

Anterior

Posterior

Superior gluteal
**L4, 5
S**1

Inferior gluteal
L5
S1, **2**

to
Gemellus
superior
and
Obturator
internus
L5
S1, **2**

Perfor-
ating
cutane-
ous
S2, **3**

Pudendal
S2, 3, 4

Common
peroneal
**L4, 5
S**1, **2**

Tibial
**L4, 5
S**1, 2, **3**

Sciatic
**L4, 5
S**1, 2, **3**

to
Gemellus
inferior
and
Quadratus
femoris
**L4, 5
S**1, (2)

Posterior
femoral
cutaneous
S1, **2**, 3

✱ to
Piriformis
(L5)
S**1, 2**

L4

L5

S1

S2

S3

S4

Spinal Segment Distribution to Nerves: Neck, Diaphragm, and Upper Extremity

NERVE	CUNNINGHAM[15]	GRAY[11]	MORRIS[16]	SPALTEHOLZ[17]	DE JONG[18]	HAYMAKER & WOODHALL[19]	COMPILATION BY KENDALLS — SPINAL SEGMENTS USED FOR SPINAL NERVE & MUSCLE CHART
Cervical Plex.							Cervical Plex. C 1,2,3,4
Brachial Plex.							Brach. Plex. C(4),5,6,7,8,T1
Phrenic							Phrenic C3,4,5
Long Thoracic							Long. Thor. C5,6,7,(8)
Dorsal Scapular							Dor. Scap. C4,5
N. to Subclavius							N. to Subclavius C5,6
Suprascapular							Suprascap. C4,5,6
Upp. Subscap.							U. Subscap. C(4),5,6,(7)
Thoracodorsal							Thoracodor. C(5),6,7,8
Low. Subscap.							L. Subscap. C5,6,(7)
Lat. Pectoral							Lat. Pect. C5,6,7
Med. Pectoral							Med. Pect. C(6),7,8,T1
Axillary							Axillary C5,6
Musculocutan.							Musculocutan. C(4),5,6,7
Radial							Radial C5,6,7,8,T1
Median							Median C5,6,7,8,T1
Ulnar							Ulnar C7,8,T1

Spinal Segment Distribution to Nerves: Trunk and Lower Extremity

NERVE	CUNNINGHAM[15]	GRAY[11]	MORRIS[16]	SPALTEHOLZ[17]	DE JONG[18]	HAYMAKER & WOODHALL[19]	COMPILATION BY KENDALLS — SPINAL SEGMENT USED FOR SPINAL NERVE AND MUSCLE CHART
Iliohypogastric							Iliohypogastric T12, L1
Ilioinguinal							Ilioinguinal T(12), L1
Lumb. Plex.							Lumb. Plex. T(12),L1,2,3,4
Femoral							Femoral L(1),2,3,4
Obturator							Obturator L(1),2,3,4
Sup. Gluteal							Sup. Glut. L4,5,S1
Inf. Gluteal							Inf. Glut. L5,S1,2
Sac. Plex.							Sac. Plex. L4,5,S1,2,3
Sciatic							Sciatic L4,5,S1,2,3
Common Peroneal							C. Peroneal L4,5,S1,2
Tibial							Tibial L4,5,S1,2,3

chapter 4

Upper Extremity and Scapular Muscles

Tests for muscles of:
  Fingers and thumb
  Wrist
  Forearm
  Elbow
  Shoulder
  Scapula
Charts of muscles grouped according to action:
  Scapular muscles

Scapular and shoulder joint muscles
Upper extremity muscles
Charts for recording muscle examinations:
  Chart for analysis of muscle imbalance, upper extremity
  Upper extremity muscle chart
(For upper extremity, nerve-muscle chart, see Chapter 3)

Adductor Pollicis

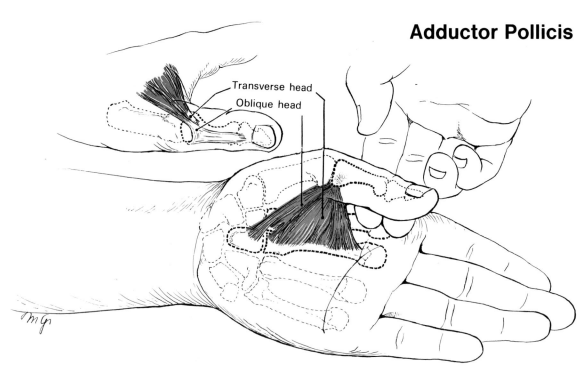

Transverse head
Oblique head

Origin of oblique fibers: Capitate bone, and bases of second and third metacarpal bones.

Origin of transverse fibers: Palmar surface of third metacarpal bone.

Insertion: Transverse head into ulnar side of base of proximal phalanx of thumb, and oblique head into extensor expansion.

Action: Adducts the carpometacarpal joint, and adducts and assists in flexion of the metacarpophalangeal joints, so that the thumb moves toward the plane of the palm. Aids in opposition of the thumb toward the little finger. By virtue of the attachment of the oblique fibers into the extensor expansion, may assist in extending the interphalangeal joint.

Nerve: Ulnar, C8, T1.

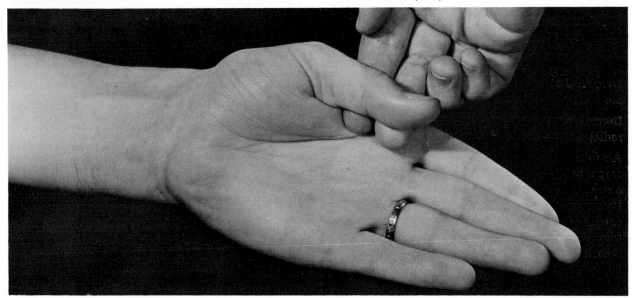

Patient: Sitting or supine.

Fixation: The hand may be stabilized by the examiner, or rest on the table for support (as illustrated).

Test: Adduction of the thumb toward the palm.

Pressure: Against the medial surface of the thumb, in the direction of abduction ventralward.

Weakness: Results in inability to clench the thumb firmly over the closed fist.

Contracture: Adduction deformity of the thumb.

Note: A test that is frequently used to determine the strength of the Adductor pollicis is the ability to hold a piece of paper between the thumb and second metacarpal. In an individual with a well-developed Adductor, the bulk of the muscle itself prevents close approximation of these parts.

Abductor Pollicis Brevis

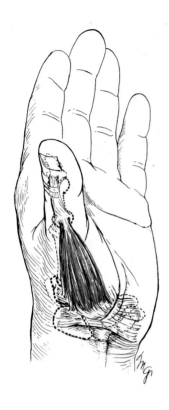

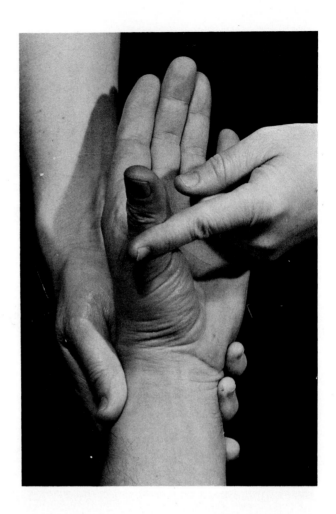

Origin: Flexor retinaculum, tubercle of trapezium bone, and tubercle of scaphoid bone.

Insertion: Base of proximal phalanx of thumb, radial side, and extensor expansion.

Action: Abducts the carpometacarpal and metacarpophalangeal joints of the thumb in a ventral direction perpendicular to the plane of the palm. By virtue of its attachment into the dorsal extensor expansion, extends the interphalangeal joint of the thumb. Assists in opposition, and may assist in flexion and medial rotation of the metacarpophalangeal joint.

Nerve: Median, C6, 7, 8, T1.

Patient: Sitting or supine.

Fixation: The examiner stabilizes the hand.

Test: Abduction of the thumb ventralward from the palm.

Pressure: Against the proximal phalanx, in the direction of adduction toward the palm.

Weakness: Decreases the ability to abduct the thumb, making it difficult to grasp a large object. An adduction deformity of the thumb may result from marked weakness.

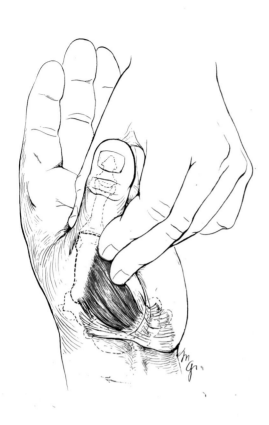

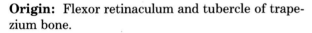

Origin: Flexor retinaculum and tubercle of trapezium bone.

Insertion: Entire length of first metacarpal bone, radial side.

Action: Opposes (i.e., flexes and abducts with slight medial rotation) the carpometacarpal joint of the thumb, placing the thumb in a position so that, by flexion of the metacarpophalangeal joint, it can oppose the fingers. (See page 25 about opposition of the thumb.)

Nerve: Median, C6, 7, 8, T1.

Note: The attachment of the Palmaris longus and the Opponens pollicis to the flexor retinaculum accounts for the fact that the Palmaris longus contracts during the Opponens test.

Patient: Sitting or supine.

Fixation: The examiner stabilizes the hand.

Test: Flexion, abduction, and slight medial rotation of the metacarpal bone so that the thumbnail shows in palmar view.

Pressure: Against the metacarpal bone, in the direction of extension and adduction with lateral rotation.

Weakness: Results in a flattening of the thenar eminence, extension and adduction of the first metacarpal, and difficulty in holding a pencil for writing, or in grasping objects firmly between the thumb and fingers.

Flexor Pollicis Longus

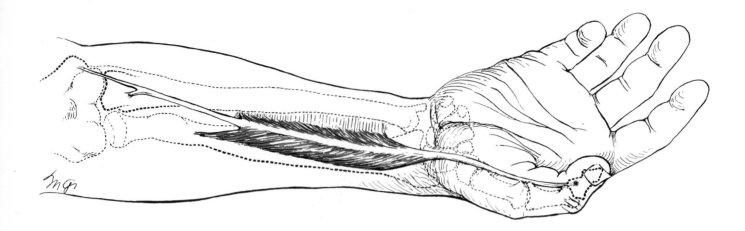

Origin: Anterior surface of body of radius below tuberosity, interosseus membrane, medial border of coronoid process of ulna, and/or medial epicondyle of humerus.

Insertion: Base of distal phalanx of thumb, palmar surface.

Action: Flexes the interphalangeal joint of the thumb, assists in flexion of the metacarpophalangeal and carpometacarpal joints, and may assist in flexion of the wrist.

Nerve: Median, C(6), 7, **8**, T1.

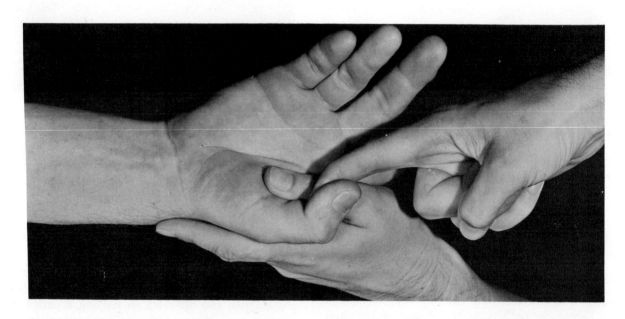

Patient: Sitting or supine.

Fixation: The hand may rest on the table for support (as illustrated) with the examiner stabilizing the metacarpal bone and proximal phalanx of the thumb in extension; or the hand may rest on its ulnar side with the wrist in slight extension and the examiner stabilizing the proximal phalanx of the thumb in extension.

Test: Flexion of the interphalangeal joint of the thumb.

Pressure: Against the palmar surface of the distal phalanx, in the direction of extension.

Weakness: Decreases the ability to flex the distal phalanx, making it difficult to hold a pencil for writing or to pick up minute objects between the thumb and fingers. Marked weakness may result in a hyperextension deformity of the interphalangeal joint.

Contracture: Flexion deformity of interphalangeal joint.

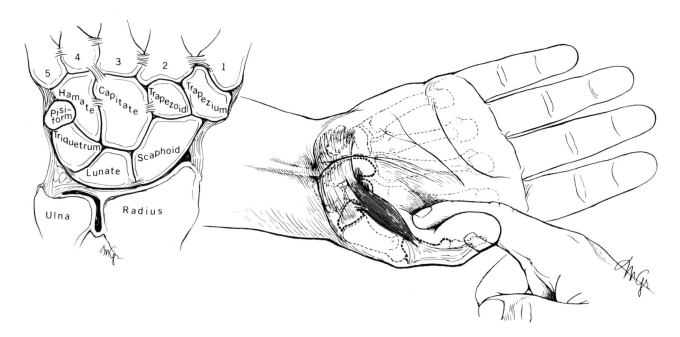

Origin of superficial head: Flexor retinaculum and trapezium bone.

Origin of deep head: Trapezoid and capitate bones.

Insertion: Base of proximal phalanx of thumb, radial side, and extensor expansion.

Action: Flexes the metacarpophalangeal and carpometacarpal joints of the thumb, and assists in opposition of the thumb toward the little finger. By virtue of its attachment into the dorsal extensor expansion, may extend the interphalangeal joint.

Nerve to superficial head: Median, C6, 7, 8, T1.

Nerve to deep head: Ulnar, C8, T1.

Patient: Sitting or supine.

Fixation: The examiner stabilizes the hand.

Test: Flexion of the metacarpophalangeal joint of the thumb without flexion of the interphalangeal joint.

Pressure: Against the palmar surface of the proximal phalanx in the direction of extension.

Weakness: Decreases the ability to flex the metacarpophalangeal joint making it difficult to grip objects firmly between the thumb and fingers. Marked weakness may result in a hyperextension deformity of the metacarpophalangeal joint.

Contracture: Flexion deformity of the metacarpophalangeal joint.

Extensor Pollicis Longus

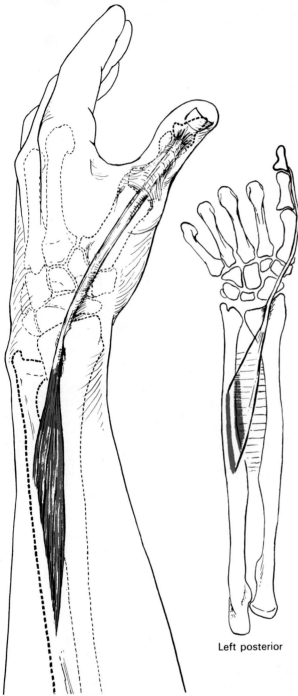

Left posterior

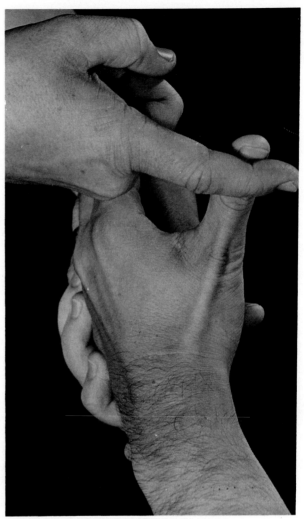

Origin: Middle one-third of posterior surface of ulna distal to origin of Abductor pollicis longus, and interosseus membrane.

Insertion: Base of distal phalanx of thumb, dorsal surface.

Action: Extends the interphalangeal joint and assists in extension of the metacarpophalangeal and carpometacarpal joints of the thumb. Assists in abduction and extension of the wrist.

Nerve: Radial, C6, **7, 8**.

Patient: Sitting or supine.

Fixation: The examiner stabilizes the hand and gives counterpressure against the palmar surface of the first metacarpal and proximal phalanx.

Test: Extension of the interphalangeal joint of the thumb.

Pressure: Against the dorsal surface of the interphalangeal joint of the thumb, in the direction of flexion.

Weakness: Decreases the ability to extend the interphalangeal joint, and may result in a flexion deformity of that joint.

Note: In a radial nerve lesion, the interphalangeal joint of the thumb may be extended by the action of the Abductor pollicis brevis, the Flexor pollicis brevis, the oblique fibers of the Adductor pollicis, or by the first Palmar interosseus by virtue of their insertions into the extensor expansion of the thumb. Interphalangeal joint extension in an otherwise complete radial nerve lesion should not be interpreted as regeneration or partial involvement if this one action only is observed.

Extensor Pollicis Brevis

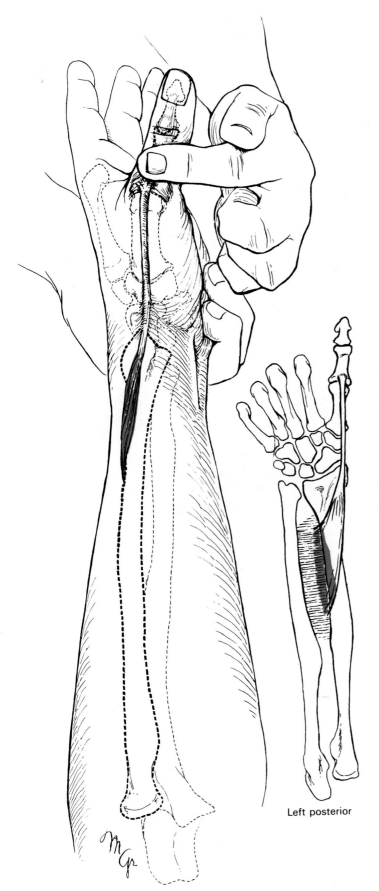

Left posterior

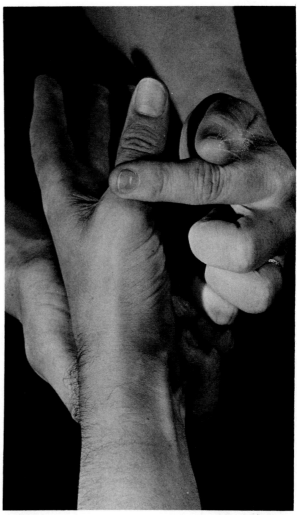

Origin: Posterior surface of body of radius distal to origin of Abductor pollicis longus, and interosseus membrane.

Insertion: Base of proximal phalanx of thumb, dorsal surface.

Action: Extends the metacarpophalangeal joint of the thumb, extends and abducts the carpometacarpal joint, and assists in abduction (radial deviation) of the wrist.

Nerve: Radial, C6, **7, 8**.

Patient: Sitting or supine.

Fixation: The examiner stabilizes the wrist.

Test: Extension of the metacarpophalangeal joint of the thumb.

Pressure: Against the dorsal surface of the proximal phalanx, in the direction of flexion.

Weakness: Decreases the ability to extend the metacarpophalangeal joint, and may result in a position of flexion of that joint.

Abductor Pollicis Longus

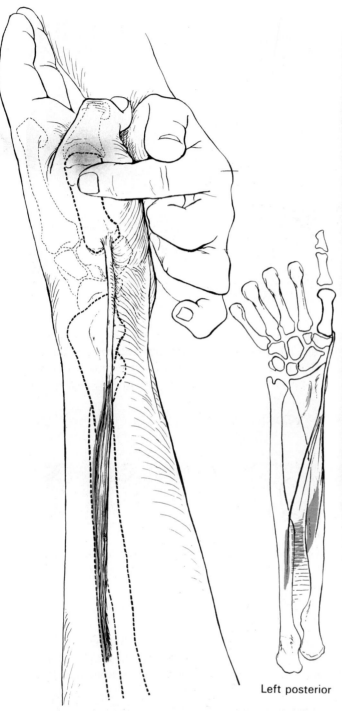

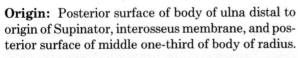

Left posterior

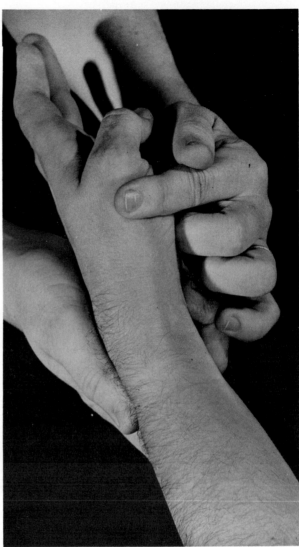

Origin: Posterior surface of body of ulna distal to origin of Supinator, interosseus membrane, and posterior surface of middle one-third of body of radius.

Insertion: Base of first metacarpal bone, radial side.

Action: Abducts and extends the carpometacarpal joint of the thumb; abducts (radial deviation) and assists in flexing the wrist.

Nerve: Radial, C6, **7**, **8**.

Patient: Sitting or supine.

Fixation: The examiner stabilizes the wrist.

Test: Abduction and slight extension of the first metacarpal bone.

Pressure: Against the lateral surface of the distal end of the first metacarpal, in the direction of adduction and flexion.

Weakness: Decreases the ability to abduct the first metacarpal, and the ability to abduct the wrist.

Contracture: Abducted and slightly extended position of the first metacarpal with slight radial deviation of the hand.

Abductor Digiti Minimi

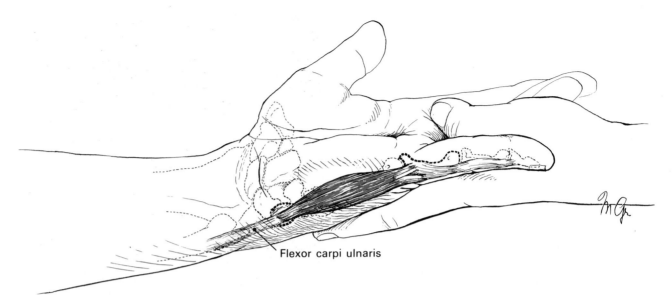

Flexor carpi ulnaris

Origin: Tendon of Flexor carpi ulnaris and pisiform bone.

Insertion: By two slips, one into base of proximal phalanx of little finger, ulnar side; the second, into the ulnar border of the extensor expansion.

Action: Abducts, assists in opposition, and may assist in flexion of the metacarpophalangeal joint of the little finger; and, by virtue of insertion into the extensor expansion, may assist in extension of interphalangeal joints.

Nerve: Ulnar, C(7), **8**, T1.

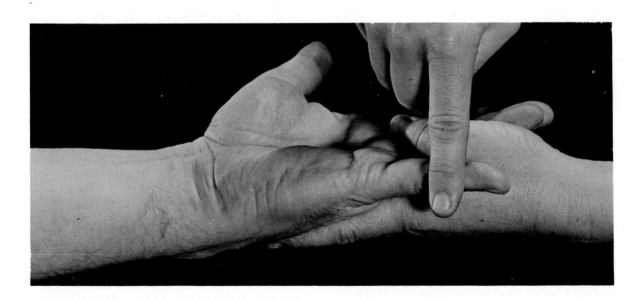

Patient: Sitting or supine.

Fixation: The hand may be stabilized by the examiner or rest on the table for support.

Test: Abduction of the little finger.

Pressure: Against the ulnar side of the little finger, in the direction of adduction toward the midline of the hand.

Weakness: Decreases the ability to abduct the little finger, and results in adduction of this digit.

Note: One should be consistent in the placing of pressure in all finger abduction and adduction tests. Pressure against the sides of the middle phalanges seems most appropriate for all these tests.

Opponens Digiti Minimi

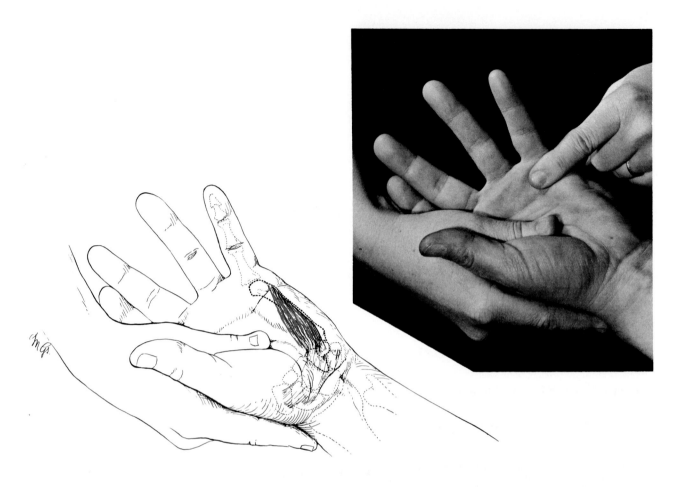

Origin: Hook of hamate bone, and flexor retinaculum.

Insertion: Entire length of fifth metacarpal bone, ulnar side.

Action: Opposes (i.e., flexes with slight rotation) the carpometacarpal joint of the little finger, lifting the ulnar border of the hand into a position so that the metacarpophalangeal flexors can oppose the little finger to the thumb. (See page 25.) Helps to cup the palm of the hand.

Nerve: Ulnar, C(7), **8**, T1.

Patient: Sitting or supine.

Fixation: The hand may be stabilized by the examiner or rest on the table for support. The first metacarpal is held firmly by the examiner.

Test: Opposition of the fifth metacarpal toward the first.

Pressure: Against the palmar surface along the fifth metacarpal in the direction of flattening the palm of the hand. The one-finger pressure was used in the illustration to avoid obscuring the belly of the muscle, but usually the thumb is used to apply pressure along the fifth metacarpal.

Weakness: Results in a flattening of the palm and makes it difficult, if not impossible, to oppose the little finger to the thumb.

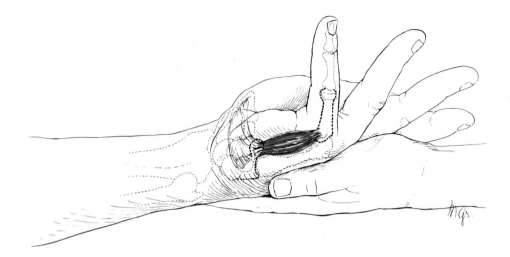

Origin: Hook of hamate bone, and flexor retinaculum.

Insertion: Base of proximal phalanx of little finger, ulnar side.

Action: Flexes the metacarpophalangeal joint of the little finger and assists in opposition of the little finger toward the thumb.

Nerve: Ulnar, C(7), **8**, T1.

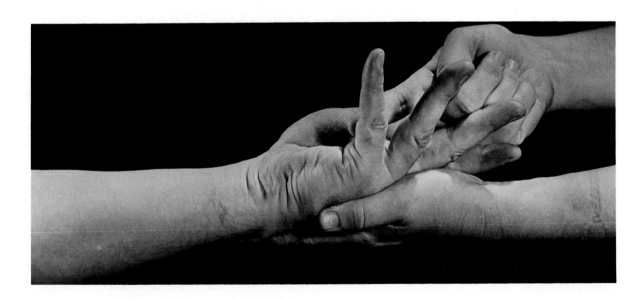

Patient: Sitting or supine.

Fixation: The hand may rest on the table for support, or be stabilized by the examiner.

Test: Flexion of the metacarpophalangeal joint with interphalangeal joints extended.

Pressure: Against the palmar surface of the proximal phalanx, in the direction of extension.

Weakness: Decreases the ability to flex and oppose the little finger.

Dorsal Interossei

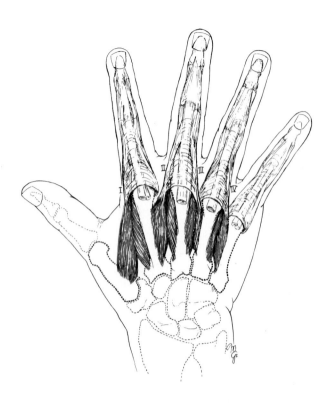

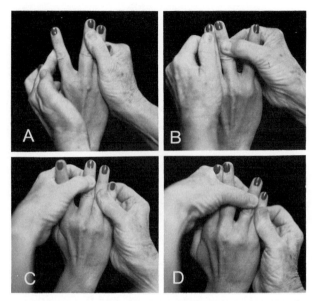

Origins

First, lateral head: Proximal one-half of ulnar border of first metacarpal bone.

First, medial head: Radial border of second metacarpal bone.

Second, third, and fourth: Adjacent sides of metacarpal bones in each interspace.

Insertions

Into extensor expansion and to base of proximal phalanx as follows:

First: Radial side of index finger, chiefly to base of proximal phalanx.

Second: Radial side of middle finger.

Third: Ulnar side of middle finger, chiefly into extensor expansion.

Fourth: Ulnar side of ring finger.

Action: Abduct the index, middle, and ring fingers from the axial line through the third digit. Assist in flexion of metacarpophalangeal joints and extension of interphalangeal joints of the same fingers. The first assists in adduction of the thumb.

Nerve: Ulnar, C8, T1.

Patient: Sitting or supine.

Fixation: In general, stabilization of adjacent digits, to give fixation of digit toward which finger is moved, and to prevent assistance from digit on other side.

Test and **pressure or traction** (against middle phalanx):

First (A), abduction of index finger toward thumb. Pressure against radial side of index finger in direction of middle finger.

Second (B), abduction of middle finger toward index finger. Hold middle finger and pull in direction of ring finger.

Third (C), abduction of middle finger toward ring finger. Hold middle finger and pull in direction of index finger.

Fourth (D), abduction of ring finger toward little finger. Hold ring finger and pull in direction of middle finger.

Weakness: Decreases the ability to abduct the index, middle, and ring fingers. Decreases the strength of extension of the interphalangeal joints and flexion of the metacarpophalangeal joints of index, middle, and ring fingers.

Contracture: Abduction of index and ring fingers. (See p. 74.)

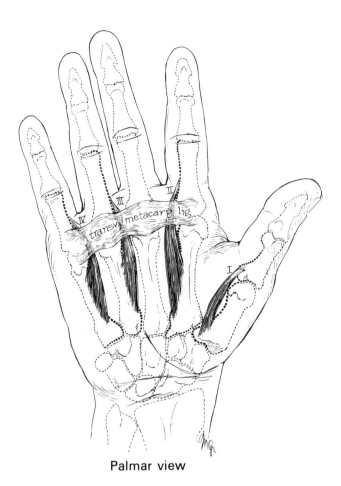

Palmar view

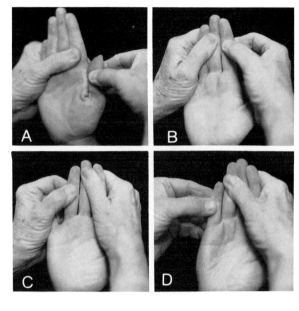

Origins

First: Base of first metacarpal bone, ulnar side.

Second: Length of second metacarpal bone, ulnar side.

Third: Length of fourth metacarpal bone, radial side.

Fourth: Length of fifth metacarpal bone, radial side.

Insertions

Chiefly, into the extensor expansion of the respective digit, with possible attachment to base of proximal phalanx as follows:

First: Ulnar side of thumb.

Second: Ulnar side of index finger.

Third: Radial side of ring finger.

Fourth: Radial side of little finger.

Action: Adduct the thumb, index, ring, and little finger toward the axial line through the third digit. Assist in flexion of metacarpophalangeal joints, and extension of interphalangeal joints of the three fingers.

Nerve: Ulnar, C8, T1.

Patient: Sitting or supine.

Fixation: In general, stabilization of adjacent digits, to give fixation of digit toward which finger is moved, and to prevent assistance from digit on other side.

Test and **traction** (against middle phalanx):

First (A), adduction of thumb toward index finger (acting with Adductor pollicis and first Dorsal interosseus). Hold thumb and pull in radial direction.

Second (B), adduction of index finger toward middle finger. Hold index finger and pull in direction of thumb.

Third (C), adduction of ring finger toward middle finger. Hold ring finger and pull in direction of little finger.

Fourth (D), adduction of little finger toward ring finger. Hold little finger and pull in ulnar direction.

Weakness: Decreases ability to adduct thumb, index, ring, and little fingers. Decreases strength in flexion of metacarpophalangeal joints and extension of interphalangeal joints of the index, ring, and little fingers.

Contracture: Fingers held in adduction. May result from wearing a cast with fingers in adduction.

Lumbricales

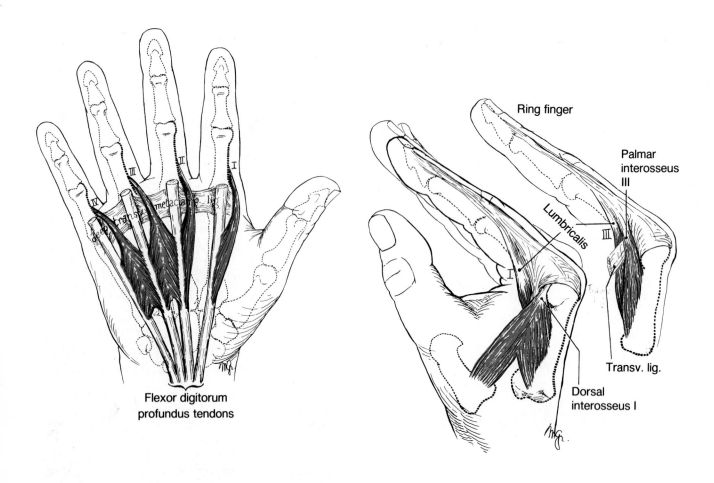

Flexor digitorum
profundus tendons

Ring finger

Palmar
interosseus
III

Lumbricalis

III

I

Transv. lig.

Dorsal
interosseus I

Origin of first and second: Radial surface of Flexor profundus tendons of index and middle fingers, respectively.

Origin of third: Adjacent sides of Flexor profundus tendons of middle and ring fingers.

Origin of fourth: Adjacent sides of Flexor profundus tendons of ring and little fingers.

Insertion: Into the radial border of the extensor expansion on the dorsum of the respective digits.

Action: Extend the interphalangeal joints and simultaneously flex the metacarpophalangeal joints of the second through fifth digits. The Lumbricales also extend the interphalangeal joints when the metacarpophalangeal joints are extended. As the fingers are extended at all joints, the Flexor digitorum profundus tendons offer a form of passive resistance to this movement. Since the Lumbricales are attached to the Flexor profundus tendons, they can diminish this resistive tension by contracting and pulling these tendons distally, and this release of tension decreases the contractile force needed by the muscles which extend the finger joints.

Nerve to Lumbricales I, II: Median, C(6), 7, **8**, **T1**.

Nerve to Lumbricales III, IV: Ulnar, C(7), **8**, **T1**.

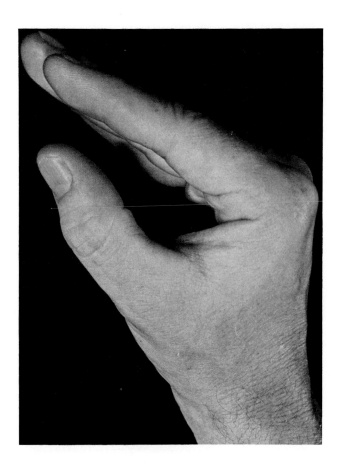

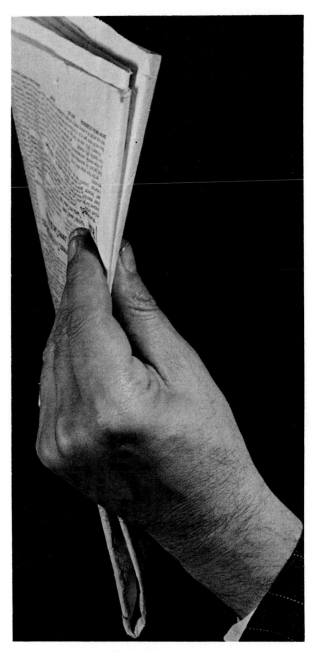

LUMBRICALES AND INTEROSSEI

Patient: Sitting or supine.

Fixation: The examiner stabilizes the wrist in slight extension if there is any weakness of wrist muscles.

Test: Extension of interphalangeal joints with simultaneous flexion of metacarpophalangeal joints.

Pressure: First, against the dorsal surface of the middle and distal phalanges, in the direction of flexion; and second, against the palmar surface of the proximal phalanges, in the direction of extension. Pressure is not illustrated in the photograph because it is applied in two stages, not both simultaneously.

Weakness: Results in claw-hand deformity.

Contracture: Metacarpophalangeal joint flexion with interphalangeal joint extension.

Shortness: See p. 76.

An important function of the Lumbricales and Interossei is illustrated by the above photograph. With marked weakness or paralysis of these muscles an individual cannot hold a newspaper or a book upright in one hand. The complaint by a patient that he cannot hold a newspaper in one hand may be the clue to this type of weakness.

Tests for Length of Lumbricales and Interossei

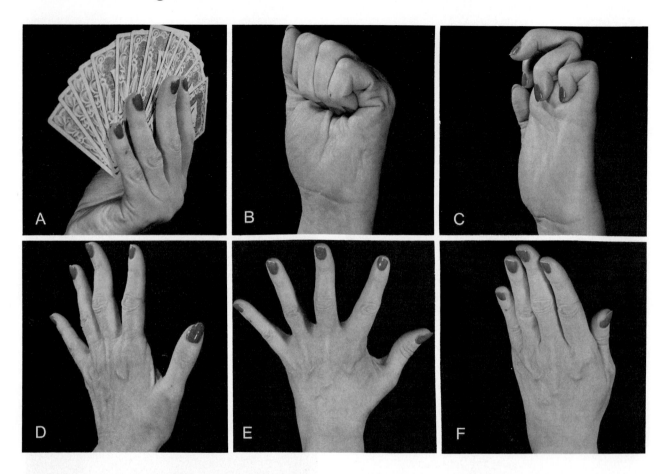

SHORTNESS OF THE INTRINSIC MUSCLES OF THE HAND

The case, illustrated by the above photographs, was that of a middle aged woman whose complaint was that her middle finger occasionally pained rather severely and there was a constant tight, "drawing" feeling along the sides of this finger. She did not feel that the pain was actually in the joints of the finger. A medical checkup had revealed no arthritis. This person was an avid card player, and the condition was present in the left hand which was the hand in which she held her cards.

Illustration A shows the position of the subject's hand in holding a hand of cards. This position is one of strong Lumbrical and Interosseus action. Just as in holding a newspaper, the middle finger is the one which strongly opposes the thumb.

On testing for length of the intrinsic muscles there was evidence of shortness chiefly in the muscles in the middle finger.

The patient could close the fingers to make a fist as in B. This was possible although some shortness existed in the Lumbricales and Interossei because the muscles were being elongated over the interphalangeal joints only, not over the metacarpophalangeal joints.

The patient could extend the fingers as in D. This was possible because the muscles were being elongated over the metacarpophalangeal joints only, not over the interphalangeal joints. (The distal phalanx of the middle finger, which opposes the thumb in holding the cards, is in slight hyperextension.)

When attempting to close the hand into a claw-hand position C, the shortness became apparent. In closing the fingers into this position, the Lumbricales and Interossei must elongate over all three joints at the same time. The middle finger shows the greatest limitation. The ring finger shows slight limitation which is demonstrated by the lack of distal joint flexion as well as by decreased hyperextension of the metacarpophalangeal joint.

The fact that the fingers could be spread apart as in E, and closed sideways as in F, suggests that the shortness may have been in the Lumbricales more than in the Interosser.

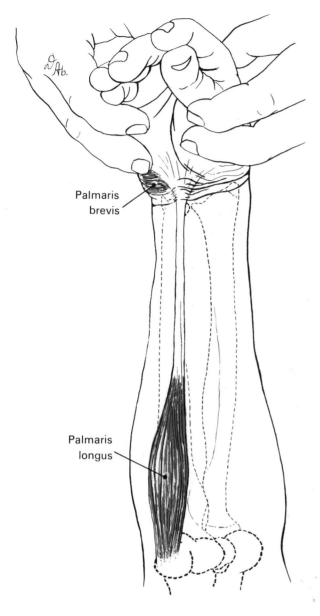

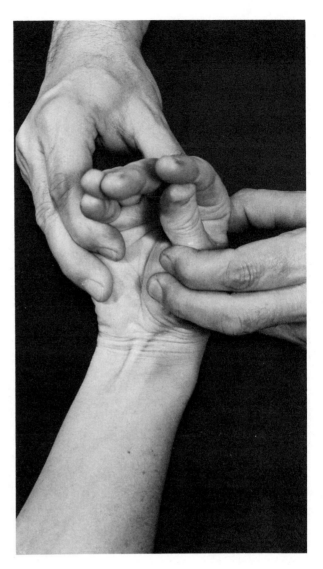

PALMARIS LONGUS

Origin: Common flexor tendon from medial epicondyle of humerus, and deep antebrachial fascia.

Insertion: Flexor retinaculum, and palmar aponeurosis.

Action: Tenses the palmar fascia, flexes the wrist, and may assist in flexion of the elbow.

Nerve: Median, C(6), **7, 8,** T1.

PALMARIS BREVIS

Origin: Ulnar border of palmar aponeurosis and palmar surface of flexor retinaculum.

Insertion: Skin on ulnar border of hand.

Action: Corrugates the skin on ulnar side of hand.

Nerve: Ulnar, C(7), **8,** T1.

PALMARIS LONGUS

Patient: Sitting or supine.

Fixation: The forearm rests on the table for support, in a position of supination.

Test: Tensing of the palmar fascia by strongly cupping the palm of the hand, and flexion of the wrist.

Pressure: Pressure is applied against the thenar and hypothenar eminences in the direction of flattening the palm of the hand, and against the hand in the direction of extending the wrist.

Weakness: Decreases the ability to cup the palm of the hand. Strength of wrist flexion is diminished.

Extensor Indicis, Extensor Digiti Minimi, and Extensor Digitorum

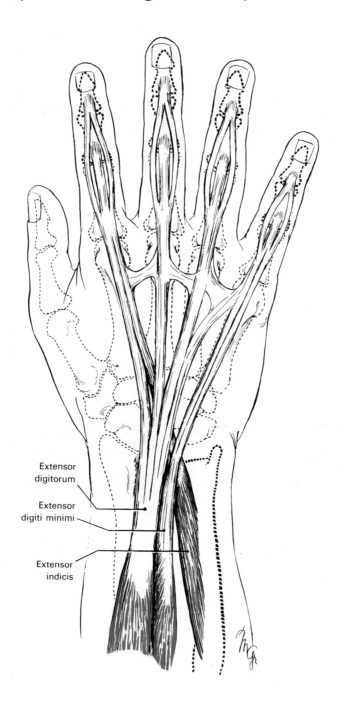

Extensor digitorum

Extensor digiti minimi

Extensor indicis

EXTENSOR INDICIS

Origin: Posterior surface of body of ulna distal to origin of Extensor pollicis longus, and interosseus membrane.

Insertion: Into extensor expansion of index finger with Extensor digitorum longus tendon.

Action: Extends the metacarpophalangeal joint and, in conjunction with the Lumbricalis and Interossei, extends the interphalangeal joints of the index finger. May assist in adduction of the index finger.

Nerve: Radial, C6, **7**, **8**.

EXTENSOR DIGITI MINIMI

Origin: Common extensor tendon from lateral epicondyle of humerus, and deep antebrachial fascia.

Insertion: Into extensor expansion of little finger with Extensor digitorum tendon.

Action: Extends the metacarpophalangeal joint and, in conjunction with the Lumbricalis and Interosseus, extends the interphalangeal joints of the little finger. Assists in abduction of the little finger.

Nerve: Radial, C6, **7**, **8**.

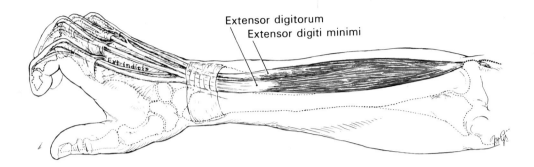

Extensor digitorum
Extensor digiti minimi

Ext. indicis

Origin: Common extensor tendon from lateral epicondyle of humerus, and deep antebrachial fascia.

Insertion: By four tendons, each penetrating a membraneous expansion on the dorsum of the second to fifth digits and dividing over the proximal phalanx into a medial and two lateral bands. The medial band inserts into the base of the middle phalanx while the lateral bands reunite over the middle phalanx and insert into the base of the distal phalanx. (See p. 78.)

Action: Extends the metacarpophalangeal joints and, in conjunction with the Lumbricales and Interossei, extends the interphalangeal joints of the second through fifth digits. Assists in abduction of the index, ring, and little fingers; and assists in extension and abduction of the wrist.

Nerve: Radial, C6, 7, 8.

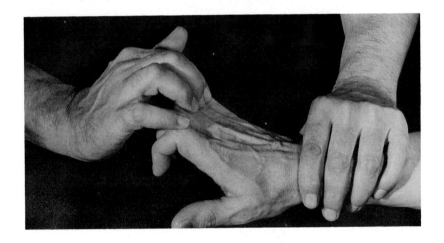

Patient: Sitting or supine.

Fixation: The examiner stabilizes the wrist, avoiding full extension.

Test: Extension of the metacarpophalangeal joints of the second through fifth digits, with interphalangeal joints relaxed.

Pressure: Against the dorsal surfaces of the proximal phalanges, in the direction of flexion.

Weakness: Decreases the ability to extend the metacarpophalangeal joints of the second through fifth digits, and may result in a position of flexion of these joints. Strength of wrist extension is diminished.

Contracture: Hyperextension deformity of the metacarpophalangeal joints.

Shortness: Hyperextension of the metacarpophalangeal joints if the wrist is flexed, or extension of the wrist if the metacarpophalangeal joints are flexed.

Flexor Digitorum Superficialis

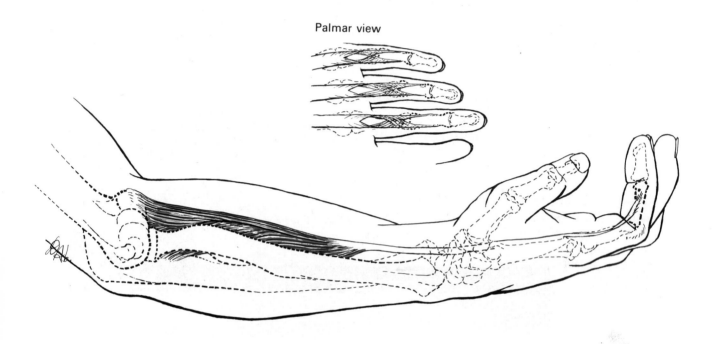

Palmar view

Origin of humeral head: Common flexor tendon from medial epicondyle of humerus, ulnar collateral ligament of elbow joint, and deep antebrachial fascia.

Origin of ulnar head: Medial side of coronoid process.

Origin of radial head: Oblique line of radius.

Insertion: By four tendons into sides of middle phalanges of second through fifth digits.

Action: Flexes the proximal interphalangeal joints of second through fifth digits, assists in flexion of the metacarpophalangeal joints and in flexion of the wrist.

Nerve: Median, C7, **8**, T1.

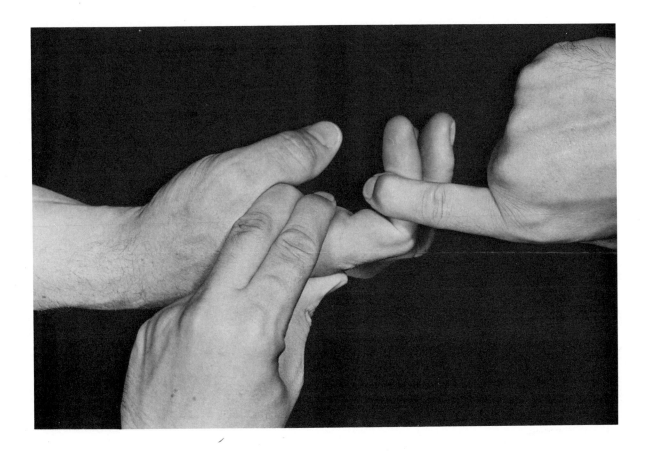

Patient: Sitting or supine.

Fixation: The examiner stabilizes the metacarpophalangeal joint, with the wrist in neutral position or in slight extension.

Test: Flexion of the proximal interphalangeal joint with the distal interphalangeal joint extended, of the second, third, fourth and fifth digits (see Note). Each finger is tested as illustrated for the index finger.

Pressure: Against the palmar surface of the middle phalanx, in the direction of extension.

Weakness: Decreases the strength of the grip and of wrist flexion. Interferes with finger function in such activities as typing, piano playing, and playing some stringed instruments in which the proximal interphalangeal joint is flexed while the distal joint is extended. Weakness causes loss of joint stability at the proximal interphalangeal joints of the fingers so that in finger extension these joints hyperextend.

Contracture: Flexion deformity of the middle phalanges of the fingers.

Shortness: Flexion of the middle phalanges of the fingers if the wrist is extended, or flexion of the wrist if the fingers are extended.

Note: It appears to be the exception rather than the rule to obtain isolated Flexor superficialis action in the fifth digit.

Flexor Digitorum Profundus

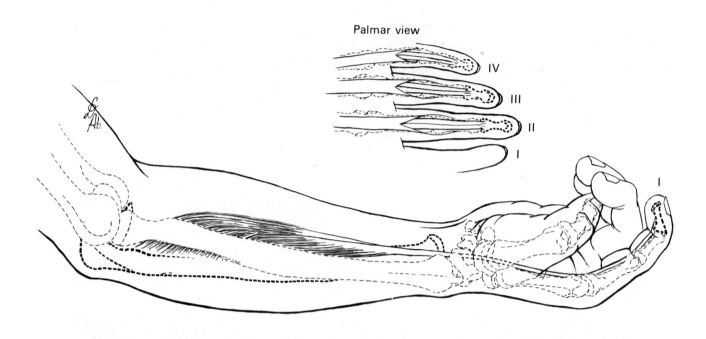

Palmar view

Origin: Anterior and medial surfaces of proximal three-fourths of ulna, interosseus membrane, and deep antebrachial fascia.

Insertion: By four tendons into bases of distal phalanges, anterior surface.

Action: Flexes distal interphalangeal joints of index, middle, ring, and little fingers, and assists in flexion of proximal interphalangeal and metacarpophalangeal joints; may assist in flexion of the wrist.

Nerve to profundus I and II: Median, C7 **8**, **T1**.

Nerve to profundus III and IV: Ulnar, C7, **8**, **T1**.

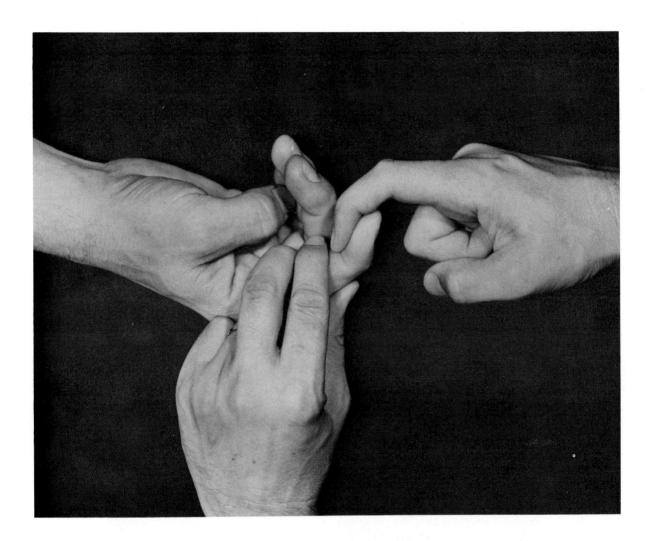

Patient: Sitting or supine.

Fixation: With the wrist in slight extension, the examiner stabilizes the proximal and middle phalanges.

Test: Flexion of the distal interphalangeal joint of the second, third, fourth, and fifth digits. Each finger is tested as illustrated above for the index finger.

Pressure: Against the palmar surface of the distal phalanx, in the direction of extension.

Weakness: Decreases the ability to flex the distal joints of the fingers in direct proportion to the extent of weakness since this is the only muscle that flexes the distal interphalangeal joints. Flexion strength of the proximal interphalangeal, metacarpophalangeal, and wrist joints may be diminished.

Contracture: Flexion deformity of the distal phalanges of the fingers.

Shortness: Flexion of the fingers if the wrist is extended, or flexion of the wrist if the fingers are extended.

Flexor Carpi Radialis

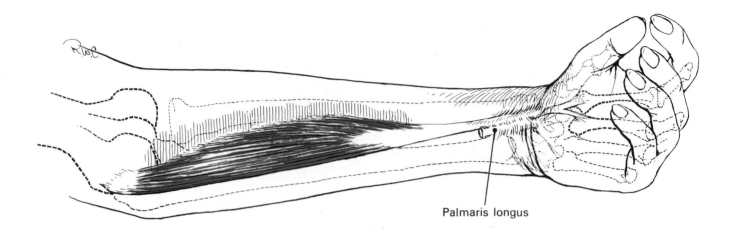

Palmaris longus

Origin: Common flexor tendon from medial epicondyle of humerus, and deep antebrachial fascia. (Fascia indicated by parallel lines.)

Insertion: Base of second metacarpal bone and a slip to base of third metacarpal bone.

Action: Flexes and abducts the wrist, and may assist in pronation of the forearm and flexion of the elbow.

Nerve: Median, C**6**, **7**, 8.

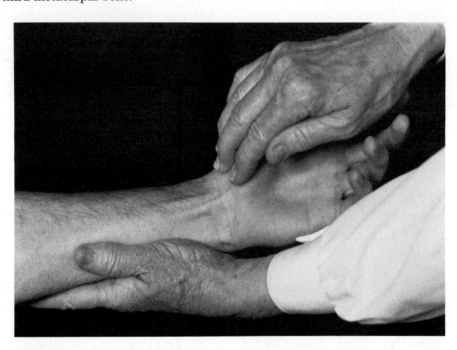

Patient: Sitting or supine.

Fixation: The forearm is in slightly less than full supination and rests on the table for support or is supported by the examiner.

Test: Flexion of the wrist toward the radial side. (See Note under Flexor carpi ulnaris.)

Pressure: Against the thenar eminence in the direction of extension toward the ulnar side.

Weakness: Decreases the strength of wrist flexion, and pronation strength may be diminished. Allows an ulnar deviation of the hand.

Contracture: Wrist flexion toward the radial side.

Note: The Palmaris longus cannot be ruled out in this test.

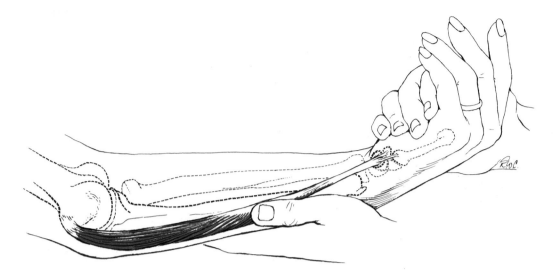

Origin of the humeral head: Common flexor tendon from medial epicondyle of humerus.

Origin of ulnar head: By aponeurosis from the medial margin of olecranon, proximal two-thirds of posterial border of ulna, and from the deep antebrachial fascia.

Insertion: Pisiform bone and, by ligaments, to hamate and fifth metacarpal bones.

Action: Flexes and adducts the wrist, and may assist in flexion of the elbow.

Nerve: Ulnar, C7, **8**, T1.

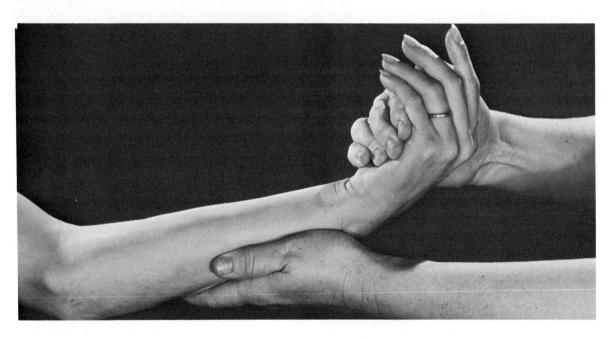

Patient: Sitting or supine.

Fixation: The forearm is in full supination and rests on the table for support or is supported by the examiner.

Test: Flexion of the wrist toward the ulnar side.

Pressure: Against the hypothenar eminence in the direction of extension toward the radial side.

Weakness: Decreases the strength of wrist flexion, and may result in a radial deviation of the hand.

Contracture: Wrist flexion toward the ulnar side.

Note: Normally, fingers will be relaxed when the wrist is flexed. If the fingers actively flex as wrist flexion is initiated, the finger flexors (profundus and superficialis) are attempting to substitute for the wrist flexors.

Extensor Carpi Radialis Longus and Brevis

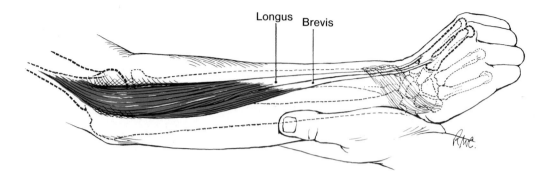

Longus Brevis

EXTENSOR CARPI RADIALIS LONGUS

Origin: Distal one-third of lateral supracondylar ridge of humerus, and lateral intermuscular septum.

Insertion: Dorsal surface of base of second metacarpal bone, radial side.

Action: Extends and abducts the wrist, and assists in flexion of the elbow.

Nerve: Radial, C5, **6, 7,** 8.

EXTENSOR CARPI RADIALIS BREVIS

Origin: Common extensor tendon from lateral epicondyle of humerus, radial collateral ligament of elbow joint, and deep antebrachial fascia.

Insertion: Dorsal surface of base of third metacarpal bone.

Action: Extends and assists in abduction of the wrist.

Nerve: Radial, C5, **6, 7,** 8.

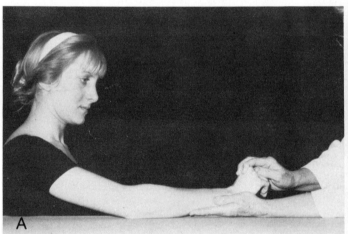
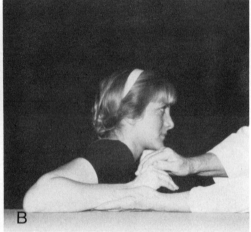

EXTENSOR CARPI RADIALIS LONGUS AND BREVIS

Patient: Sitting with elbow about 30° from zero extension.

Fixation: The forearm is in slightly less than full pronation and rests on the table for support.

Test: Extension of the wrist toward the radial side. (Fingers should be allowed to flex as the wrist is extended.)

Pressure: Against the dorsum of the hand along the second and third metacarpal bones, in the direction of flexion toward the ulnar side.

Weakness: Decreases the strength of wrist extension, and allows an ulnar deviation of the hand.

Contracture: Wrist extension with radial deviation.

Note: (See note under Extensor carpi ulnaris.)

EXTENSOR CARPI RADIALIS BREVIS

Patient: Sitting with elbow fully flexed. (Have subject lean forward to flex elbow.)

Fixation: The forearm is in slightly less than full pronation and rests on the table for support.

Test: Extension of the wrist toward the radial side. Elbow flexion makes the Extensor carpi radialis longus less effective by being in a shortened position.

Pressure: Against the dorsum of the hand along the second and third metacarpal bones, in the direction of flexion toward the ulnar side.

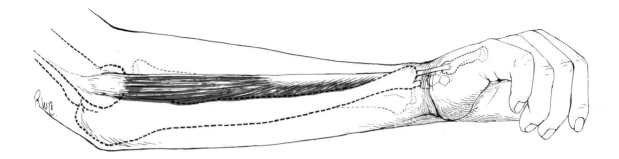

Origin: Common extensor tendon from lateral epicondyle of humerus, by aponeurosis from posterior border of ulna, and deep antebrachial fascia.

Insertion: Base of fifth metacarpal bone, ulnar side.

Action: Extends and adducts the wrist.

Nerve: Radial, C6, **7**, **8**.

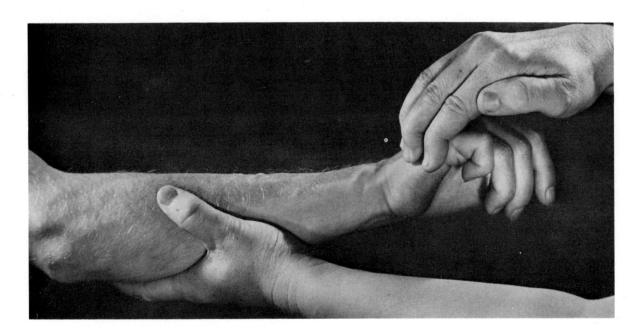

Patient: Sitting or supine.

Fixation: The forearm is in full pronation and rests on the table for support or is supported by the examiner.

Test: Extension of the wrist toward the ulnar side.

Pressure: Against the dorsum of the hand along the fifth metacarpal bone, in the direction of flexion toward the radial side.

Weakness: Decreases the strength of wrist extension, and may result in a radial deviation of the hand.

Contracture: Ulnar deviation of the hand with slight extension.

Note: Normally, fingers will be in a position of passive flexion when the wrist is extended. If the fingers actively extend as wrist extension is initiated, the finger extensors (digitorum, indicis, and digiti minimi) are attempting to substitute for the wrist extensors.

Pronator Teres and Pronator Quadratus

PRONATOR TERES

Origin of humeral head: Immediately above medial epicondyle of humerus, common flexor tendon, and deep antebrachial fascia.

Origin of ulnar head: Medial side of coronoid process of ulna.

Insertion: Middle of lateral surface of radius.

Action: Pronates the forearm and assists in flexing the elbow joint.

Nerve: Median, C**6**, **7**.

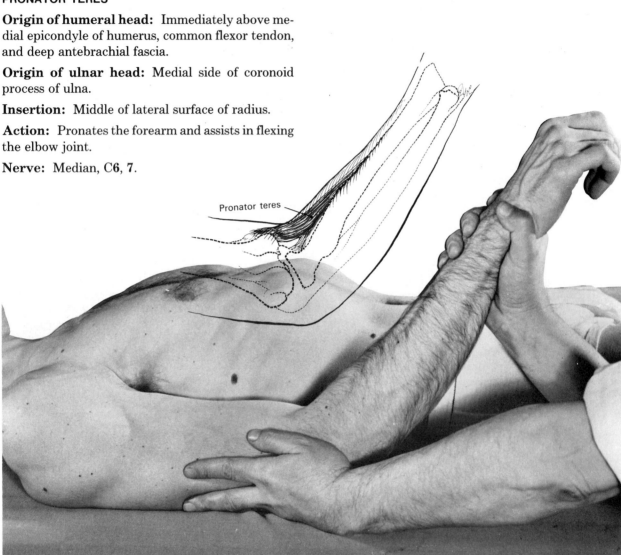

Pronator teres

PRONATORS TERES AND QUADRATUS

Patient: Supine or sitting.

Fixation: The elbow should be held against the patient's side, or be stabilized by the examiner to avoid any shoulder abduction movement.

Test: Pronation of the forearm with the elbow partially flexed.

Pressure: The examiner's hand is placed at the lower forearm above the wrist (to avoid twisting the wrist), and pressure is applied in the direction of supinating the forearm.

Weakness: Allows a supinated position of the forearm; interferes with many everyday functions such as turning a doorknob, using a knife to cut meats, or turning the hand downward in picking up a cup or other object.

Contracture: With the forearm held in a position of pronation, there is a marked interference with many normal functions of the hand and forearm that require moving from pronation to supination.

Origin: Medial side, anterior surface of distal one-fourth of ulna.

Insertion: Lateral side, anterior surface of distal one-fourth of radius.

Action: Pronates the forearm.

Nerve: Median, C7, **8,** T1.

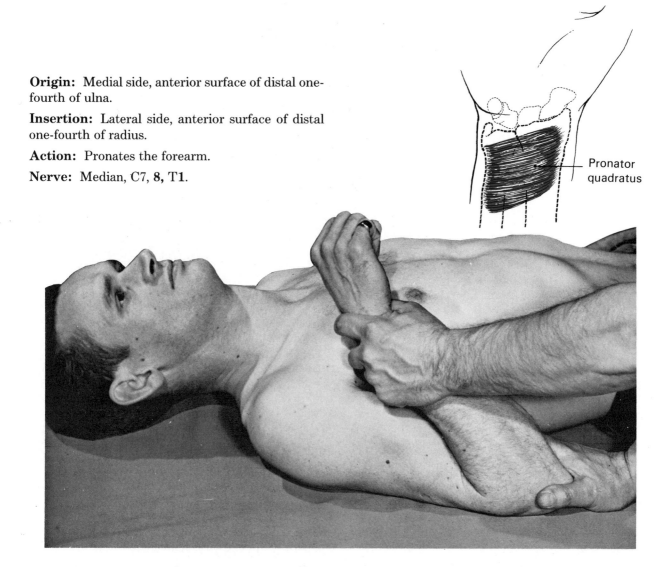

Pronator quadratus

Patient: Supine or sitting.

Fixation: The elbow should be held against the patient's side (by either the patient or the examiner) to avoid shoulder abduction.

Test: Pronation of the forearm with the elbow completely flexed in order to make the humeral head of the Pronator teres less effective by being in a shortened position.

Pressure: The examiner's hand is placed at the lower forearm above the wrist (to avoid twisting the wrist), and pressure is applied in the direction of supinating the forearm.

Supinator and Biceps

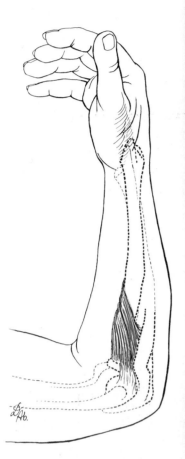

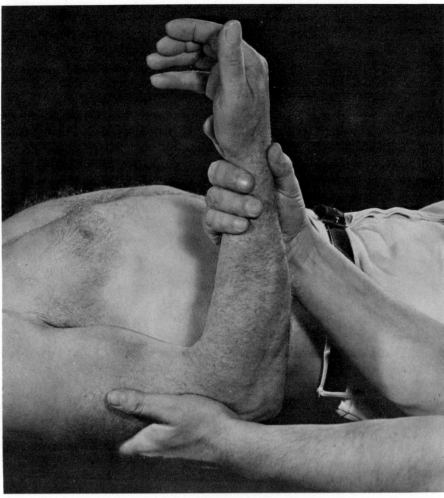

SUPINATOR

Origin: Lateral epicondyle of humerus, radial collateral ligament of elbow joint, annular ligament of radius, and supinator crest of ulna.

Insertion: Lateral surface of upper one-third of body of radius covering part of anterior and posterior surfaces.

Action: Supinates the forearm.

Nerve: Radial, C5, **6**, (7).

SUPINATOR AND BICEPS

Patient: Supine.

Fixation: The elbow should be held against the patient's side to avoid shoulder movement.

Test: Supination of the forearm with elbow at right angle or slightly below.

Pressure: At the distal end of the forearm above the wrist (to avoid twisting the wrist), and pressure is applied in the direction of pronating the forearm.

Weakness: Allows the forearm to remain in a pronated position. Interferes with many functions of the extremity, particularly those involved in feeding oneself.

Contracture: Elbow flexion with forearm supination. Results in marked interference with functions of the extremity that involve the change from supinated to pronated position of the forearm.

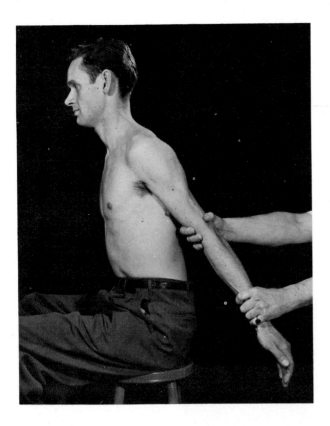

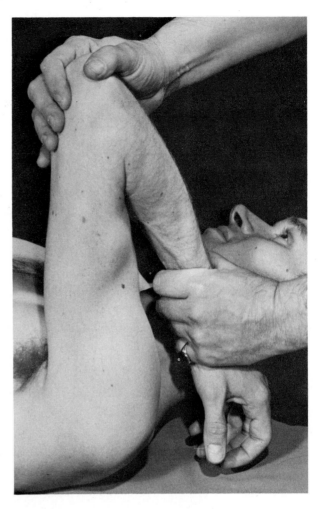

SUPINATOR (Tested with Biceps elongated.)

Patient: Sitting or standing.

Fixation: The examiner holds the shoulder and elbow in extension.

Test: Supination of the forearm.

Pressure: At the distal end of the forearm above the wrist, in the direction of pronation. The subject may attempt to rotate the humerus laterally to make it appear that the forearm remains in supination as pressure is applied and the forearm starts to pronate.

SUPINATOR (Tested with Biceps in shortened position.)

Patient: Supine.

Fixation: The examiner holds the shoulder in flexion with the elbow completely flexed. It is usually advisable to have the subject close the fingers in order to keep them from touching the table, which he may do in an effort to brace the forearm in test position.

Test: Supination of the forearm.

Pressure: At the distal end of the forearm above the wrist, in the direction of pronation. Care should be taken to *avoid* maximum pressure because, as strong pressure is applied, the Biceps comes into action and, in this shortened position, goes into a "cramp." A severe cramp may leave the muscle sore for several days. This test should be used merely as a differential diagnostic aid.

Note: In a radial nerve lesion involving the supinator, the test position cannot be maintained. The forearm will fail to hold the fully supinated position even though the Biceps is normal.

Brachioradialis

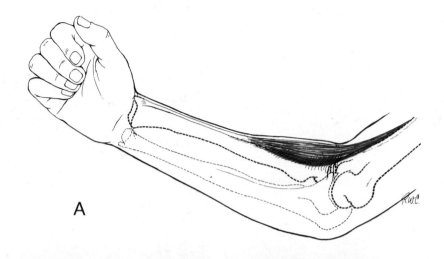

A

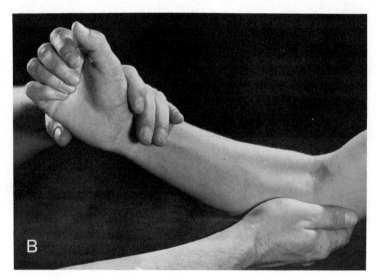

B

Origin: Proximal two-thirds of lateral supracondylar ridge of humerus, and lateral intermuscular septum.

Insertion: Lateral side of base of styloid process of radius.

Action: Flexes the elbow joint, and assists in pronating and supinating the forearm when these movements are resisted.

Nerve: Radial: C**5, 6**.

Patient: Supine or sitting.

Fixation: The examiner places one hand under the elbow to cushion it from table pressure.

Test: Flexion of the elbow with the forearm neutral between pronation and supination. The belly of the brachioradialis must be seen and felt during this test because the movement can be produced by other muscles which flex the elbow.

Pressure: Against the lower forearm, in the direction of extension.

Weakness: Decreases the strength of elbow flexion and of resisted supination or pronation to midline.

Coracobrachialis

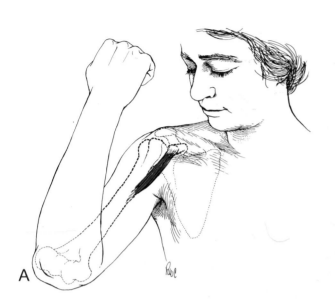

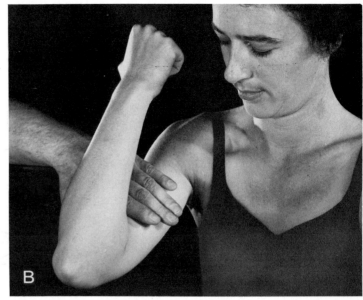

Origin: Apex of coracoid process of scapula.

Insertion: Medial surface of middle of shaft of humerus, opposite deltoid tuberosity.

Action: Flexes and adducts the shoulder joint.

Nerve: Musculocutaneous, C6, 7.

Patient: Sitting or supine.

Fixation: If trunk is stable, no fixation by examiner should be necessary.

Test: Shoulder flexion in lateral rotation with the elbow completely flexed and forearm supinated. Assistance from the biceps in shoulder flexion is decreased in this test position because the complete elbow flexion and forearm supination place the muscle in too short a position to be effective in shoulder flexion.

Pressure: Against the anteromedial surface of the lower third of the humerus in the direction of extension and slight abduction.

Weakness: Decreases the strength of shoulder flexion particularly in movements which involve complete elbow flexion and supination, for example, combing the hair.

Shortness: The coracoid process is depressed anteriorly when the arm is down at the side.

Biceps Brachii and Brachialis

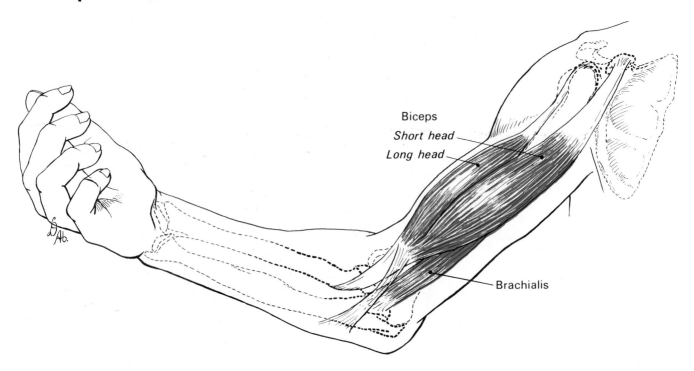

Biceps
Short head
Long head

Brachialis

BICEPS BRACHII

Origin of short head: Apex of coracoid process of scapula.

Origin of long head: Supraglenoid tubercle of scapula.

Insertion: Tuberosity of radius, and aponeurosis of Biceps brachii (lacertus fibrosus).

Action: Flexes the shoulder joint, and the long head may assist with abduction if the humerus is laterally rotated. *With the origin fixed*, flexes the elbow joint moving the forearm toward the humerus, and supinates the forearm. *With the insertion fixed*, flexes the elbow joint moving the humerus toward the forearm as in pull-up or chinning exercises.

Nerve: Musculocutaneous, C**5,6**.

BRACHIALIS

Origin: Distal one-half of anterior surface of humerus, and medial and lateral intermuscular septa.

Insertion: Tuberosity and coronoid process of ulna.

Action: *With the origin fixed*, flexes the elbow joint moving the forearm toward the humerus. *With the insertion fixed*, flexes the elbow joint moving the humerus toward the forearm as in pull-up or chinning exercises.

Nerve: Musculocutaneous, and small branch from radial, C**5,6**.

Patient: Supine or sitting.

Fixation: The examiner places one hand under the elbow to cushion it from table pressure.

Test: Elbow flexion slightly less than or at right angle, with forearm in supination.

Pressure: Against the lower forearm, in the direction of extension.

Weakness: Decreases the ability to flex the forearm against gravity. There is marked interference with such daily activities as feeding oneself or combing the hair.

Contracture: Flexion deformity of the elbow.

Note: If the Biceps and Brachialis are weak as in a musculocutaneous lesion, the patient will pronate the forearm before he flexes the elbow using Brachioradialis, Extensor carpi radialis longus, Pronator teres, and wrist flexors.

The lower figure on the facing page illustrates that, against resistance, the Biceps acts in flexion even though the forearm is in pronation. Since the Brachialis is inserted on the ulna, the position of the forearm, whether in supination or pronation, does not affect the action of this muscle in elbow flexion. The Brachioradialis appears to have a slightly stronger action in the pronated position of the forearm in the elbow flexion test than in the supinated position, although its strongest action in flexion is with the forearm in midposition.

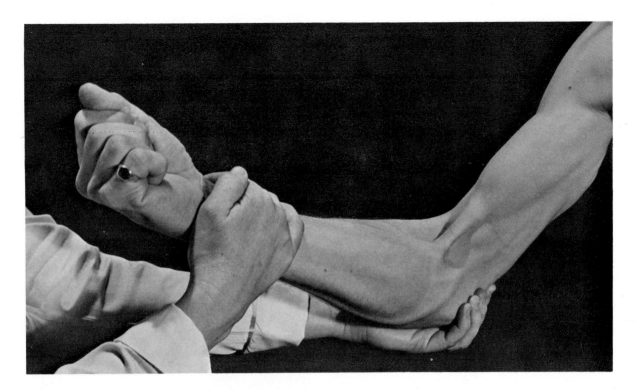

Elbow flexion with forearm supinated.

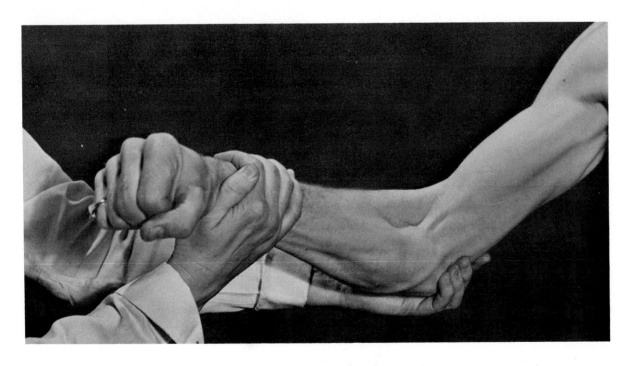

Elbow flexion with forearm pronated.

Triceps Brachii and Anconeus

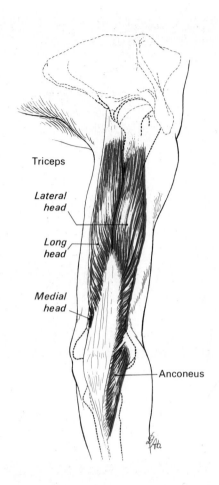

Triceps

Lateral head

Long head

Medial head

Anconeus

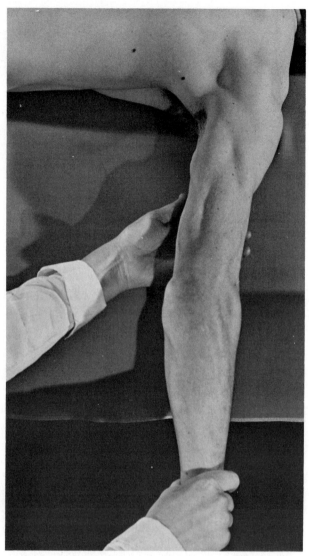

ANCONEUS

Origin: Lateral epicondyle of humerus, posterior surface.

Insertion: Lateral side of olecranon process, and upper one fourth of posterior surface of body of ulna.

Action: Extends the elbow joint, and may stabilize the ulna during pronation and supination.

Nerve: Radial, C7,8.

TRICEPS BRACHII AND ANCONEUS

Patient: Prone.

Fixation: The shoulder is at 90° abduction, neutral with regard to rotation, and with the arm supported between the shoulder and the elbow by the table. The examiner places one hand under the arm near the elbow to cushion the arm from table pressure.

Test: Extension of the elbow joint (to slightly less than full extension).

Pressure: Against the forearm, in the direction of flexion.

TRICEPS BRACHII

Origin of long head: Infraglenoid tubercle of scapula.

Origin of lateral head: Lateral and posterior surfaces of proximal one-half of body of humerus, and lateral intermuscular septum.

Origin of medial head: Distal two-thirds of medial and posterior surfaces of humerus below the radial groove, and from medial intermuscular septum.

Insertion: Posterior surface of olecranon process of ulna and antebrachial fascia.

Action: Extends the elbow joint. In addition, the long head assists in adduction and extension of the shoulder joint.

Nerve: Radial, C6,7,8 T1

TRICEPS BRACHII AND ANCONEUS

Patient: Supine.

Fixation: The shoulder is at approximately 90° flexion, with the arm supported in a position perpendicular to the table.

Test: Extension of the elbow (to slightly less than full extension).

Pressure: Against the forearm, in the direction of flexion.

Weakness: Results in the inability to extend the forearm against gravity. There is interference with everyday functions which involve elbow extension as in reaching upward toward a high shelf. There is loss of ability to throw objects or push with the extended elbow. An individual is handicapped in using crutches or a cane since he cannot extend his elbow and transfer weight to his hand.

Contracture: Extension deformity of the elbow. Marked interference with everyday functions that involve elbow flexion.

Note: When the shoulder is horizontally abducted (see facing page), the long head of the Triceps is shortened over both the shoulder and elbow joints. When the shoulder is flexed (horizontally adducted), the long head of the Triceps is shortened over the elbow joint while elongated over the shoulder joint. Because of this two-joint action of the long head, it is made less effective in the prone position by being shortened over both joints, with the result that the Triceps withstands less pressure when tested in the prone position than in the supine position. While the Triceps and Anconeus act together in extending the elbow joint, it may be useful to differentiate these two muscles. Since the belly of the Anconeus muscle is below the elbow joint, it can be distinguished from the Triceps by palpation. The branch of the radial nerve to the Anconeus arises near the midhumeral level and is quite long. It is possible for a lesion to involve only this branch leaving the Triceps unaffected. Paralysis of the Anconeus materially reduces the strength of elbow extension. One may find that a grade of "good" elbow extension strength is actually the result of a normal Triceps and a zero Anconeus.

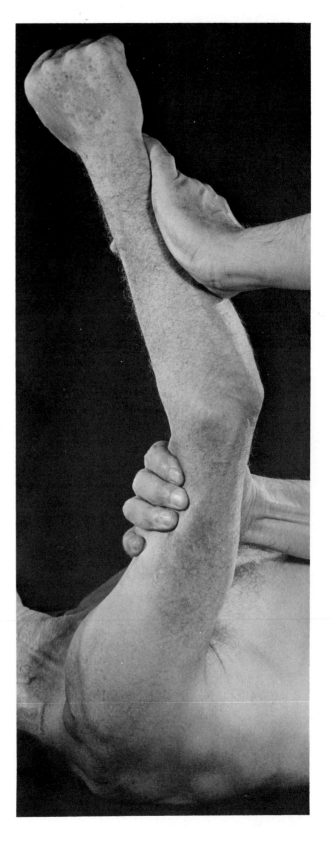

Supraspinatus

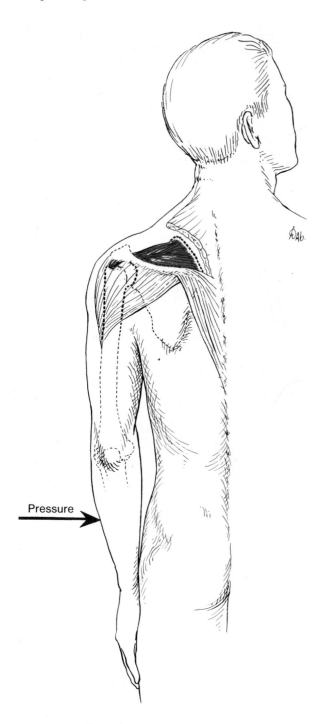

Pressure →

Origin: Medial two-thirds of supraspinous fossa of scapula.

Insertion: Superior facet of greater tubercle of humerus, and shoulder joint capsule.

Action: Abducts the shoulder joint, and stabilizes the head of the humerus in the glenoid cavity during movements of this joint.

Nerve: Suprascapular, C4,**5**,6.

Patient: Sitting or standing with arm at side, head and neck extended and laterally flexed to same side and the face rotated toward the opposite side.

Fixation: None necessary since maximum pressure is not required.

Note: No effort is made to distinguish the Supraspinatus from the Deltoid in the strength test for the purpose of grading since these muscles act simultaneously in abducting the shoulder. However, the Supraspinatus can be palpated to determine whether it is active.

Since the Supraspinatus is completely covered by the upper and middle fibers of the Trapezius, to palpate this muscle the Trapezius should be as relaxed as possible. This is accomplished by extending and laterally flexing the head and neck so that the face is rotated toward the opposite side, as illustrated; and by testing the activity of the Supraspinatus at the beginning of the abduction movement when the activity of the Trapezius is at a low level. The Deltoid and the Supraspinatus act together in initiating abduction, and this test is not to be construed to mean that the Supraspinatus is responsible for the first few degrees of abduction.

Test: Initiation of abduction of the humerus.

Pressure: Against the forearm in the direction of adduction.

Weakness: The tendon of the Supraspinatus is firmly attached to the superior surface of the capsule of the shoulder joint. Weakness of the muscle or a rupture of the tendon decreases shoulder joint stability, allowing the head of the humerus to alter its relationship with the glenoid cavity.

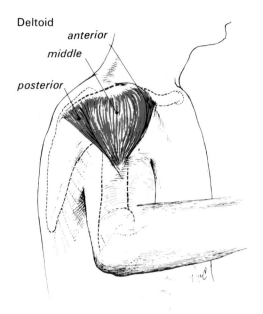

Deltoid
anterior
middle
posterior

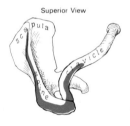

Superior View

DELTOID

Origin of anterior fibers: Anterior border, superior surface, lateral one-third of clavicle.

Origin of middle fibers: Lateral margin and superior surface of acromion.

Origin of posterior fibers: Inferior lip of posterior border of spine of scapula.

Insertion: Deltoid tuberosity of humerus.

Action: Abduction of the shoulder joint, performed chiefly by the middle fibers with stabilization by the anterior and posterior fibers. In addition, the anterior fibers flex and, in the supine position, medially rotate the shoulder joint; the posterior fibers extend and, in the prone position, laterally rotate.

Nerve: Axillary, C**5,6**.

Patient: Sitting.

Fixation: The position of the trunk in relation to the arm in this test is such that a stable trunk will need no further stabilization by the examiner. If the scapular fixation muscles are weak, the examiner must stabilize the scapula.

Test: Shoulder abduction without rotation. When placing the shoulder in test position, the elbow should be flexed to indicate the neutral position of rotation, but may be extended after the shoulder position is established in order to use the extended extremity for a longer lever. The examiner should be consistent in his technique for subsequent tests.

Pressure: Against the dorsal surface of the distal end of the humerus if the elbow is flexed, or against the forearm if the elbow is extended.

Weakness: Results in the inability to lift the arm in abduction against gravity. In the presence of paralysis of the entire Deltoid and Supraspinatus, the humerus tends to subluxate downward if the arm remains unsupported in a hanging position. The capsule of the shoulder joint permits almost an inch of separation of the head of the humerus from the glenoid cavity. In cases of axillary nerve involvement in which the Deltoid is weak while the Supraspinatus is not affected, the relaxation of the joint is not as marked, but tends to progress if the Deltoid strength does not return.

Deltoid, Anterior and Posterior (sitting)

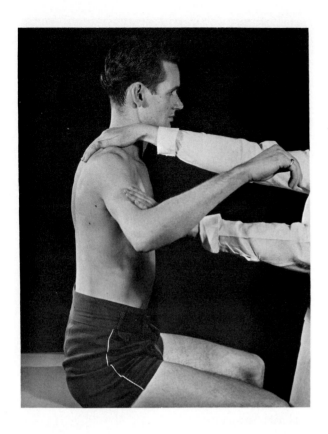 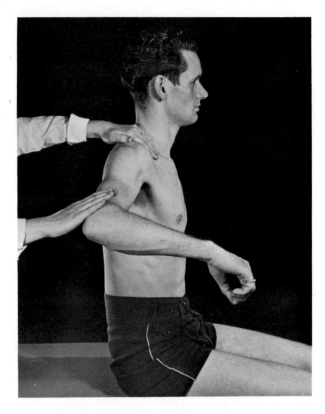

ANTERIOR DELTOID

Patient: Sitting.

Fixation: If scapular fixation muscles are weak, the scapula must be stabilized by the examiner. As pressure is applied on the arm, counterpressure is applied posteriorly to the shoulder girdle.

Test: Shoulder abduction in slight flexion, with the humerus in slight lateral rotation. In the erect sitting position it is necessary to place the humerus in slight lateral rotation to increase the effect of gravity on the anterior fibers. (The anatomical action of the anterior Deltoid, which entails slight medial rotation, is part of the test of the anterior Deltoid in the supine position.) (See p. 101.)

Pressure: Against the anteromedial surface of the arm, in the direction of adduction and slight extension.

POSTERIOR DELTOID

Patient: Sitting.

Fixation: If scapular fixation muscles are weak, the scapula must be stabilized by the examiner. As pressure is applied on the arm, counterpressure is applied anteriorly on the shoulder girdle.

Test: Shoulder abduction in slight extension, with the humerus in slight medial rotation. In the erect sitting position it is necessary to place the humerus in slight medial rotation in order to have the posterior fibers in an antigravity position. (The anatomical action of the posterior Deltoid, which entails slight lateral rotation, is part of the posterior Deltoid test in the prone position.) (See p. 101.)

Pressure: Against the posterolateral surface of the arm, above the elbow, in the direction of adduction and slight flexion.

Deltoid, Anterior (supine) and Posterior (prone)

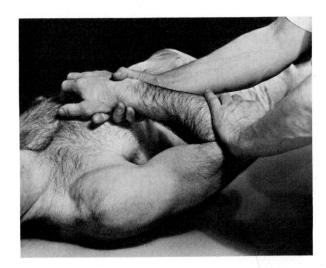
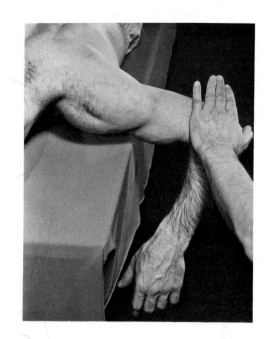

ANTERIOR DELTOID

Patient: Supine.

Fixation: The Trapezius and Serratus anterior should stabilize the scapula in all the Deltoid tests, and if these muscles are weak the examiner should stabilize the scapula.

Test: Shoulder abduction in the position of slight flexion and medial rotation. One hand of the examiner is placed under the patient's wrist to make sure that he does not lift the elbow by reverse action of the wrist extensors which action may occur if the patient is allowed to press the hand down on the chest.

Pressure: Against the anterior surface of the arm just above the elbow, in the direction of adduction toward the side of the body.

POSTERIOR DELTOID

Patient: Prone.

Fixation: The scapula must be held stable by scapular muscles or by the examiner.

Test: Horizontal abduction of the shoulder with slight lateral rotation.

Pressure: Against the posterolateral surface of the arm, in a direction obliquely downward midway between adduction and horizontal adduction.

Latissimus Dorsi

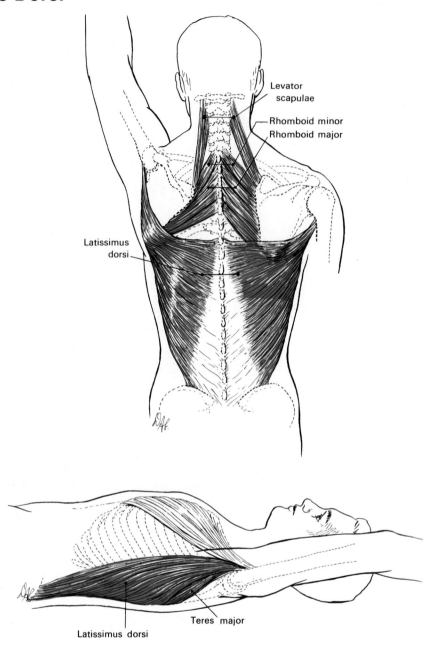

Levator
scapulae

Rhomboid minor
Rhomboid major

Latissimus
dorsi

Teres major

Latissimus dorsi

Origin: Spinous processes of last six thoracic vertebrae, last three or four ribs, through the thoracolumbar fascia from the lumbar and sacral vertebrae and posterior one-third of external lip of iliac crest, a slip from the inferior angle of the scapula.

Insertion: Intertubercular groove of humerus.

Action: *With the origin fixed*, medially rotates, adducts, and extends the shoulder joint. By continued action, depresses the shoulder girdle, and assists in lateral flexion of the trunk. (See p. 222.) *With the insertion fixed*, assists in tilting the pelvis anteriorly and laterally. Acting bilaterally, this muscle assists in hyperextending the spine and anteriorly tilting the pelvis, or in flexing the spine depending upon its relation to the axes of motion.

This muscle is important in relation to movements such as climbing, walking with crutches, or hoisting the body up on parallel bars, in which the muscles act to lift the body toward the fixed arms. The strength of the Latissimus dorsi is a factor in such forceful arm movements as swimming, rowing, and chopping. All adductors and medial rotators act in these strong movements but the Latissimus dorsi may be of major importance.

The Latissimus dorsi may act as an accessory muscle of respiration.

Nerve: Thoracodorsal, **C6,7,8**.

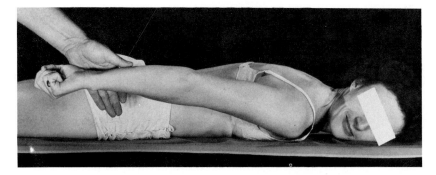

Patient: Prone.

Fixation: None necessary.

Test: Adduction of the arm, with extension, in the medially rotated position.

Pressure: Against the forearm, in the direction of abduction and slight flexion of the arm.

Weakness: Weakness interferes with activities which involve adduction of the arm toward the body or the body toward the arm. The strength of lateral trunk flexion is diminished.

Shortness: Results in a limitation of elevation of the arm in flexion or abduction. Tends to depress the shoulder girdle down and forward. In a right C-curve of the spine the anterior fibers of the left Latissimus dorsi usually are shortened, as they are bilaterally in a marked kyphosis.

Shortness of the Latissimus dorsi may be found in individuals who have walked with crutches for a prolonged period of time, as, for example, the paraplegic who uses a swing-through gait.

Tests for length of shoulder adductors.

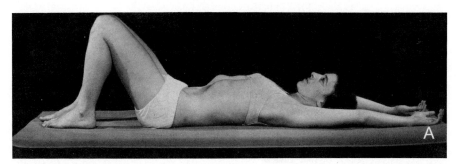

With no shortness of shoulder adductors, the shoulder joint can be completely flexed while the low back is flat on the table, as illustrated above.

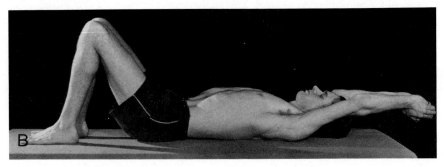

With shortness of the shoulder adductors, the shoulder joint cannot be completely flexed while the low back is held flat. The presence of shortness in the Latissimus dorsi and Teres major is obvious in this subject.

Pectoralis Major

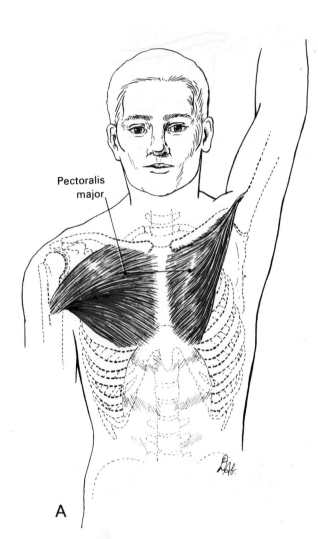

Pectoralis
major

Origin of upper fibers (clavicular portion): Anterior surface of sternal one-half of clavicle.

Origin of lower fibers (sternocostal portion): Anterior surface of sternum, cartilages of first six or seven ribs, and aponeurosis of the External oblique.

Insertion of upper and lower fibers: Crest of greater tubercle of humerus. Upper fibers are more anterior and caudal on the crest than the lower fibers which twist on themselves and are more posterior and cranial.

Action of muscle as a whole: *With the origin fixed*, it adducts and medially rotates the humerus. With the *insertion fixed,* the Pectoralis major may assist in elevating the thorax as in forced inspiration. In crutch-walking or in parallel-bar work, it will assist in supporting the weight of the body.

Action of upper fibers: Flex and medially rotate the shoulder joint, and horizontally adduct the humerus toward the opposite shoulder.

Nerve to upper fibers: Lateral Pectoral C**5,6,7**.

Action of lower fibers: Depress the shoulder girdle by virtue of attachment on the humerus, and obliquely adduct the humerus toward the opposite iliac crest.

Nerves to lower fibers: Lateral and medial pectoral, C**6,7,8, T1.**

A

Tests for length of Pectoralis major

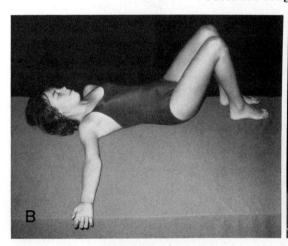

B

Normal length of upper fibers.

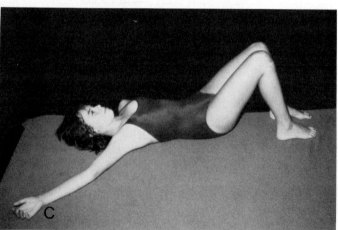

C

Normal length of lower fibers.

Pectoralis Major, Upper

Pectoralis Major, Lower

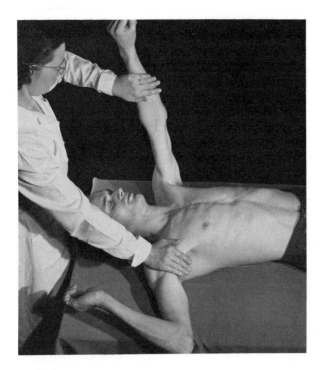

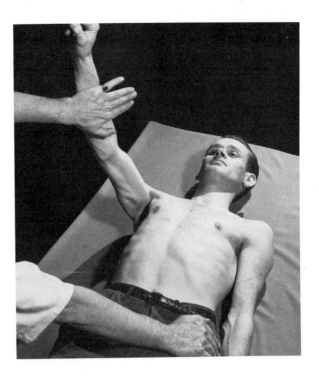

Patient: Supine.

Fixation: The examiner holds the opposite shoulder firmly on the table. The Triceps maintains the elbow in extension.

Test: Starting with the elbow extended, and the shoulder in 90° flexion and slight medial rotation, the humerus is horizontally adducted toward the sternal end of the clavicle.

Pressure: Against the forearm, in the direction of horizontal abduction.

Weakness: Decreases the ability to draw the arm in horizontal adduction across the chest, making it difficult to touch the hand to the opposite shoulder. Decreases strength of shoulder flexion and medial rotation.

Shortness: The range of motion in horizontal abduction and lateral rotation of the shoulder is decreased. A shortness of the Pectoralis major holds the humerus in medial rotation and adduction, and, secondarily results in abduction of the scapula from the spine.

Note: The authors have seen one patient with rupture and another with weakness of the lower part of the Pectoralis major resulting from Indian wrestling. The arm was in a position of lateral rotation and abduction when a forceful effort was made to medially rotate and adduct it.

Patient: Supine.

Fixation: The examiner places one hand on opposite iliac crest to hold the pelvis firmly on the table. The anterior part of the External and Internal oblique muscles stabilize the thorax on the pelvis. In cases of abdominal weakness, the thorax instead of the pelvis, must be stabilized. The Triceps maintains the elbow in extension.

Test: Starting with the elbow extended, and the shoulder in flexion and slight medial rotation, adduction of the arm obliquely toward the opposite iliac crest.

Pressure: Against the forearm, obliquely in a lateral and cranial direction.

Weakness: Decreases the strength of adduction obliquely toward the opposite hip. There is a loss of continuity of muscle action from the Pectoralis major to External oblique and Internal oblique on the opposite side with the result that chopping or striking movements are difficult. From a supine position, if the subject's arm is placed diagonally overhead, he will find it difficult to lift the arm from the table. He will also have difficulty holding any large or heavy object in both hands at or near waist level.

Shortness: There is a forward depression of the shoulder girdle by the pull of the Pectoralis major on the humerus which often accompanies the pull of a tight Pectoralis minor on the scapula. Flexion and abduction ranges of motion overhead are limited.

Pectoralis Minor

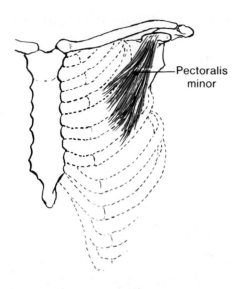

Pectoralis minor

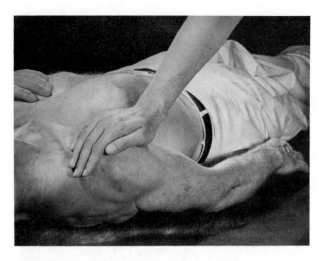

Origin: Superior margins, outer surfaces of third, fourth, and fifth ribs near the cartilages; and from fascia over corresponding intercostal muscles.

Insertion: Medial border, superior surface of coracoid process of scapula.

Action: *With the origin fixed*, tilts the scapula anteriorly, i.e., rotates the scapula about a coronal axis so that the coracoid process moves anteriorly and caudally, while the inferior angle moves posteriorly and medially. With the scapula stabilized *to fix the insertion*, the Pectoralis minor assists in forced inspiration.

Nerve: Medial pectoral with fibers from a communicating branch of the lateral pectoral, C(6), **7,8,**T1. (For explanation, see p. 52)

Test for shortness of Pectoralis minor

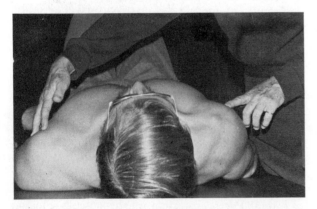

Left, normal length; right, short, holding shoulder forward.

Patient: Supine.

Fixation: None by the examiner unless the abdominal muscles are weak, in which case the rib cage on the same side should be held firmly down.

Test: Forward thrust of the shoulder with the arm at the side. The subject must exert no downward pressure on the hand to force the shoulder forward. (If necessary raise the subject's hand and elbow off the table.)

Pressure: Against the anterior aspect of the shoulder, downward toward the table.

Weakness: Strong extension of the humerus is dependent upon fixation of the scapula by the Rhomboids and Levator scapulae posteriorly and the Pectoralis minor anteriorly. With weakness of the Pectoralis minor, the strength of arm extension is diminished.

With the scapula stabilized in a position of good alignment, the Pectoralis minor acts as an accessory muscle of inspiration. Weakness of this muscle will increase respiratory difficulty in patients already suffering from involvement of respiratory muscles.

Contracture: With the origin of this muscle on the ribs and the insertion on the coracoid process of the scapula, a contracture of this muscle tends to depress the coracoid process of the scapula forward and downward. Such muscle contracture is an important contributing factor in many cases of arm pain. With the cords of the brachial plexus and the axillary blood vessels lying between the coracoid process and rib cage, contracture of the Pectoralis minor may produce an impingement on these large vessels and nerves.

A contracted Pectoralis minor restricts flexion of the shoulder joint by limiting scapular rotation and preventing the glenoid cavity from attaining the cranial orientation necessary for complete flexion of the joint.

Teres Major

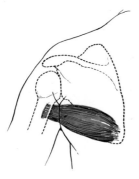

Origin: Dorsal surfaces of inferior angle and lower third of lateral border of scapula.

Insertion: Crest of lesser tubercle of humerus.

Action: Medially rotates, adducts, and extends the shoulder joint.

Nerve: Lower subscapular, C5,**6**,7.

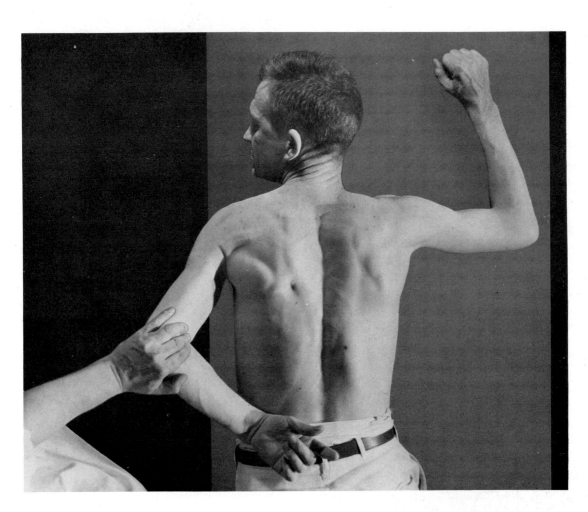

Patient: Prone.

Fixation: Usually none is necessary because the weight of the trunk is sufficient fixation. If necessary, the opposite shoulder may be held down on the table.

Test: Extension and adduction of the humerus in the medially rotated position, with the hand resting on the posterior iliac crest.

Pressure: Against the arm above the elbow, in the direction of abduction and flexion.

Weakness: Diminishes the strength of medial rotation, adduction and extension of the humerus.

Shortness: Prevents full range of lateral rotation and abduction of the humerus. With tightness of the Teres major the scapula will begin to rotate laterally almost simultaneously with flexion or abduction. Scapular movements which accompany shoulder flexion and abduction are influenced by the degree of muscle shortness of the Teres major and Subscapularis.

Shoulder Medial Rotators

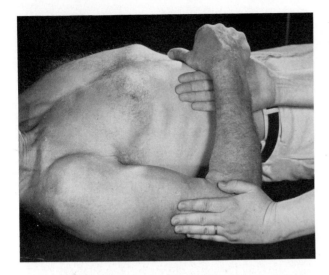 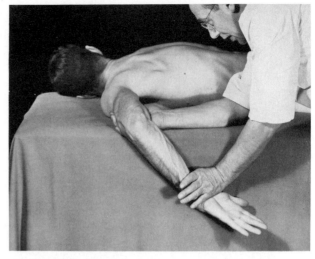

The chief muscles acting in this shoulder medial rotation test are Latissimus dorsi, Pectoralis major, Subscapularis, and Teres major.

Patient: Supine.

Fixation: Counterpressure is applied by the examiner against the outer aspect of the distal end of the humerus in order to insure a rotation motion.

Test: Medial rotation of the humerus with arm at side and elbow held at right angle.

Pressure: Using the forearm as a lever, pressure is applied in the direction of laterally rotating the humerus.

Note: For the purpose of objectively grading a weak medial rotator group against gravity, the test in the prone position (see above right) is preferred over the test in supine position. For a maximum strength test, the test in supine position is preferred because less scapular fixation is required.

Patient: Prone.

Fixation: The arm rests on the table. The examiner's hand, near the elbow, cushions against table pressure and stabilizes the humerus to insure a rotation action by preventing any adduction or abduction. The Rhomboids give fixation of the scapula.

Test: Medial rotation of the humerus with the elbow held at right angle.

Pressure: Using the forearm as a lever, pressure is applied in the direction of laterally rotating the humerus.

Weakness: Inasmuch as the medial rotators are also strong adductors, the ability to perform both medial rotation and adduction is decreased.

Shortness: Range of shoulder flexion overhead and lateral rotation are limited.

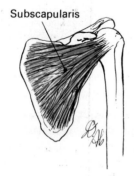

Subscapularis

SUBSCAPULARIS

Origin: Subscapular fossa of scapula.

Insertion: Lesser tubercle of humerus and shoulder joint capsule.

Action: Medially rotates the shoulder joint, and stabilizes the head of the humerus in the glenoid cavity during movements of this joint.

Nerve: Upper and lower subscapular, C**5,6,**7.

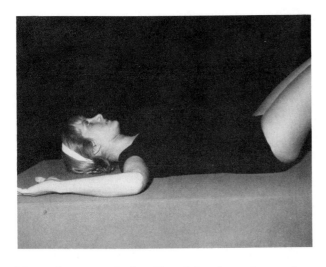

Normal range of shoulder joint lateral rotation (90°).

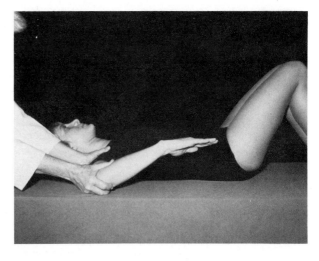

Normal range of shoulder joint medial rotation (about 70°). Shoulder is held down to prevent shoulder girdle movement.

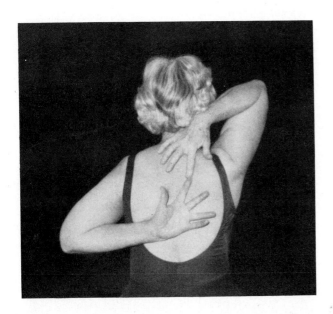

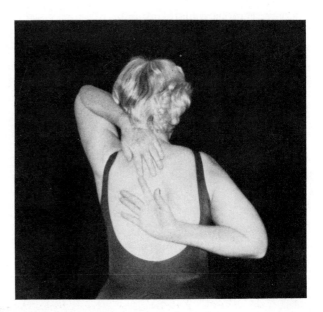

Placing hands behind back, as illustrated, requires normal range of shoulder joint motion without abnormal shoulder girdle movement.

If shoulder joint medial rotation were limited, there would be an effort to compensate by substituting shoulder girdle movement. This substitution would be seen as depression of the shoulder girdle anteriorly, and winging of the scapula.

Shoulder Lateral Rotators (prone)

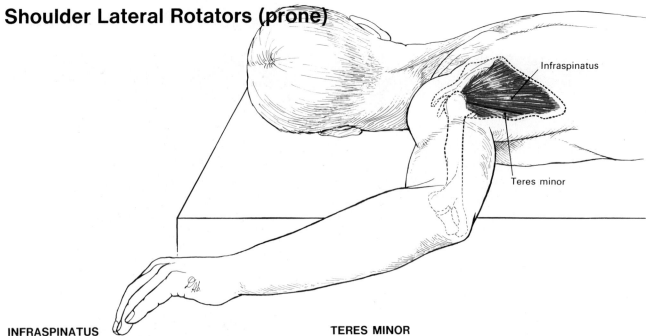

Infraspinatus

Teres minor

INFRASPINATUS

Origin: Medial two-thirds of infraspinous fossa of scapula.

Insertion: Middle facet of greater tubercle of humerus, and shoulder joint capsule.

Action: Laterally rotates the shoulder joint and stabilizes the head of the humerus in the glenoid cavity during movements of this joint.

Nerve: Suprascapular, C(4),**5,6.**

TERES MINOR

Origin: Upper two-thirds, dorsal surface of lateral border of scapula.

Insertion: Lowest facet of greater tubercule of humerus, and shoulder joint capsule.

Action: Laterally rotates the shoulder joint, and stabilizes the head of the humerus in the glenoid cavity during movements of this joint.

Nerve: Axillary, C5,6.

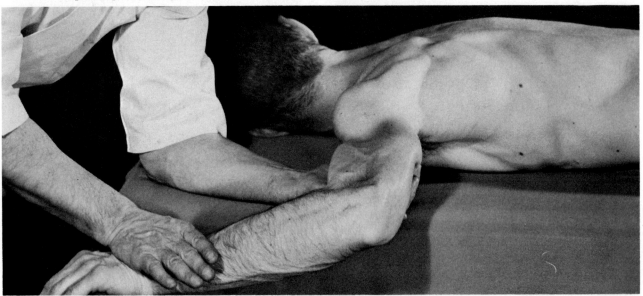

Patient: Prone.

Fixation: The arm rests on the table. The examiner places one hand under the arm near the elbow and stabilizes the humerus. He ensures a rotation action by preventing adduction or abduction motion. His hand cushions against the table pressure. This test requires strong fixation by the scapular muscles, particularly the middle and lower Trapezius, and in using this test one must observe whether the lateral rotators of the scapula or the lateral rotators of the shoulder "give" when pressure is applied.

Test: Lateral rotation of the humerus with the elbow held at right angle.

Pressure: Using the forearm as a lever, pressure is applied in the direction of medially rotating the humerus.

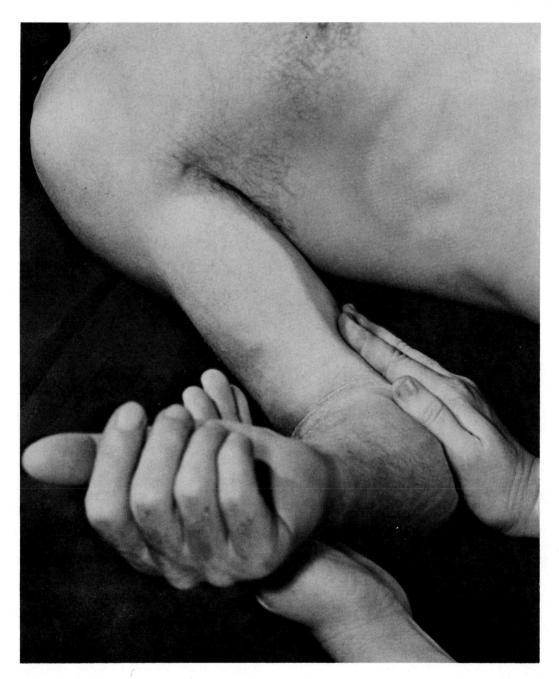

Patient: Supine.

Fixation: Counterpressure is applied by the examiner against the inner aspect of the distal end of the humerus in order to ensure a rotation motion.

Test: Lateral rotation of the humerus with the elbow held at right angle.

Pressure: Using the forearm as a lever, pressure is applied in the direction of medially rotating the humerus.

Weakness: The humerus assumes a position of medial rotation. Lateral rotation, in antigravity positions, is difficult or impossible.

For the purpose of objectively grading a weak lateral rotator group against gravity and for palpation of the rotator muscles, the test in prone position is preferred over the Teres minor and Infraspinatus test in supine position. For action of these two rotators without much assistance from the posterior Deltoid, and without the necessity of maximal Trapezius fixation, the test in supine position is preferred.

Rhomboids, Levator Scapulae, and Trapezius

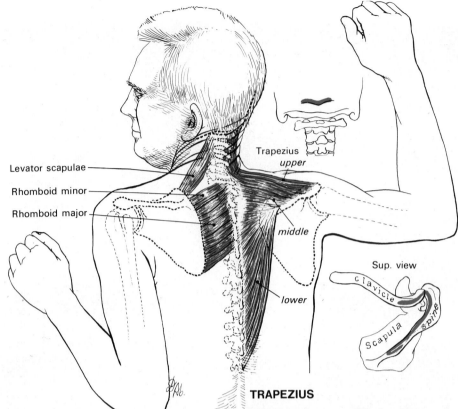

RHOMBOIDS

Origin of major: Spinous processes of second through fifth thoracic vertebrae.

Insertion of major: By fibrous attachment to medial border of scapula between spine and inferior angle.

Origin of minor: Ligamentum nuchae, spinous processes of seventh cervical and first thoracic vertebrae.

Insertion of minor: Medial border at root of spine of scapula.

Action: Adduct and elevate the scapula, and rotate it so the glenoid cavity faces caudally.

Nerve: Dorsal scapular, C**4,5**.

LEVATOR SCAPULAE

Origin: Transverse processes of first four cervical vertebrae.

Insertion: Medial border of scapula between superior angle and root of spine.

Action: *With the origin fixed*, elevates the scapula and assists in rotation so the glenoid cavity faces caudally. *With the insertion fixed*, and acting *unilaterally*, laterally flexes the cervical vertebrae and rotates their spines toward the same side. Acting *bilaterally*, the Levator scapulae may assist in extension of the cervical spine.

Nerve: Cervical **3,4** and Dorsal scapular C**4,5**.

TRAPEZIUS

Origin of upper fibers: External occipital protuberance, medial one-third of superior nuchal line, ligamentum nuchae, and spinous process of seventh cervical vertebra.

Origin of middle fibers: Spinous processes of first through fifth thoracic vertebrae.

Origin of lower fibers: Spinous processes of sixth through twelfth thoracic vertebrae.

Insert of upper fibers: Lateral one-third of clavicle and acromion process of scapula.

Insert of middle fibers: Medial margin of acromion and superior lip of spine of scapula.

Insert of lower fibers: Tubercle at apex of spine of scapula.

Action: *With the origin fixed*, adduction of the scapula, performed chiefly by the middle fibers with stabilization by the upper and lower fibers. Rotation of the scapula so the glenoid cavity faces cranially, performed chiefly by the upper and lower fibers with stabilization by the middle fibers. In addition, the upper fibers elevate and the lower fibers depress the scapula. *With the insertion fixed*, and acting *unilaterally*, the upper fibers extend, laterally flex, and rotate the head and joints of the cervical vertebrae so that the face turns toward the opposite side; and acting *bilaterally*, the upper trapezius extends the neck.

Nerve: Spinal portion of cranial nerve XI (accessory), and ventral ramus, C**2,3,4**.

Rhomboids and Levator Scapulae

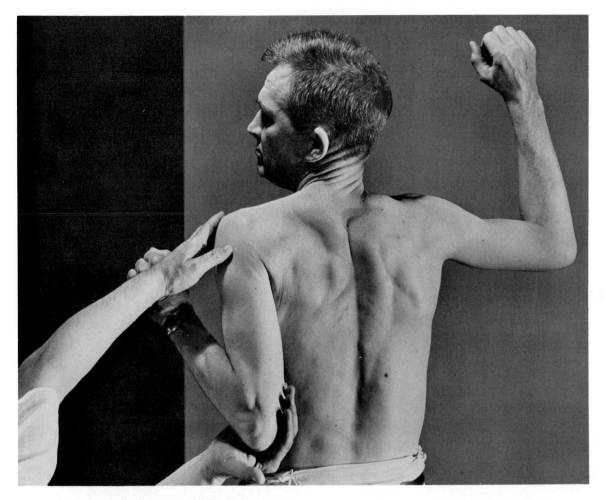

Patient: Prone.

Fixation: None is necessary on the part of the examiner, but it is assumed that the adductors of the shoulder joint have been tested and found to be strong enough to hold the arm for use as a lever in this test.

Test: Adduction and elevation of the scapula with medial rotation of the inferior angle. To obtain this position of the scapula, and to obtain leverage for pressure in the test, the arm is placed in the position as illustrated. With the elbow flexed, the humerus is adducted toward the side of the body, and in slight extension and slight lateral rotation.

The test is to determine the ability of the Rhomboids to hold the scapula in test position as pressure is applied against the arm.

Pressure: The examiner applies pressure with one hand against the patient's arm in the direction of abducting the scapula and rotating the inferior angle laterally; and against the patient's shoulder with the other hand in the direction of depression.

Weakness: The scapula abducts and the inferior angle rotates outward. The strength of adduction and extension of the humerus is diminished by loss of Rhomboid fixation of the scapula. Ordinary function of the arm is affected less by loss of Rhomboids than by loss of either Trapezius or Serratus anterior.

Contracture: The scapula is drawn into a position of adduction and elevation. Contracture tends to accompany paralysis or weakness of Serratus anterior because the Rhomboids are direct opponents of the Serratus.

Modified Test: If the shoulder muscles are weak, the examiner places the scapula in the test position and attempts to depress and derotate the scapula.

Note: The accompanying photograph shows the Rhomboids in a state of contraction. (See p. 102, for right Rhomboids in neutral position and left, in elongated position.)

Middle Trapezius

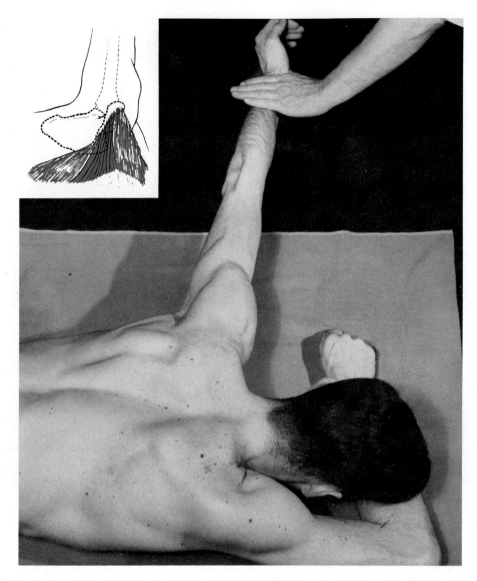

Patient: Prone.

Fixation: None necessary by the examiner, but the elbow extensors and the posterior shoulder muscles (posterior Deltoid, Teres minor, Infraspinatus) must give the necessary fixation in order to use the arm as a lever.

Test: Adduction of the scapula from a position of rotation in which the inferior angle is rotated laterally. To obtain this position of the scapula and to obtain leverage for the test, the elbow is extended, the shoulder is placed in 90° abduction and lateral rotation. This rotation of the shoulder is denoted by the position of the hand with the palm facing cranially. Note that the shoulder girdle is *not* elevated.

Pressure: Against the forearm in a downward direction toward the table.

Modification of test when posterior shoulder joint muscles are weak.

Patient: Prone with shoulder at edge of table, arm hanging down over side of table.

Fixation: None.

Test: Supporting the weight of the arm, the examiner places the scapula in a position of adduction, with some lateral rotation of the inferior angle, and without elevation of the shoulder girdle.

Pressure: As support of the arm is released, the weight of the suspended arm will exert a force that tends to abduct the scapula. A very weak Trapezius will not hold the scapula adducted against this force. If the middle Trapezius can hold against the weight of the suspended arm, then pressure may be applied against the scapula in the direction of abducting it. When recording the grade of strength, it must be noted that the test was "without arm", meaning without using the arm as a lever.

114

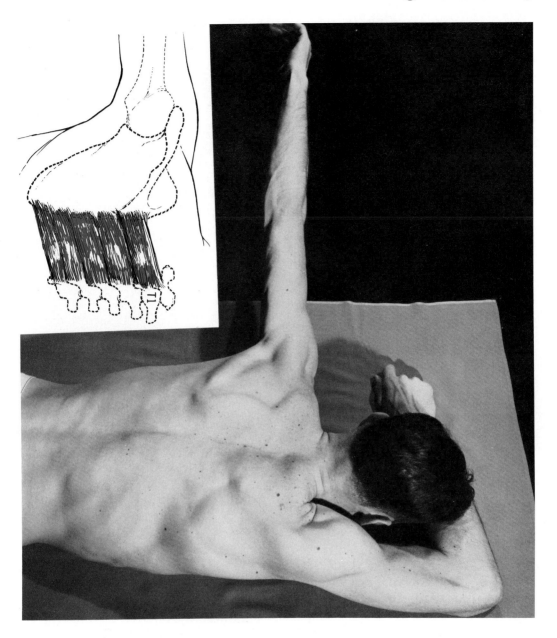

If a position of medial rotation of the humerus and elevation of the scapula is permitted in testing the middle Trapezius, it ceases to be a Trapezius test. As seen in this illustration, the humerus is medially rotated, the scapula is elevated, depressed anteriorly, and adducted by Rhomboid action rather than by middle Trapezius action. A comparison of this photograph with the one on the facing page gives an example of what is meant by obtaining the specific action in which a muscle is the prime mover.

Lower Trapezius

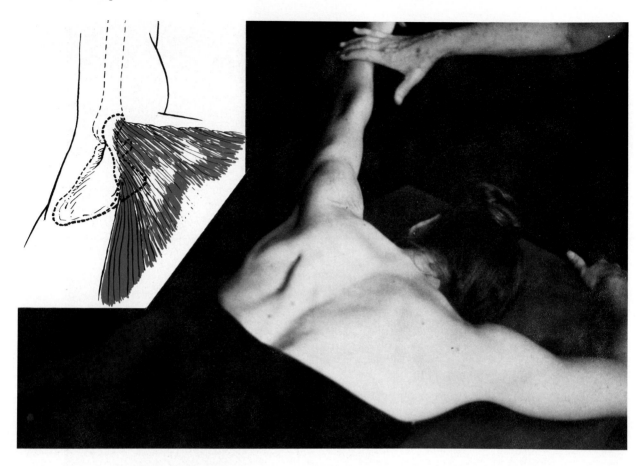

Patient: Prone.

Fixation: None is necessary on the part of the examiner, but the elbow extensors and shoulder muscles, particularly posterior Deltoid, must give necessary fixation to use the arm as a lever in this test.

Test: Depression, lateral rotation of the inferior angle, and adduction of the scapula. To obtain this position of the scapula in order to place emphasis on the action of the lower fibers, and to obtain leverage for the test, the arm is placed diagonally overhead with the shoulder laterally rotated (as denoted by the position of the hand).

Pressure: Against forearm in a downward direction toward table.

Modification of test when posterior shoulder joint muscles are weak:

Patient: Prone with shoulder at edge of table, arm hanging down over side of table.

Fixation: None.

Test: The examiner places the scapula in a position of adduction and depression with the inferior angle rotated laterally.

Pressure: Against the scapula in the direction of elevating it, and somewhat abducting it. When recording the grade of strength, it must be noted that the test was "without arm", meaning without using the arm as a lever.

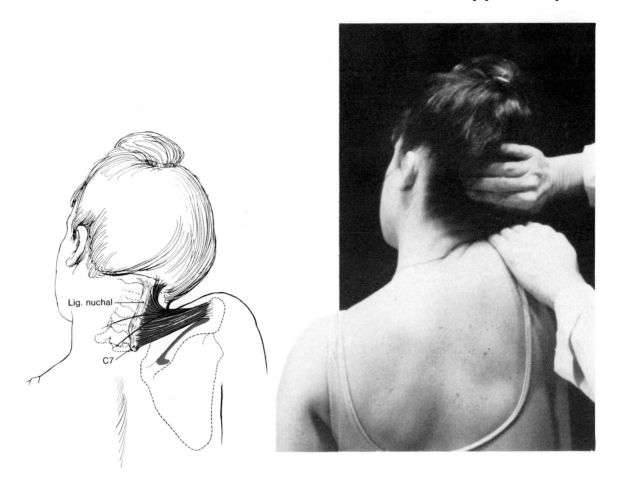

Patient: Sitting.

Fixation: None necessary.

Test: Elevation of the acromial end of the clavicle and scapula; extension and rotation of the head and neck toward the elevated shoulder with the face rotated in the opposite direction.

Pressure: Against the shoulder in the direction of depression, and against the head in the direction of flexion anterolaterally.

Weakness: Decreases the ability to approximate the acromial end of the scapula and the occiput. Decreases the ability to raise the head from a prone position. Interferes with abduction and flexion of the humerus above shoulder level. (See p. 122 for posture of shoulder when entire Trapezius is paralyzed.)

Shortness: Results in a position of elevation of the shoulder girdle (commonly seen in prize fighters and swimmers). In a faulty posture with forward head and kyphosis, the cervical spine is in extension and the upper Trapezius muscles are in a shortened position.

Contracture: Unilateral contracture frequently seen in torticollis cases. For example, the left upper Trapezius is usually contracted along with a contracture of the left Sternocleidomastoid and Scaleni. (See also p. 261.)

Serratus Anterior

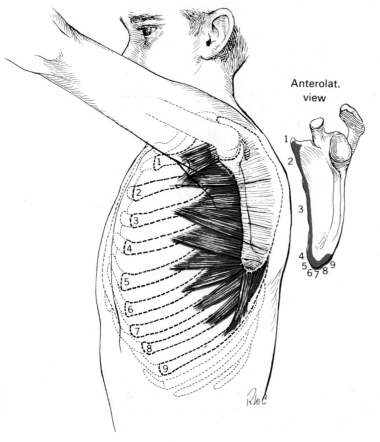

Anterolat. view

Origin: Outer surfaces and superior borders of upper eight or nine ribs.

Insertion: Costal surface of medial border of scapula.

Action: *With the origin fixed*, abducts the scapula, rotates it so the glenoid cavity faces cranially, and holds the medial border of the scapula firmly against the thorax. In addition, the lower fibers may depress the scapula, and the upper fibers may elevate it slightly. Starting from a position with the humerus fixed in flexion and the hands against a wall (see the standing Serratus test, p. 120), the Serratus acts to displace the thorax posteriorly as the effort is made to push the body away from the wall. Another example of this type of action is in a properly executed push-up.

With the scapula stabilized in adduction by the Rhomboids, thereby *fixing the insertion*, the Serratus may act in forced inspiration.

Nerve: Long thoracic, C**5,6,7,**8.

Patient: Supine.

Fixation: None necessary unless the shoulder or elbow muscles are weak, in which case the examiner will support the extremity in the perpendicular position as the test is done.

Test: Abduction of the scapula projecting the upper extremity anteriorly (upward from the table). Movement of the scapula must be observed and the inferior angle palpated to ensure that the scapula is abducting. Projection of the extremity can be accomplished by the action of the Pectoralis minor (aided by the Levator and Rhomboids) when the Serratus is weak, in which case, the scapula tilts forward at the coracoid process, and the inferior angle moves posteriorly and in the direction of medial rotation. Since this type of substitution can occur during this test, the test in which the patient is seated is preferred.

Pressure: Against the subject's fist, transmitting the pressure downward through the extremity to the scapula and forcing it in the direction of adduction. Some pressure may be applied against the lateral border of the scapula as well as against the fist.

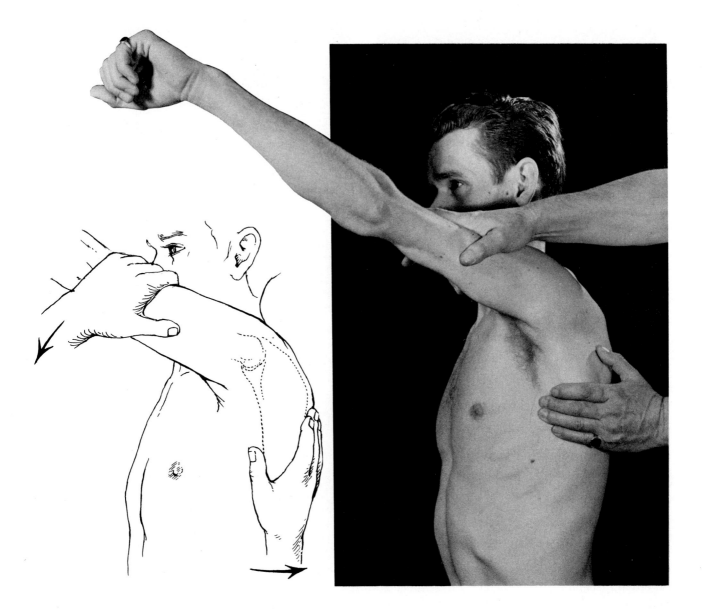

Patient: Sitting.

Fixation: None should be necessary by the examiner if the trunk is stable, but the shoulder flexors must be strong in order to use the arm as a lever in this test.

Test: Stabilization of the scapula in abduction with lateral rotation of the inferior angle in order to maintain the humerus between approximately 120° and 130° flexion. This test emphasizes the upward rotation action of the Serratus in the abducted position as compared to the emphasis on the abduction action shown in the supine and standing tests.

Pressure: Against the dorsal surface of the arm between shoulder and elbow, in the direction of extension, and some against the lateral border of the scapula in the direction of rotating the inferior angle medially. For purposes of photography, the examiner stood behind the subject and applied pressure with the finger tips on the scapula as illustrated. In practice, it is preferable to stand beside the subject and apply pressure as illustrated by the inset. It is not advisable to use a long lever by applying pressure on the forearm or at the wrist because normal shoulder flexors will often "give" before the Serratus even if the Serratus grades only good minus (70%).

Weakness: Makes it difficult to raise the arm in flexion or abduction. Results in "winging" of scapula.

Serratus Anterior

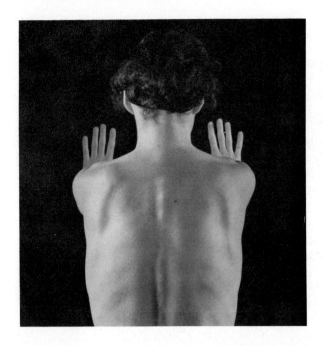

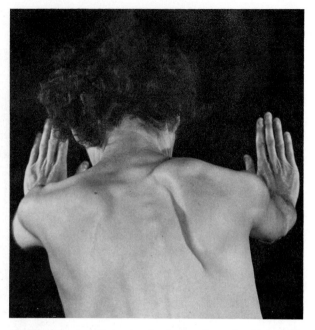

The above photograph shows the winging of the right scapula during a test of the right Serratus anterior.

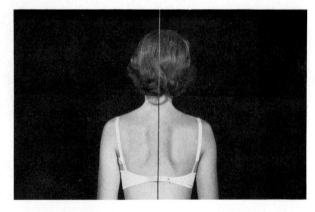

Patient: Standing.

Fixation: None necessary.

Test: Facing a wall and with the elbows straight, the subject places his hands against the wall at shoulder level or slightly above. To begin, the thorax is allowed to "sag" forward so that the scapulae are in a position of some adduction. The subject then pushes hard against the wall, displacing the thorax backward until the scapulae are in a position of abduction.

Resistance: The thorax acts as resistance in this test movement. By fixation of the hands and extended elbows, the scapulae become relatively fixed and the anterolateral rib cage is drawn backward toward the scapulae. (In contrast, the scapula is pulled forward toward the fixed rib cage during the forward thrust of the arm in the supine test shown on p. 118.) Because the resistance of displacing the weight of the thorax makes this a strenuous test, it will differentiate only between strong and weak for purposes of grading.

The above photograph illustrates the posture of the shoulders and scapulae as seen in some cases of mild Serratus anterior weakness. There is slight winging of the scapulae made readily visible when the upper back is very straight. When the upper back is rounded, the scapulae may be adducted and more elevated by the Rhomboids which are direct opponents of the Serratus. Mild Serratus weakness is more prevalent than generally realized. When weakness exists, it can be aggravated by attempting such strenuous exercises as push-ups.

The above photograph (the same subject is on the facing page) shows the extent to which the right arm could be elevated overhead with the subject in standing position. In the absence of Serratus anterior strength, the Trapezius compensated to rotate the scapula as can be seen in the illustration. The upper and lower fibers, particularly, stand out clearly. In repeating the movement five or six times, however, the muscle fatigued and her ability to raise the arm above shoulder level decreased.

In subjects without any paralysis there is wide range of strength in the lower and middle Trapezius.

This variation in strength is associated with postural or occupational stress on these muscles. The grade of strength will range from Fair (50%) to Normal (100%). Because of these wide variations, there also will be wide variations in the ability to raise an arm overhead among those who develop an isolated paralysis of the Serratus. If an individual already has marked weakness of the Trapezius of a postural or an occupational nature, and he subsequently incurs a Serratus paralysis, he will not be able to raise the arm overhead as in the above illustration.

Paralysis of Right Trapezius and Serratus Anterior

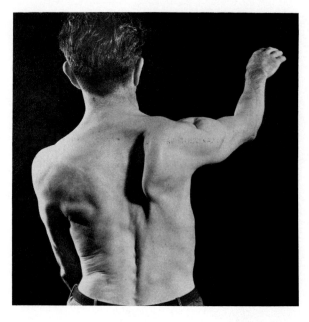

The above photograph shows the subject's inability to raise the arm overhead when both the Serratus and Trapezius are paralyzed. The winging of the medial border of the scapula makes it appear that the Rhomboids were weak, but they were not. (See photo at right.)

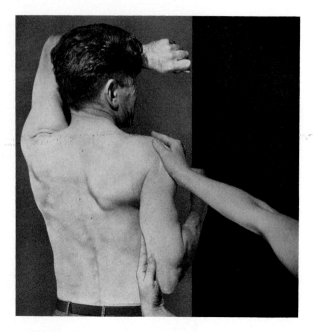

The subject in the photograph above is the same as the one at the left, being tested for strength of the Rhomboids. The subject showed good strength in adducting and stabilizing the scapula. If one did not test the Rhomboid but relied on the appearance during the arm raising, one would easily be lead to the erroneous conclusion that the Rhomboids were weak.

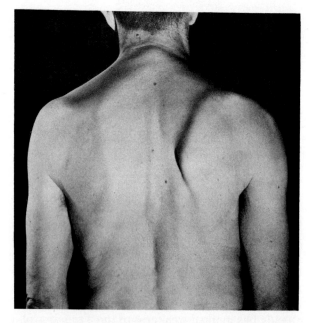

The above photograph shows the abnormal position of the right scapula that results from a paralysis of both the Trapezius and Serratus anterior. The acromial end is abducted and depressed. The inferior angle is rotated medially and elevated.

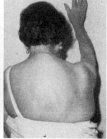

With paralysis of the Trapezius, the ability to elevate the arm in abduction and flexion is diminished. Abduction is more affected than flexion when the Trapezius is paralyzed. Flexion is more affected than abduction when the Serratus is paralyzed.

Muscles Listed According to Spinal Segment Innervation and Acting in Scapular Movements

Scapular Muscles	Spinal Segment								Elevation	Adduction	Downward or Med. Rotat.	Upward or Lat. Rotat.	Depression	Abduction	Anterior Tilt
	Cervical							Th							
	2	3	4	5	6	7	8	1							
Trapezius	2	3	4						Upp. Trap.	Trapezius		Trap.	Low. Trap.		
Levator scapulae		3	4	5					Lev. scap.		Lev. scap.				
Rhomboids, mj. & mi.			4	5					Rhomboids	Rhomboids	Rhomboids				
Serratus anterior				5	6	7	8		Upp. Serratus ant.			Serr. ant.	Low. Serratus ant.	Serr. ant.	
Pectoralis minor					(6)	7	8	1							Pect. mi.

Muscles acting in combined shoulder joint and scapular movements		
Movement	Shoulder muscles	Scapular muscles
Full flexion (to 180°)	Flexors: Anterior deltoid, Biceps, Pectoralis major, upper, Coracobrachialis. Lateral rotators: Infraspinatus, Teres minor, Posterior deltoid	Abductor: Serratus anterior. Lateral rotators: Serratus anterior, Trapezius
Full abduction (to 180°)	Abductors: Deltoid, Supraspinatus, Biceps, long head. Lateral rotators: Infraspinatus, Teres minor, Posterior deltoid	Adductor: Trapezius, acting to stabilize scapula in adduction. Lateral rotators: Trapezius, Serratus anterior
Full extension (to 45°)	Extensors: Posterior deltoid, Teres major, Latissimus dorsi, Triceps, long head	Adductors, medial rotators & elevators: Rhomboids, Levator scapulae. Ant. tilt of scapula by: Pectoralis minor
Full adduction to side against resistance	Adductors: Pectoralis major, Teres major, Latissimus dorsi, Triceps, long head	Adductors, Rhomboids, Trapezius

123

UPPER EXTREMITY MUSCLES, Listed According to Spinal Segment Innervation and Grouped According to Joint Action

Spinal Segment

Cervical 4	5	6	7	8	Th 1	MUSCLE	SHOULDER Abduction	Lat. Rotat.	Flexion	Med. Rotat.	Extension	Adduction	ELBOW Flexion	Extension	FOREARM Supination	Pronation
4	5	6				Supraspinatus	Supraspin.									
(4)	5	6				Infraspinatus		Infraspin.								
	5	6				Teres minor		Teres mi.								
	5	6				Deltoid	Deltoid	Delt., post.	Delt., ant.	Delt., ant.	Delt., post.					
	5	6				Biceps	Biceps, l.h.		Biceps			Biceps, s.h.	Biceps		Biceps	
	5	6				Brachialis							Brachialis			
	5	6				Brachioradialis							Brachiorad.		Brachiorad.	Brachiorad.
	5	6	7			Pectoralis maj., Upp.			Pect. mj., u.	Pect. mj., u.		Pect. mj., u.				
	5	6	7			Subscapularis				Subscap.						
	5	6	(7)			Supinator									Supinator	
	5	6	7			Teres major				Teres mj.	Teres mj.	Teres mj.				
	5	6	7	8		Ext. carpi rad. l. & b.							Ext. c. r. l.			
		6	7			Coracobrachialis			Coracobr.			Coracobr.				
		6	7			Pronator teres							Pron. teres			Pron. teres
		6	7	8		Flex. carpi rad.							Fl. c. rad.			Fl. c. rad.
		6	7	8		Latissimus dorsi				Lat. dorsi	Lat. dorsi	Lat. dorsi				
		6	7	8		Ext. digitorum										
		6	7	8		Ext. digit. min.										
		6	7	8		Ext. carpi ulnaris										
		6	7	8		Abd. poll. long.										
		6	7	8		Ext. poll. brev.										
		6	7	8		Ext. poll. long.										
		6	7	8		Ext. indicis										
		6	7	8	1	Pect. maj., lower						Pect. mj., l.				
		6	7	8	1	Triceps					Tri., l.h.	Tri., l.h.		Triceps		
		(6)	7	8	1	Palmaris long.							Palm. l.			
		(6)	7	8	1	Flex. poll. long.										
		(6)	7	8	1	Lumb. I & II										
		6	7	8	1	Abd. poll. brev.										
		6	7	8	1	Opponens poll.										
		6	7	8	1	Flex. poll br. (s. h.)										
			7	8		Anconeus								Anconeus		
			7	8	1	Flex. carpi ulnaris							Fl. c. ul.			
			7	8	1	Flex. digit. super.										
			7	8	1	Flex. digit. prof.										
			7	8	1	Pronator quad.										Pron. quad.
			(7)	8	1	Abd. digiti min.										
			(7)	8	1	Opp. digiti min.										
			(7)	8	1	Flex. digiti min.										
			(7)	8	1	Lumb. III & IV										
				8	1	Dor. interossei										
				8	1	Palm. interossei										
				8	1	Flex. poll. br. (d.h.)										
				8	1	Add. pollicis										

UPPER EXTREMITY MUSCLES, Listed According to Spinal Segment Innervation and Grouped According to Joint Action (Continued)

| WRIST | | | | CARPOMETACARPAL OF THUMB & LITTLE FINGER AND METACARPOPHALANGEAL JOINTS | | | | | DIG. 2–5 PROX. INTERPHAL. JTS. | | DIG. 1–5 DISTAL INTERPHAL. JTS. | |
Extension	Flexion	Abduction	Adduction	Extension	Abduction	Flexion	Opposition	Adduction	Extension	Flexion	Extension	Flexion
Ext. c. r. l & b		Ext. c. r. l & b										
	Fl. c. rad.	Fl. c. rad.										
Ext. dig.		Ext. dig.		Ext. dig.	Ext. dig.				Ext. dig.		Ext. dig.	
				Ext. dig. min.	Ext. dig. min.				Ext. dig. min.		Ext. dig. min.	
Ext. c. ul.			Ext. c. ul.									
	Abd. poll. l.	Abd. poll. l.		Abd. poll. l.	Abd. poll. l.				/////	/////		
	Ext. poll. b.			Ext. poll. b.	Ext. poll. b.				/////	/////		
Ext. poll. l.	Ext. poll. l			Ext. poll. l.					/////	/////	Ext. poll. l.	
				Ext. ind.				Ext. Ind.	Ext. ind.		Ext. ind.	
	Palm. l.											
	Fl. poll. l.					Fl. poll. l.			/////	/////		Fl. poll. l.
						Lumb. l, ll			Lumb. l, ll		Lumb. l, ll	
					Abd. poll. b.	Abd. poll. b.	Abd. poll. b.		/////	/////	Abd. poll. b.	
							Opp. poll.					
						Fl. poll. b. (s)	Fl. poll. br. (s)		/////	/////	Fl. poll. br. (s)	
	Fl. c. ul.		Fl. c. ul.									
	Fl. dig. sup.					Fl. dig. sup.				Fl. dig. sup.		
	Fl. dig. pro.					Fl. dig. pro.				Fl. dig. pro.		Fl. dig. pro.
					Abd. d. min.	Abd. d. min.			Abd. d. min.		Abd. d. min.	
							Opp. d. min.					
					Fl. d. min.	Fl. d. min.						
						Lumb. ll, lll			Lumb. lll, lV		Lumb. lll, lV	
					Dor. int.				Dor. int.		Dor. int.	
								Palm. int.	Palm. int.		Palm. int.	
						Fl. poll. b. (d)	Fl. poll. b. (d)		/////	/////		
				Add. poll.		Add. poll.	Add. poll.	Add. poll.	/////	/////		

125

UPPER EXTREMITY MUSCLE CHART

PATIENT'S NAME _____   CLINIC No. _____

LEFT   RIGHT

					Examiner / Date						
					Trapezius, upper						
					Trapezius, middle						
					Trapezius, lower						
					Serratus anterior						
					Rhomboids						
					Pectoralis minor						
					Pectoralis major						
					Latissimus dorsi						
					Shoulder medial rotators						
					Shoulder lateral rotators						
					Deltoid, anterior						
					Deltoid, middle						
					Deltoid, posterior						
					Biceps						
					Triceps						
					Brachioradialis						
					Supinators						
					Pronators						
					Flexor carpi radialis						
					Flexor carpi ulnaris						
					Extensor carpi radialis						
					Extensor carpi ulnaris						
					1 Flexor digitorum profundus 1						
					2 Flexor digitorum profundus 2						
					3 Flexor digitorum profundus 3						
					4 Flexor digitorum profundus 4						
					1 Flexor digit. superficialis 1						
					2 Flexor digit. superficialis 2						
					3 Flexor digit. superficialis 3						
					4 Flexor digit. superficialis 4						
					1 Extensor digitorum 1						
					2 Extensor digitorum 2						
					3 Extensor digitorum 3						
					4 Extensor digitorum 4						
					1 Lumbricalis 1						
					2 Lumbricalis 2						
					3 Lumbricalis 3						
					4 Lumbricalis 4						
					1 Dorsal interosseus 1						
					2 Dorsal interosseus 2						
					3 Dorsal interosseus 3						
					4 Dorsal interosseus 4						
					1 Palmar interosseus 1						
					2 Palmar interosseus 2						
					3 Palmar interosseus 3						
					4 Palmar interosseus 4						
					Flexor pollicis longus						
					Flexor pollicis brevis						
					Extensor pollicis longus						
					Extensor pollicis brevis						
					Abductor pollicis longus						
					Abductor pollicis brevis						
					Adductor pollicis						
					Opponens pollicis						
					Flexor digiti minimi						
					Abductor digiti minimi						
					Opponens digiti minimi						

NOTES: _____

_____

_____

CHART FOR ANALYSIS OF MUSCLE IMBALANCE
UPPER EXTREMITY

Name: .... Date: 1st. Ex.- .... 2nd. Ex.- ....

Diagnosis: .... Onset: .... Exam. of .... extremity

		2nd. EX.	1st. EX.	1st. EX.	2nd. EX.		
	FLEXOR POLLICIS BREVIS					EXTENSOR POLLICIS BREVIS	
	FLEXOR POLLICIS LONGUS					EXTENSOR POLLICIS LONGUS	
	OPPONENS POLLICIS					ADDUCTOR POLLICIS	
	ABDUCTOR POLLICIS LONGUS					1 PALMAR INTEROSSEUS	
	ABDUCTOR POLLICIS BREVIS					1 DORSAL INTER. (THUMB ADD.)	
	PALMAR INTEROSSEUS 2					1 DORSAL INTER. (INDEX ABD.)	
	(DORSAL INTEROSSEUS 3)					2 DORSAL INTEROSSEUS	
	(DORSAL INTEROSSEUS 2)					3 DORSAL INTEROSSEUS	
	PALMAR INTEROSSEUS 3					4 DORSAL INTEROSSEUS	
	PALMAR INTEROSSEUS 4					ABDUCTOR DIGITI MINIMI	
	FLEXOR DIGITORUM PROFUNDUS 1					1	
	FLEXOR DIGITORUM PROFUNDUS 2					2 DISTAL INTER-PHALANGEAL	
	FLEXOR DIGITORUM PROFUNDUS 3					3 JOINT EXTENSORS	
	FLEXOR DIGITORUM PROFUNDUS 4					4	
	FLEXOR DIGITORUM SUPERFICIALIS 1					1	
	FLEXOR DIGITORUM SUPERFICIALIS 2					2 PROXIMAL INTER-PHALANGEAL	
	FLEXOR DIGITORUM SUPERFICIALIS 3					3 JOINT EXTENSORS	
	FLEXOR DIGITORUM SUPERFICIALIS 4					4	
	LUMBRICALES & INTEROSSEI 1					1 EXT. DIGIT. & INDICIS	
	LUMBRICALES & INTEROSSEI 2					2 EXT. DIGIT.	
	LUMBRICALES & INTEROSSEI 3					3 EXT. DIGIT.	
	& FLEXOR DIGITI MINIMI 4					4 EXT. DIGIT. COM. & DIG. MIN.	
	OPPONENS DIGITI MINIMI						
	PALMARIS BREVIS						
	PALMARIS LONGUS					EXTENSOR CARPI RADIALIS LONGUS & BREVIS	
	FLEXOR CARPI ULNARIS						
	FLEXOR CARPI RADIALIS					EXTENSOR CARPI ULNARIS	
	BICEPS .... } SUPINATORS					PRONATORS { .... QUADRATUS	
	SUPINATOR ...... }					PRONATORS { .... TERES	
	BRACHIORADIALIS ..... } ELBOW					ELBOW { .... TRICEPS	
	BRACHIALIS .... } FLEXORS					EXTENSORS { .... ANCONEUS	
	BICEPS .... }						
	CORACOBRACHIALIS						
	ANTERIOR DELTOID						
	MIDDLE DELTOID					LATISSIMUS DORSI	
	POSTERIOR DELTOID					CLAV. PECTORALIS MAJOR	
	SUPRASPINATUS					STER. PECTORALIS MAJOR	
	TERES MINOR & INFRASPINATUS					TERES MAJOR & SUBSCAPULARIS	
	SERRATUS ANTERIOR					RHOMBOIDS & LEV. SCAP.	
	UPPER TRAPEZIUS						
	MIDDLE TRAPEZIUS						
	LOWER TRAPEZIUS					PECTORALIS MINOR	

chapter 5

Lower Extremity Muscles

Tests for muscles of:
  Toes
  Ankle
  Knee
  Hip
Chart of muscles grouped according to joint action, lower extremity

Charts for recording muscle examinations
  Chart for analysis of muscle imbalance, lower extremity
  Lower extremity muscle chart
  (For lower extremity nerve-muscle chart, see Chapter 3)

Abductor Hallucis and Adductor Hallucis

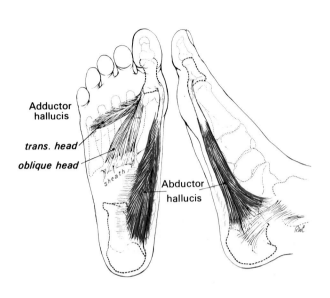

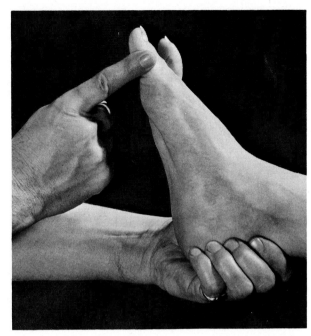

ABDUCTOR HALLUCIS

Origin: Medial process of tuberosity of calcaneus, flexor retinaculum, plantar aponeurosis, and adjacent intermuscular septum.

Insertion: Medial side of base of proximal phalanx of great toe. Some fibers are attached to the medial sesamoid bone, and a tendinous slip may extend to the base of the proximal phalanx of the great toe.

Action: Abducts and assists in flexion of the metatarsophalangeal joint of the great toe, and assists with adduction of the forefoot.

Nerve: Tibial, L4, **5**, **S1**.

Patient: Supine or sitting.

Fixation: The examiner grips the heel firmly.

Test: If possible, abduction of the big toe from the axial line of the foot. This is difficult for the average individual, and the action may be demonstrated by having the patient pull the forefoot in adduction against pressure by the examiner.

Pressure: Against the medial side of the first metatarsal and proximal phalanx. The muscle can be palpated and often seen along the medial border of the foot.

Weakness: Allows forefoot valgus, hallux valgus, and medial displacement of the navicular.

Contracture: Pulls the foot into forefoot varus with the big toe abducted.

ADDUCTOR HALLUCIS (No test illustrated.)

Origin: *Oblique head* from bases of second, third, and fourth metatarsal bones, and sheath of tendon of Peroneus longus. *Transverse head* from plantar metatarsophalangeal ligaments of third, fourth, and fifth digits and deep transverse metatarsal ligament.

Insertion: Lateral side of base of proximal phalanx of great toe.

Action: Adducts and assists in flexing the metatarsophalangeal joint of the great toe.

Nerve: Tibial, S1, **2**.

Contracture: Adduction deformity of great toe. (Hallux valgus.)

Flexor Hallucis Brevis

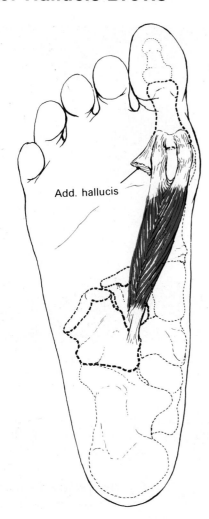

Add. hallucis

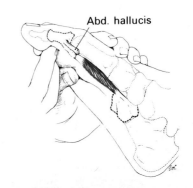

Abd. hallucis

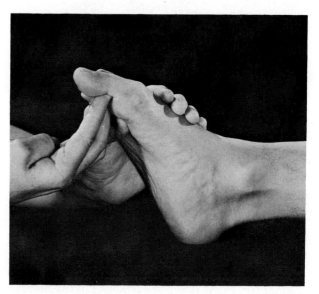

Origin: Medial part of plantar surface of cuboid bone, adjacent part of lateral cuneiform bone, and from prolongation of tendon of Tibialis posterior.

Insertion: Medial and lateral sides of base of proximal phalanx of great toe.

Action: Flexes the metatarsophalangeal joint of the great toe.

Nerve: Tibial L4, **5**, S1.

Patient: Supine or sitting.

Fixation: The examiner stabilizes the foot proximal to the metatarsophalangeal joint, and maintains a neutral position of the foot and ankle. (Plantar flexion of the foot may cause restriction of the test movement by the tension of the opposing long toe extensor muscles.)

Test: Flexion of the metatarsophalangeal joint of the great toe.

Pressure: Against the plantar surface of the proximal phalanx, in the direction of extension.

Note: When the Flexor hallucis longus is paralyzed and the brevis is active, the action of the brevis is clear because the toe flexes at the metatarsophalangeal joint without any flexion of the interphalangeal joint. When the Flexor hallucis brevis is paralyzed and the longus is active, the metatarsophalangeal joint hyperextends, and the interphalangeal joint flexes.

Weakness: Allows hammer toe position of great toe. Lessens stability of the longitudinal arch.

Contracture: The proximal phalanx is held in flexion.

Flexor Digitorum Brevis

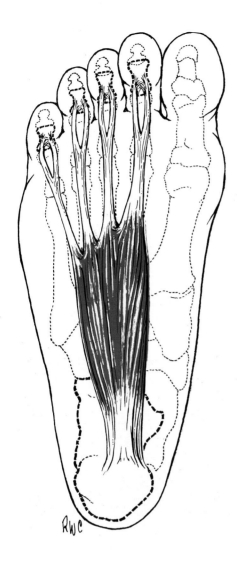

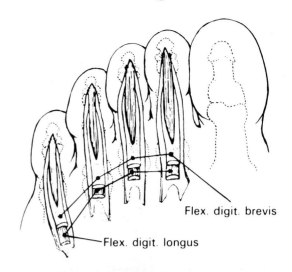

Flex. digit. brevis
Flex. digit. longus

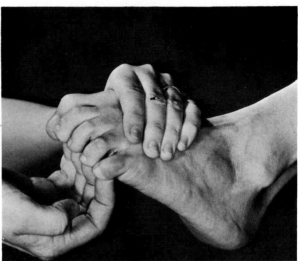

Origin: Medial process of tuberosity of calcaneus, central part of plantar aponeurosis, and adjacent intermuscular septa.

Insertion: Middle phalanx of second through fifth digits.

Action: Flexes the proximal interphalangeal joints, and assists in flexion of the metatarsophalangeal joints of the second through fifth digits.

Nerve: Tibial, L4, **5**, S1.

Patient: Supine or sitting.

Fixation: The examiner stabilizes the proximal phalanges, and maintains a neutral position of the foot and ankle. If the Gastrocnemius and Soleus are paralyzed, the examiner must stabilize the calcaneus, which is the bone of origin, during the toe flexor test.

Test: Flexion of the proximal interphalangeal joints of the second, third, fourth, and fifth digits.

Pressure: Against the plantar surface of the middle phalanx of the four toes, in the direction of extension.

Note: When the Flexor digitorum longus is paralyzed and the brevis is active, the toes flex at the middle phalanx while the distal phalanx remains extended.

Weakness: The ability to flex the proximal interphalangeal joints of the four lateral toes is decreased, and the muscular support of the longitudinal and transverse arches is diminished.

Contracture: Restriction of extension of the toes. The middle phalanges flex, and there is a tendency toward a cavus if the Gastrocnemius and Soleus are weak.

Flexor Hallucis Longus

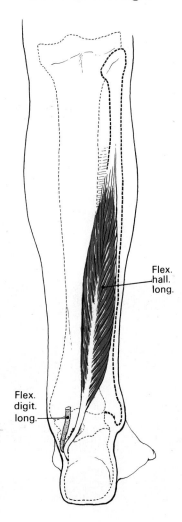

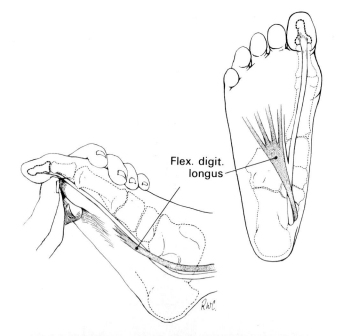

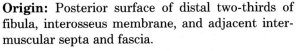

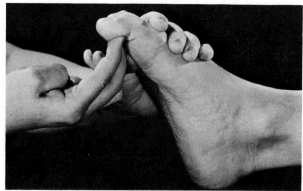

Origin: Posterior surface of distal two-thirds of fibula, interosseus membrane, and adjacent intermuscular septa and fascia.

Insertion: Base of distal phalanx of great toe, plantar surface.

Note: The Flexor hallucis longus is connected to the Flexor digitorum longus by a strong tendinous slip.

Action: Flexes the interphalangeal joint of the great toe, and assists in flexion of the metatarsophalangeal joint, plantar flexion of the ankle joint, and inversion of the foot.

Nerve: Tibial, L5, S1, 2.

Patient: Supine or sitting.

Fixation: The examiner stabilizes the metatarsophalangeal joint in neutral position and maintains the ankle joint approximately midway between dorsal and plantar flexion. (Full dorsiflexion may pro-

duce passive flexion of the interphalangeal joint, and full plantar flexion would allow the muscle to shorten too much to exert its maximum force.) If the Flexor hallucis brevis is very strong and the Flexor hallucis longus weak, it is necessary to restrict the tendency for the metatarsophalangeal joint to flex by holding the proximal phalanx in slight extension.

Test: Flexion of the interphalangeal joint of the great toe.

Pressure: Against the plantar surface of the distal phalanx in the direction of extension.

Weakness: Results in tendency toward hyperextension of interphalangeal joint. Decreases the strength of inversion of the foot and plantar flexion of the ankle. In weight-bearing, permits a tendency toward a pronation of the foot.

Contracture: Hammer-toe deformity of great toe.

Flexor Digitorum Longus and Quadratus Plantae

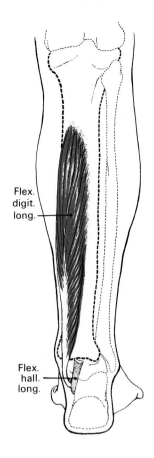

Flex. digit. long.

Flex. hall. long.

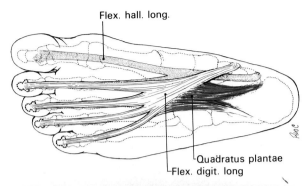

Flex. hall. long.

Quadratus plantae

Flex. digit. long

FLEXOR DIGITORUM LONGUS

Origin: Middle three-fifths of posterior surface of body of tibia, and from fascia covering the Tibialis posterior.

Insertion: Bases of distal phalanges of second through fifth digits.

Action: Flexes proximal and distal interphalangeal and metatarsophalangeal joints of the second through fifth digits. Assists in plantar flexion of the ankle joint and inversion of the foot.

Nerve: Tibial, L5, S1, (2).

Patient: Supine or sitting. In the presence of Gastrocnemius tightness, the knee should be flexed to permit a neutral position of the foot.

Fixation: The examiner stabilizes the metatarsals and maintains a neutral position of the foot and ankle.

Test: Flexion of the distal interphalangeal joints of the second, third, fourth, and fifth digits. The Flexor digitorum is assisted by Quadratus plantae.

Pressure: Against the plantar surface of the distal phalanges of the four toes in the direction of extension.

Weakness: Results in tendency toward hyperextension of distal interphalangeal joints of the four toes. Decreases the ability to invert the foot and plantar flex the ankle. In weight-bearing, weakness permits a tendency toward a pronation of the foot.

Contracture: Flexion deformity of distal phalanges of four lateral toes, with restriction of dorsiflexion and eversion of foot.

QUADRATUS PLANTAE (FLEXOR ACCESSORIUS)

Origin of medial head: Medial surface of calcaneus, and medial border of long plantar ligament.

Origin of lateral head: Lateral border of plantar surface of calcaneus, and lateral border of long plantar ligament.

Insertion: Lateral margin, and dorsal and plantar surfaces of tendon of Flexor digitorum longus.

Action: Modifies the line of pull of the Flexor digitorum longus tendons, and assists in flexing the second through fifth digits.

Nerve: Tibial, S1, 2.

Note: NO TEST ILLUSTRATED.

Lumbricales and Interossei

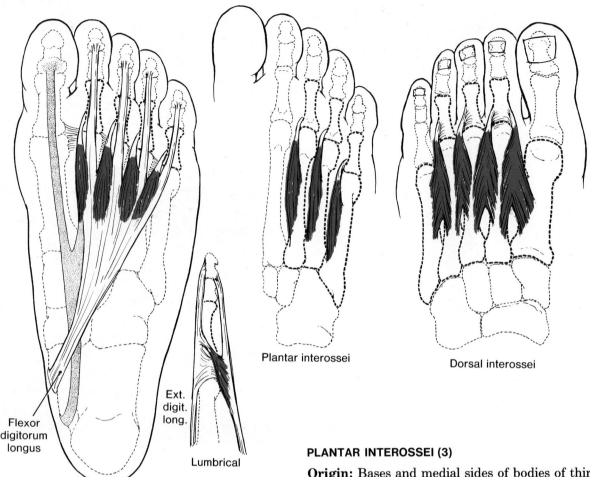

Flexor digitorum longus

Ext. digit. long.

Lumbrical

Lumbricales

Plantar interossei

Dorsal interossei

LUMBRICALES

Origin: First from medial side of first Flexor digitorum longus tendon, second from adjacent sides of first and second Flexor digitorum longus tendons, third from adjacent sides of second and third Flexor digitorum longus tendons, fourth from adjacent sides of third and fourth Flexor digitorum longus tendons.

Insertions: Medial side of proximal phalanx and dorsal expansion of the Extensor digitorum longus tendon of the second through fifth digits.

Action: Flex the metatarsophalangeal joints and assists in extending the interphalangeal joints of the second through fifth digits.

Nerve to Lumbricalis I: Tibial, L4, **5**, S1.

Nerve to Lumbricales II, III, IV: Tibial, L (4), (5), S1, 2.

PLANTAR INTEROSSEI (3)

Origin: Bases and medial sides of bodies of third, fourth, and fifth metarsal bones.

Insertion: Medial sides of bases of proximal phalanges of same digits.

Action: Adduct the third, fourth, and fifth digits toward the axial line through the second digit. Assist in flexion of the metatarsophalangeal joints, and may assist in extension of interphalangeal joints of third, fourth, and fifth digit.

Nerve: Tibial, S1, **2**.

DORSAL INTEROSSEI (4)

Origin: Each by two heads from adjacent sides of the metatarsal bones.

Insertion: Side of proximal phalanx and capsule of metatarsophalangeal joint. First, to medial side of second digit; other three to lateral sides of second, third, and fourth digits.

Action: Abducts the second, third, and fourth digits from the axial line through the second digit. Assists in flexion of the metatarsophalangeal joints, and may assist in extension of interphalangeal joints of second, third, and fourth digits.

Nerve: Tibial, S1, **2**.

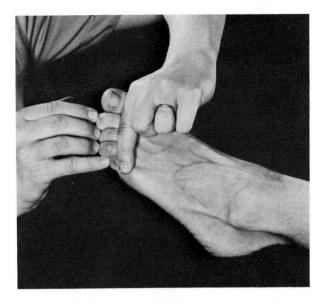

Patient: Supine or sitting.

Fixation: Examiner stabilizes the mid-tarsal region and maintains a neutral position of the foot and ankle.

Test: Flexion of the metatarsophalangeal joints of the second, third, fourth, and fifth digits with an effort to avoid flexion of the interphalanageal joints.

Pressure: Against the plantar surface of the proximal phalanges of the four lateral toes.

Weakness: When these muscles are weak, and the Flexor digitorum longus is active, hyperextension occurs at the metatarsophalangeal joints. The distal joints flex causing a hammer toe position of the four lateral toes. Muscular support of the transverse arch is decreased.

Patient: Supine or sitting.

Fixation: The examiner stabilizes the metatarsophalangeal joints, and maintains the foot and ankle in approximately 20° to 30° plantar flexion.

Test: Extension of the interphalangeal joints of the four lateral toes. (A separate test for the adduction and abduction actions of the interossei is not practical since most individuals cannot perform these movements of the toes.)

Pressure: Against the dorsal surface of the distal phalanges, in the direction of flexion.

Note: Testing the strength of the distal joint extensors is important in determining the imbalance which lends itself to a hammer-toe condition.

Deformities of the foot and ankle. In the following list, foot deformities are defined in terms of the positions of the involved joints. In severe deformities the position of the joint is beyond the normal range of joint motion.

Talipes valgus: Foot everted and accompanied by flattening of the longitudinal arch.

Talipes varus: Foot inverted and accompanied by an increase in the height of the longitudinal arch.

Talipes equinus: Ankle joint plantar flexed.

Talipes equinovalgus: Ankle joint plantar flexed and foot everted.

Talipes equinovarus: Ankle joint plantar flexed and foot inverted. (Club foot.)

Talipes calcaneus: Ankle joint dorsiflexed.

Talipes calcaneovalgus: Ankle joint dorsiflexed and foot everted.

Talipes calcaneovarus: Ankle joint dorsiflexed and foot inverted.

Talipes cavus: Ankle joint dorsiflexed, forefoot plantar flexed, resulting in a high longitudinal arch. With the change in position of the calcaneus, the posterior prominence of the heel tends to be obliterated, and weight-bearing on the calcaneus shifts posteriorly.

Extensor Digitorum Longus and Brevis, and Peroneus Tertius

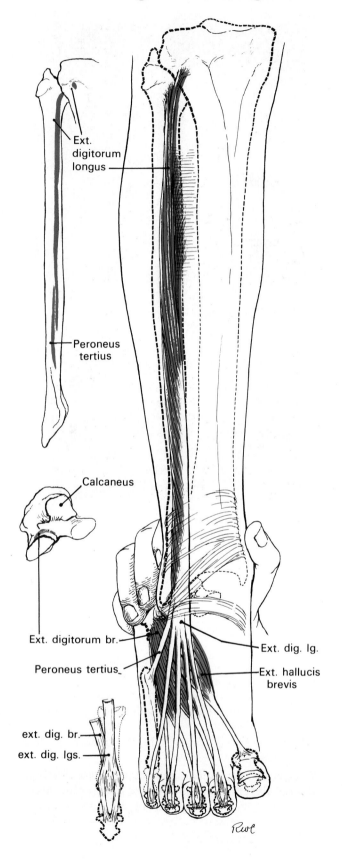

Ext. digitorum longus

Peroneus tertius

Calcaneus

Ext. digitorum br.

Peroneus tertius

ext. dig. br.

ext. dig. lgs.

Ext. dig. lg.

Ext. hallucis brevis

EXTENSOR DIGITORUM LONGUS

Origin: Lateral condyle of tibia, proximal three fourths of anterior surface of body of fibula, proximal part of interosseus membrane, adjacent intermuscular septa, and deep fascia.

Insertion: By four tendons to the second through fifth digits. Each tendon forms an expansion on the dorsal surface of the toe, divides into an intermediate slip attached to base of middle phalanx, and two lateral slips attached to base of distal phalanx.

Action: Extends the metatarsophalangeal joints and assists in extending the interphalangeal joints of the second through fifth digits. Assists in dorsiflexion of the ankle joint and eversion of the foot.

Nerve: Peroneal, L4, **5,** S1.

EXTENSOR DIGITORUM BREVIS

Origin: Distal part of superior and lateral surfaces of calcaneus, lateral talocalcaneal ligament, and apex of inferior extensor retinaculum.

Insertion: By four tendons to first through fourth digits. The most medial slip, also known as the Extensor hallucis brevis, inserts into dorsal surface of base of proximal phalanx of great toe. Other three tendons join lateral sides of tendons of Extensor digitorum longus to second, third, and fourth digits.

Action: Extends the metatarsophalangeal joints of the first through fourth digits, and assists in extending the interphalangeal joints of second, third, and fourth digits.

Nerve: Deep peroneal, L4, **5,** S1.

Note: Since the Extensor digitorum brevis tendons fuse with the tendons of the Extensor longus to the second, third, and fourth digits, the brevis as well as the longus will extend all joints of these toes. Without an Extensor longus, however, there will be no extension of the fifth digit at the metatarsophalangeal joint. To differentiate, palpate the tendon of the longus and the belly of the brevis; and try to detect any difference in movement of the toes.

PERONEUS TERTIUS

Origin: Distal one-third of anterior surface of fibula, interosseus membrane and adjacent intermuscular septum.

Insertion: Dorsal surface, base of fifth metatarsal.

Action: Dorsiflexes ankle joint and everts foot.

Nerve: Deep peroneal, L4, **5,** S1.

Extensor Digitorum Longus and Brevis, and Peroneus Tertius

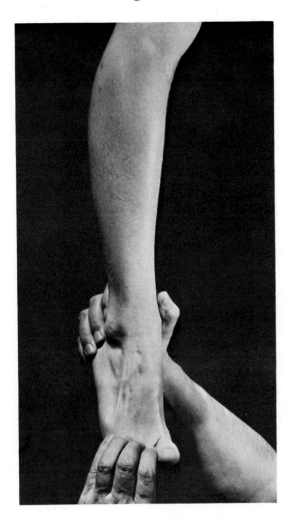 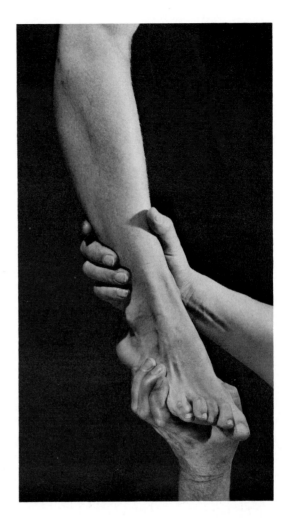

EXTENSOR DIGITORUM LONGUS AND BREVIS

Patient: Supine or sitting.

Fixation: The examiner stabilizes the foot in slight plantar flexion.

Test: Extension of all joints of the second, third, fourth, and fifth digits.

Pressure: Against the dorsal surface of the toes in the direction of flexion.

Weakness: Allows a tendency toward drop-foot and forefoot varus. Diminishes the ability to dorsi-flex the ankle joint and evert the foot. In many cases of flat feet (collapse of the long arch) there is an accompanying weakness of toe extensors.

Contracture: Hyperextension of metatarsopha-langeal joints.

PERONEUS TERTIUS

Patient: Supine or sitting.

Fixation: The examiner supports the leg above the ankle joint.

Test: Dorsiflexion of the ankle joint, with eversion of the foot.

Note: The Peroneus tertius is assisted in this test by the Extensor digitorum longus of which it is a part.

Pressure: Against the lateral side, dorsal surface of the foot in the direction of plantar flexion and inversion.

Weakness: Decreases the ability to evert the foot and dorsiflex the ankle joint.

Contracture: Dorsiflexion of the ankle joint and eversion of the foot.

Extensor Hallucis Longus and Brevis

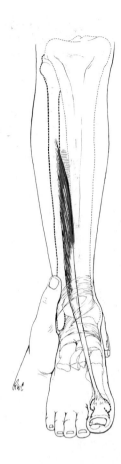

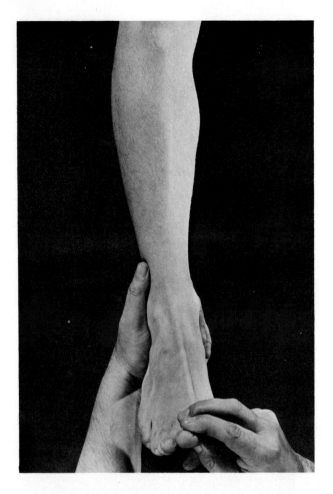

EXTENSOR HALLUCIS LONGUS

Origin: Middle two-quarters of anterior surface of fibula and adjacent interosseous membrane.

Insertion: Base of distal phalanx of great toe.

Action: Extends the metatarsophalangeal and interphalangeal joints of the great toe. Assists in inversion of the foot and dorsiflexion of the ankle joint.

Nerve: Deep peroneal, L4, **5**, S1.

EXTENSOR HALLUCIS BREVIS (same as medial slip of Extensor digitorum brevis)

Origin: Distal part of superior and lateral surfaces of calcaneus, lateral talocalcaneal ligament, and apex of inferior extensor retinaculum.

Insertion: Dorsal surface of base of proximal phalanx of great toe.

Action: Extends the metatarsophalangeal joint of great toe.

Nerve: Deep peroneal L4, **5**, S1.

Patient: Supine or sitting.

Fixation: The examiner stabilizes the foot in slight plantar flexion.

Test: Extension of metatarsophalangeal and interphalangeal joints of the great toe.

Pressure: Against the dorsal surface of the distal and proximal phalanges of the great toe, in the direction of flexion.

Weakness: Decreases the ability to extend the great toe, and allows a position of flexion. The ability to dorsiflex the ankle joint is decreased.

Contracture: Extension of the great toe, with the head of the first metatarsal driven downward.

Note: The paralysis of an Extensor hallucis brevis (first slip of the Extensor digitorum brevis) cannot be determined accurately in the presence of a strong Extensor hallucis longus. However, in paralysis of the longus the action of the brevis is clear. The distal phalanx does not extend, and the proximal phalanx extends in the direction of adduction (toward the axial line of the foot).

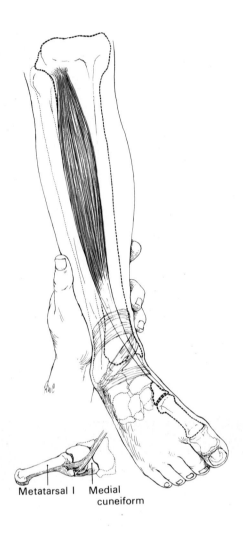

Metatarsal I   Medial cuneiform

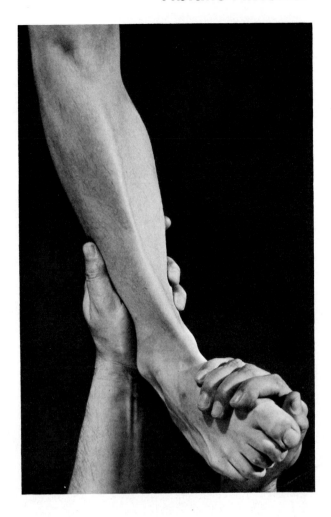

Origin: Lateral condyle and proximal one-half of lateral surface of tibia, interosseus membrane, deep fascia, and lateral intermuscular septum.

Insertion: Medial and plantar surface of medial cuneiform bone, base of first metatarsal bone.

Action: Dorsiflexes the ankle joint and assists in inversion of the foot.

Nerve: Deep peroneal, L4, **5**, S1.

Patient: Supine or sitting (with knee flexed if any gastrocnemius tightness is present).

Fixation: The examiner supports the leg just above the ankle joint.

Test: Dorsiflexion of the ankle joint and inversion of the foot, without extension of the great toe.

Pressure: Against the medial side, dorsal surface of the foot, in the direction of plantar flexion of the ankle joint and eversion of the foot.

Weakness: Decreases the ability to dorsiflex the ankle joint and allows a tendency toward eversion of the foot. This may be seen as a partial drop-foot and tendency toward pronation.

Contracture: Dorsiflexion of ankle joint with inversion of the foot, that is, calcaneo-varus position of the foot.

Note: Although Tibialis anterior weakness may be found in conjunction with a pronated foot, such weakness is seldom found in a congenital flat foot.

Tibialis Posterior

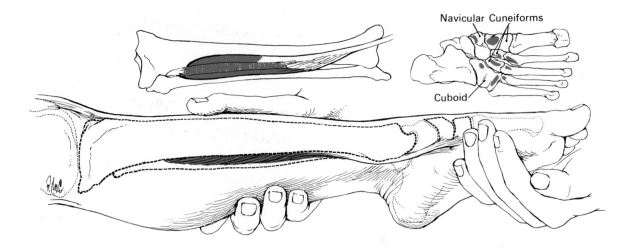

Navicular Cuneiforms

Cuboid

Origin: Most of interosseus membrane, lateral portion of posterior surface of tibia, proximal two-thirds, of medial surface of fibula, adjacent intermuscular septa, and deep fascia.

Nerve: Tibial L(4), **5**, **S1.**

Insertion: Tuberosity of navicular bone and by fibrous expansions to the sustentaculum tali, three cuneiforms, cuboid, and bases of second, third, and fourth metatarsal bones.

Action: Inverts the foot and assists in plantar flexion of the ankle joint.

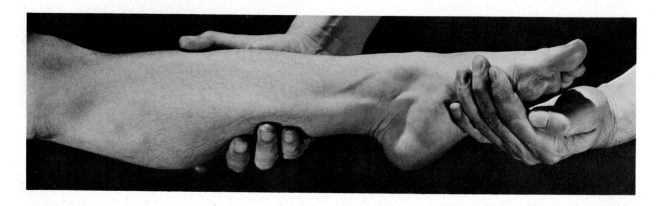

Patient: Supine with extremity in lateral rotation.

Fixation: The examiner supports the leg above the ankle joint.

Test: Inversion of the foot with plantar flexion of the ankle joint.

Pressure: Against the medial side and plantar surface of the foot, in the direction of dorsiflexion of the ankle joint and eversion of the foot.

Note: If the Flexor hallucis longus and Flexor digitorum longus are being substituted for the Tibialis

posterior, the toes will be strongly flexed as pressure is applied.

Weakness: Decreases the ability to invert the foot and plantar flex the ankle joint. Results in pronation of the foot and decreased support of the longitudinal arch. Interferes with the ability to rise on toes, and inclines toward what is commonly called a Gastroncnemius limp.

Contracture: Equino-varus position in non-weight bearing, and a supinated position of the heel with forefoot varus in weight-bearing.

Peroneus Longus and Brevis

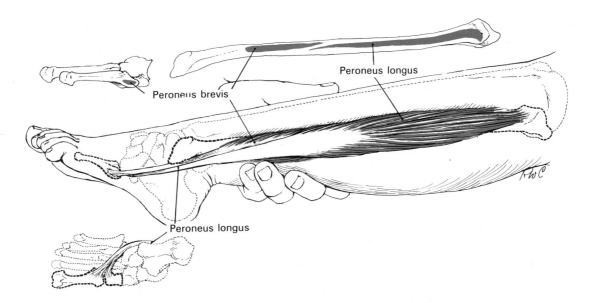

PERONEUS LONGUS

Origin: Lateral condyle of tibia, head and proximal two-thirds of lateral surface of fibula, intermuscular septa, and adjacent deep fascia.

Insertion: Lateral side of base of first metatarsal and of medial cuneiform bone.

Action: Everts foot, assists in plantar flexion of ankle joint, and depresses head of first metatarsal.

Nerve: Superficial peroneal, L4, **5**, **S1.**

PERONEUS BREVIS

Origin: Distal two-thirds of lateral surface of fibula and adjacent intermuscular septa.

Insertion: Tuberosity at base of fifth metatarsal bone, lateral side.

Action: Everts foot, and assists in plantar flexion of ankle.

Nerve: Superficial peroneal, L4, **5**, **S1.**

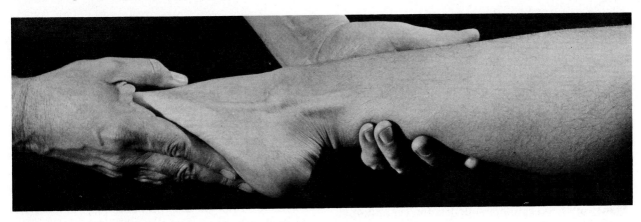

Patient: Supine with extremity medially rotated, or sidelying (on opposite side).

Fixation: The examiner supports the leg above the ankle joint.

Test: Eversion of the foot with plantar flexion of the ankle joint.

Pressure: Against the lateral border and sole of the foot, in the direction of inversion of the foot and dorsiflexion of the ankle joint.

Weakness: Decreases the strength of eversion of the foot and plantar flexion of the ankle joint. Allows a varus position of the foot and lessens the ability to rise on the toes. Lateral stability of the ankle is decreased.

Contracture: Results in an everted or valgus position of the foot.

Note: In weight-bearing, with a strong pull on its insertion at the base of the first metatarsal, the Peroneus longus causes the head of the first metatarsal to be pressed downward into the supporting surface.

Soleus

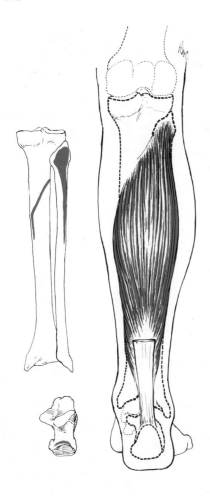

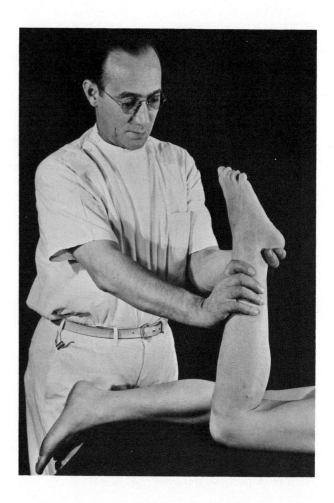

Origin: Posterior surfaces of head of fibula and proximal one-third of its body, soleal line and middle one-third of medial border of tibia, and tendinous arch between tibia and fibula.

Insertion: With tendon of Gastrocnemius into posterior surface of calcaneus.

Action: Plantar flexes the ankle joint.

Nerve: Tibial, L5, S1, 2.

Patient: Prone with the knee flexed 90° or more.

Fixation: The examiner supports the leg proximal to the ankle.

Test: Plantar flexion of the ankle joint without inversion or eversion of the foot.

Pressure: Against the calcaneus (as illustrated) in the direction of pulling the heel plantarward. When there is marked weakness, the patient may not be able to hold against pressure at the heel. When weakness is not marked, more leverage is necessary and is obtained by applying pressure simultaneously against the sole of the foot. (See p. 146).

Note: Inversion of the foot shows substitution by Tibialis posterior and toe flexors. Eversion shows substitution by the Peroneals. Extension of the knee is evidence of attempting to assist with Gastrocnemius, that is, the Gastrocnemius is at a disadvantage with the knee flexed 90° or more, and, to bring it into a stronger action, the patient will attempt to extend the knee.

Weakness: Permits a calcaneus position of the foot and predisposes toward a cavus. Results in inability to rise on toes. In standing, the insertion of the Soleus muscle on the calcaneus becomes the fixed point for action of this muscle in maintaining normal alignment of the leg in relation to the foot. The deviation that results from weakness of the Soleus may appear as a slight knee flexion fault in posture, but more often results in an anterior displacement of the body weight from the normal plumb-line distribution, as seen when the plumbline is hung slightly anterior to the outer malleolus. (See p. 279. Shortness and contracture: Continued on p. 146.)

Gastrocnemius and Plantaris

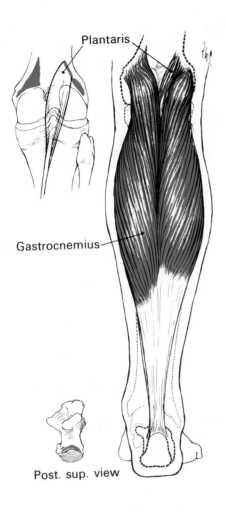

Plantaris

Gastrocnemius

Post. sup. view

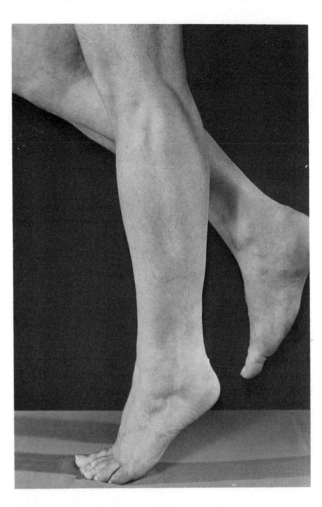

GASTROCNEMIUS

Origin of medial head: Proximal and posterior part of medial condyle and adjacent part of femur, capsule of knee joint.

Origin of lateral head: Lateral condyle and posterior surface of femur, capsule of knee joint.

Insertion: Middle part of posterior surface of calcaneus.

Nerve: Tibial, S1, **2**.

PLANTARIS

Origin: Distal part of lateral supracondylar line of femur and adjacent part of its popliteal surface, oblique popliteal ligament of knee joint.

Insertion: Posterior part of calcaneus.

Nerve: Tibial, L4, **5**, S1, (2).

Action: The Gastrocnemius and the Plantaris plantar flex the ankle joint and assist in flexion of the knee joint.

ANKLE PLANTAR FLEXORS

Patient: Standing. (Patient may steady himself with a hand on the table, but should not take any weight on the hand.)

Test: Patient rises on toes, pushing the body weight directly upward.

Note: Inclining the body forward and flexing the knee is evidence of weakness; the patient dorsiflexes the ankle joint attempting to clear the heel from the floor by tension of the plantar flexors as the body weight is thrown forward.

Shortness: Constant wearing of high-heeled shoes by women tends to develop a shortness of the Gastrocnemius and Soleus muscles.

Muscles which act in plantar flexion:

Soleus	Ankle joint plantar flex-
Gastrocnemius	ors. (Tendo calcaneus
Plantaris	group)
Tibialis posterior	Forefoot, and ankle joint
Peroneus longus	plantar flexors.
Peroneus brevis	
Flexor hallucis longus	Toe, forefoot, and ankle
Flexor digit, longus	joint plantar flexors.

Ankle Plantar Flexors

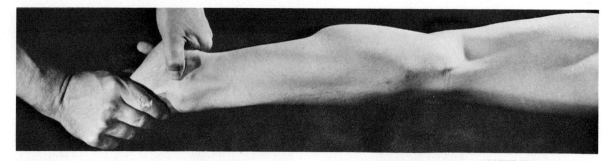

Patient: Prone with knee extended, and the foot projecting over the end of the table.

Fixation: The weight of the extremity resting on a firm table should be sufficient fixation of the part.

Test: Plantar flexion of the foot with emphasis on pulling the heel upward more than pushing the forefoot downward. This test movement does not attempt to isolate the Gastrocnemius action from other plantar flexors, but the presence or absence of a Gastrocnemius can be determined by careful observation during the test.

Pressure: For maximum pressure in this position, it is necessary to apply pressure against the forefoot as well as against the calcaneus. If the muscle is very weak, pressure against the calcaneus is sufficient.

The Gastrocnemius usually can be seen and always can be palpated if it is contracting during the plantar flexion test. Movements of the toes and forefoot should be observed carefully during the test to detect substitutions. The patient may be able to flex the anterior part of the foot by toe flexors, Tibialis posterior, and Peroneus longus without a direct upward pull on the heel by the Tendo calcaneus. If the Gastrocnemius and Soleus are weak, the heel will be "pushed" up secondary to flexion of the anterior part of the foot rather than "pulled" up simultaneously with the flexion of the forepart of the foot. If pressure is applied to the heel rather than to the ball of the foot, it is possible to isolate, partially, the combined action of the Gastrocnemius and Soleus from the other plantar flexors. Movement of the foot toward eversion or inversion will show imbalance in opposing lateral and medial muscles and, if pronounced, will show an attempt to substitute the Peroneals or Tibialis posterior for the Gastrocnemius and Soleus.

Action of the Gastrocnemius often can be demonstrated in the knee flexion test when the Hamstrings are weak. In the prone position with the knee fully extended, the patient is asked to bend the knee against resistance. If the Gastrocnemius is strong, there will be plantar flexion at the ankle as the Gastrocnemius acts to initiate knee flexion. When the Gastrocnemius is weak and stretched, the foot is pulled into dorsiflexion to place slight tension on the Gastrocnemius as it acts to flex the knee.

Weakness: Permits a calcaneus position of the foot if the Gastrocnemius and Soleus are weak. In standing, results in hyperextension of the knee and inability to rise on toes. In walking, the inability to transfer weight normally results in a "Gastrocnemius limp."

Contracture: Equinus position of the foot and flexion of the knee.

Shortness: Restriction of dorsiflexion of the ankle when the knee is extended, and restriction of knee extension when the ankle is dorsiflexed. In the transfer of weight during the stance phase in walking, shortness limits the normal dorsiflexion of the ankle joint.

(SOLEUS con't. from p. 144.)

A nonparalytic type of weakness may result from sudden trauma to the muscle as in landing from a jump in a position of ankle dorsiflexion and knee flexion; or gradual trauma from repeated deep knee bending in which the ankle is fully dorsiflexed. The Gastrocnemius escapes the stretch because of the knee flexion.

Contracture: Equinus position of the foot both in weight-bearing and nonweight-bearing.

Shortness: A tendency toward hyperextension of the knee in the standing position. When walking barefoot the shortness is compensated for by toeing out, thereby transferring the weight from posterolateral heel to anteromedial forefoot. In heels, the shortness may go unnoticed.

Weakness of Soleus and Gastrocnemius in Standing

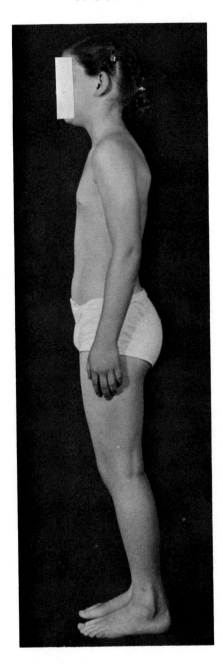 

In standing, the knee joints will tend to *flex* and the ankle joints dorsiflex if the *Soleus* muscles are weak.

Because the Soleus acts to plantar flex the foot, in standing it acts to hold the leg back in normal alignment with the foot. With weakness of the muscle, the ankle joint dorsiflexes. In standing, this generally is accompanied by some knee flexion or may be accompanied by a forward inclination of the body as a whole from the ankles up. A strong Soleus may help compensate for a weak Quadriceps by pulling the leg back, thus passively extending the knee.

In standing, the knee joints will tend to *hyperextend* and the ankle joints plantar flex if the *Gastrocnemius* muscles are weak.

Unlike the Soleus, the Gastrocnemius passes over the knee joint. It is a flexor of the knee joint and as such is a stabilizer in helping to prevent hyperextension. While the action of the Gastrocnemius is the same as the Soleus on the ankle joint, and conceivably weakness could result in the same change of alignment as does weakness of the Soleus, this does not occur. Instead, with Gastrocnemius weakness, the knee tends to hyperextend in standing.

Tests for Length of Hamstring Muscles

Supine with the low back in a position of normal anterior curve and with the pelvis in neutral position, normal Hamstring length will permit approximately 70° of hip joint flexion. (See fig. A.) The position of anterior curve is not a stable one for testing and, in the supine position, there are variations in the amount of anterior curve among different subjects. In order to standardize testing as much as possible, it is necessary to have the low back flat on the table. To help ensure that the back is flat, it is necessary, also, that the subject be on a firm surface, not on a soft, padded table top for the test.

In figs. B, C, and D, the pelvis is in 10° posterior tilt and the low back is flat on the table. The left leg is extended while the right is tested. (The hip-flexor and low-back muscles are normal in length.)

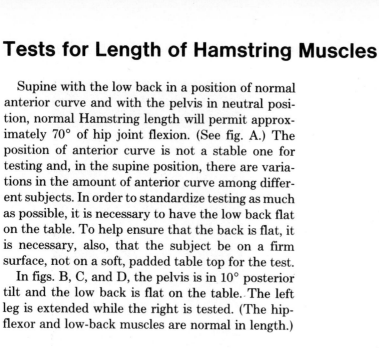

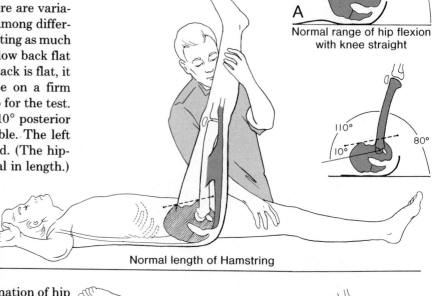

A

Normal range of hip flexion with knee straight

B   Normal length of Hamstring

The straight-leg-raising test is a combination of hip flexion and flexion (flattening) of the low back. The 70° hip flexion range and the 10° posterior pelvic tilt permit the range of 80° upward from the table in the straight-leg-raising test as seen in fig. B. This 80° range is considered normal.

Either limited or excessive lumbar spine flexion adversely affects the test for Hamstring length. (See illustrations on facing page and photographs on pp. 152 and 153.)

Fig. C shows excessive Hamstring length, and fig. D shows short Hamstrings.

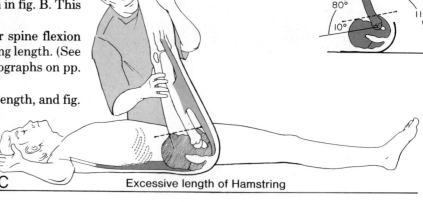

C   Excessive length of Hamstring

Short Hamstrings exert a downward pull on the ischium in the direction of posteriorly tilting the pelvis as the straight leg is raised. To prevent excessive posterior pelvic tilt and excessive flexion of the back, it is necessary to stabilize the pelvis with the low back in the flat back position by holding the opposite leg firmly down.

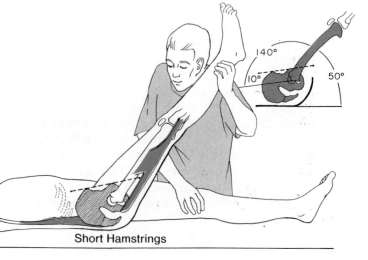

D   Short Hamstrings

Tests for Length of Hamstring Muscles

In standing, or in supine position with legs extended, shortness of hip flexor muscles pulls the low back forward into hyperextension. Fig. A' illustrates the shortness that is present in the left hip flexors in figs. B' and C'. This hip flexor shortness adversely affects the test for Hamstring length and may lead to a misdiagnosis regarding the length of these muscles. Since accuracy in testing for Hamstring length requires that the low back be flat, there must be a modification of the position of the extended leg to allow the back to flatten when hip flexor shortness is present. (See p. 152, *lower left fig.*)

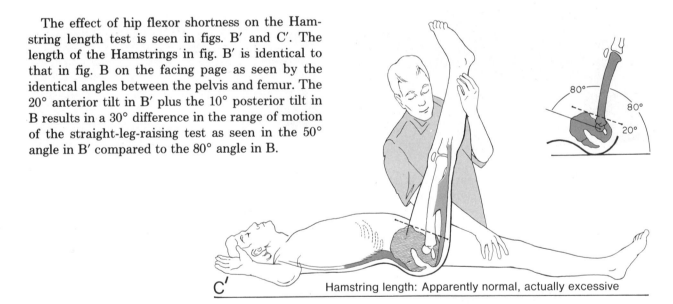

A'

Short hip flexors hold back in hyperextension

B'

Hamstring length: Apparently short, actually normal

The effect of hip flexor shortness on the Hamstring length test is seen in figs. B' and C'. The length of the Hamstrings in fig. B' is identical to that in fig. B on the facing page as seen by the identical angles between the pelvis and femur. The 20° anterior tilt in B' plus the 10° posterior tilt in B results in a 30° difference in the range of motion of the straight-leg-raising test as seen in the 50° angle in B' compared to the 80° angle in B.

C'

Hamstring length: Apparently normal, actually excessive

The Hamstring length in fig. C' is the same as in C on the facing page. The arc of motion in relation to the table makes it appear as if the Hamstrings in fig. C' are normal when, in fact, they are excessive in length.

Fig. D' shows excessive posterior tilt of the pelvis which allows the leg to be raised slightly higher than in fig. D on the facing page although the Hamstring length is the same in both instances. With the opposite leg held firmly down, excessive posterior tilt will not occur except in subjects who have excessive length in hip flexors.

D'

Hamstring length: Apparent length greater than actual

Tests for Length of Hamstring Muscles

The straight-leg-raising test, with low back flat on the table, shows normal length of Hamstring muscles, which permits flexion of the thigh toward the pelvis (hip joint flexion) to an angle of about 80° up from the table.

In forward bending, normal Hamstring length permits flexion of the *pelvis toward the thigh* (hip joint flexion) as illustrated.

When the low back is hyperextended (arched up from the table), the normal Hamstrings appear to be short.

When the hip and knee are flexed on one side as the opposite straight leg is raised, the back flexes too much and the Hamstrings appear to be longer than normal. (This position should not be used for testing or stretching Hamstrings.)

The subject shows good postural alignment in standing, and strength and length tests showed good muscle balance.

Excessive Hamstring length permits excessive flexion of the *thigh toward the pelvis* (hip joint flexion).

In forward bending, excessive Hamstring length permits excessive flexion of the *pelvis toward the thigh* (hip joint flexion.) This subject also has excessive flexion in the midback (thoracolumbar) area.

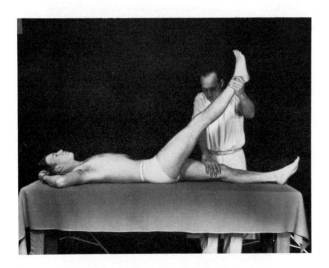

Short Hamstrings limit flexion of the *thigh toward the pelvis* (hip joint flexion).

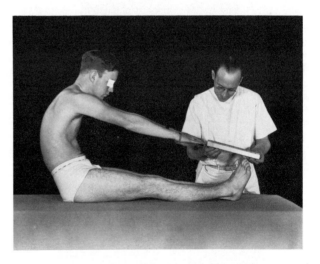

In forward bending, short Hamstrings limit flexion of the *pelvis toward the thigh* (hip joint flexion). Note that the angle of the posterior surface of the pelvis with the table is the same as the angle of the posterior thigh with the table in figure at left.

Tests for Length of Hamstring Muscles

The Hamstrings appear to be short, but the test is not accurate because the low back is not flat on the table. Shortness of the hip flexors on the side of the extended leg holds the back in hyperextension.

Flexion of the *pelvis toward the thigh* (hip flexion) appears to be almost normal in forward bending. (Since both hips are in flexion in forward bending, hip flexor shortness does not interfere with the movement of the pelvis toward the thigh as occurs when one leg is extended.)

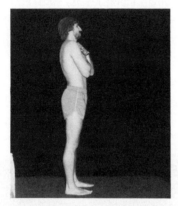

The lordosis in standing is evidence of the hip flexor shortness in this subject.

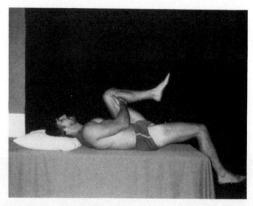

A test for length of hip flexors confirms that there is shortness of these muscles.

To accommodate for the hip flexor shortness and allow the low back to flatten, the thigh is *passively* flexed by a pillow under the knee, *not actively* held in flexion by the subject. With the back flat, the test accurately shows Hamstrings to be almost normal in length.

In testing for Hamstring length and in exercising to stretch short Hamstrings, *avoid* placing one hip and knee in flexed position, as illustrated, while raising the other. The flexibility of the low back rather than of the Hamstrings is adding to the range of motion. Not infrequently, there is excessive back flexibility along with Hamstring shortness. To be therapeutic, the stretching needs to be localized to the area where it is needed.

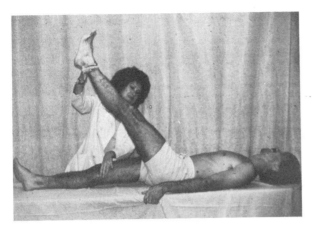

Short Hamstrings limit the flexion of the *thigh toward the pelvis* (hip joint flexion).

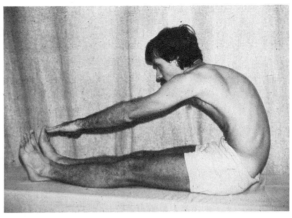

Short Hamstrings limit the flexion of the *pelvis toward the thigh*, but the excessive flexion of the low back allows this subject to almost touch his toes in forward bending.

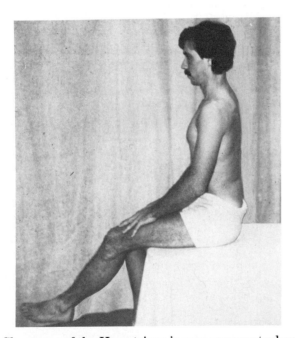

Shortness of the Hamstrings is very apparent when the subject tries to extend the knee while sitting with back held straight.

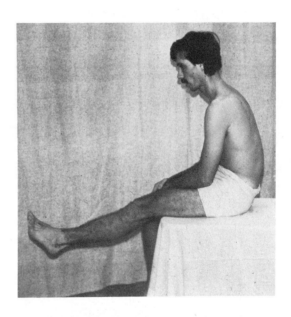

When the subject is allowed to sit with the low back in flexion, Hamstring shortness is not apparent. As an exercise to stretch Hamstrings in the sitting position, the back must be kept straight.

Medial Hamstrings: Semitendinosus and Semimembranosus

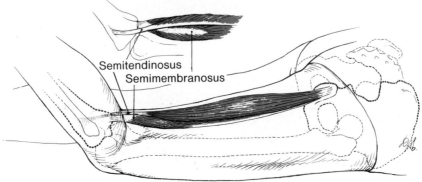

Lt. extremity, medial view

Semitendinosus
Semimembranosus

Rt. extremity, posterolateral view

SEMITENDINOSUS

Origin: Tuberosity of ischium by tendon common with long head of Biceps femoris.

Insertion: Proximal part of medial surface of body of tibia, and deep fascia of leg.

Action: Flexes and medially rotates the knee joint. Extends and assists in medial rotation of the hip joint.

Nerve: Sciatic (tibial br.), L4, **5**, S1, **2**.

SEMIMEMBRANOSUS

Origin: Tuberosity of ischium, proximal and lateral to Biceps femoris and Semitendinosus.

Insertion: Posteromedial aspect of medial condyle of tibia.

Action: Flexes and medially rotates the knee joint. Extends and assists in medial rotation of the hip joint.

Nerve: Sciatic (tibial br.), L4, **5**, S1, **2**.

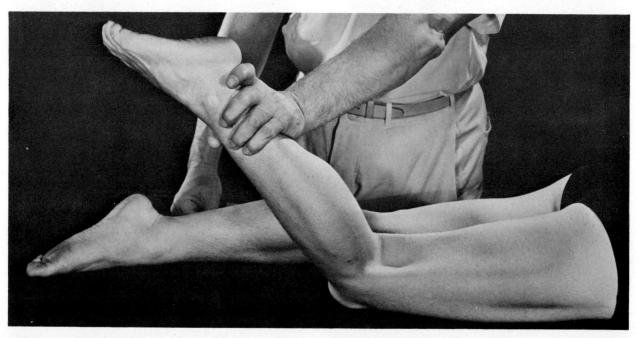

Patient: Prone.

Fixation: The examiner should hold the thigh down firmly on the table. (To avoid covering the muscle belly of the medial Hamstrings, fixation is not illustrated.)

Test: Flexion of the knee to less than 90°, with the thigh in medial rotation, and the leg medially rotated on the thigh.

Pressure: Against the leg proximal to the ankle, in the direction of knee extension.

Lateral Hamstrings: Biceps Femoris

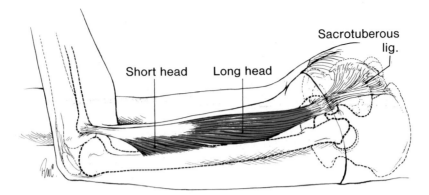

Short head Long head Sacrotuberous lig.

Origin of long head: Distal part of sacrotuberous ligament, and posterior part of tuberosity of ischium.

Origin of short head: Lateral lip of linea aspera, proximal two-thirds of supracondylar line, and lateral intermuscular septum.

Insertion: Lateral side of head of fibula, lateral condyle of tibia, deep fascia on lateral side of leg.

Action: The long and short heads of the Biceps femoris flex and laterally rotate the knee joint. In addition, the long head extends and assists in lateral rotation of the hip joint.

Nerve to long head: Sciatic (tibial br.), L5, S1, 2, 3.

Nerve to short head: Sciatic (peroneal br.), L5, S1, 2.

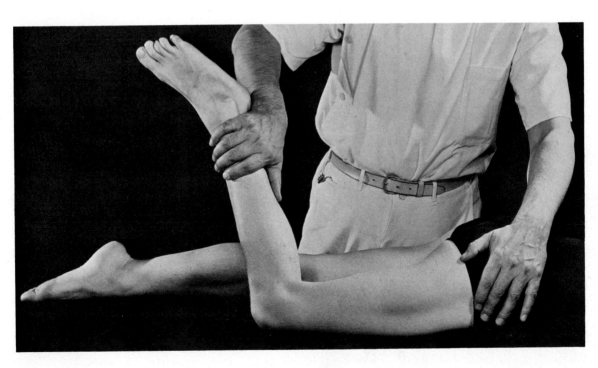

Patient: Prone.

Fixation: The examiner should hold the thigh firmly down on the table. (Not illustrated in order to avoid covering the muscles.)

Test: Flexion of the knee to less than 90° with the thigh in slight lateral rotation, and the leg in slight lateral rotation on the thigh.

Pressure: Against the leg proximal to the ankle, in the direction of knee extension.

Hamstrings and Gracilis

Weakness: Weakness of both the medial and lateral Hamstrings permits hyperextension of the knee. When this weakness is bilateral, the pelvis may tilt anteriorly and the lumbar spine may assume a lordotic position. If the weakness is unilateral, a pelvic rotation may result. Weakness of lateral Hamstrings causes a tendency toward loss of lateral stability of the knee, allowing a thrust in the direction of bow-leg position in weight-bearing. Weakness of medial Hamstrings decreases the medial stability of the knee joint, and permits a knock-knee position with a tendency toward lateral rotation of the leg on the femur.

Contracture: Contracture of both the medial and lateral Hamstrings results in a position of knee flexion, and, if the contracture is extreme, it will be accompanied by a posterior tilting of the pelvis and a flattening of the lumbar spine.

Shortness: Restriction of knee extension when the hip is flexed, or restriction of hip flexion when the knee is extended. (See pp. 148 and 153.) Shortness of the Hamstrings will permit standing erect, but the posture will be characterized by a posterior tilting of the pelvis and a decrease of the lumbar curvature.

Note: Ordinarily, the hip flexors act to *safeguard* the Hamstrings during knee flexion. Do not expect the subject to hold full-knee flexion nor to hold against the same amount of pressure with the hip extended in the prone position that he could resist with the hip flexed in sitting. The frequent occurrence of muscle "cramping" during the Hamstring test on nonparalytic individuals is evidence of too much knee flexion in proportion to the amount of pressure. In other words, the degree of knee flexion should be less if very strong pressure is applied. For maximum pressure, hip flexion should be permitted by testing in the sitting position.

Weakness of Popliteus and Gastrocnemius may interfere with initiating knee flexion. Substitution of Sartorius action will appear in the form of hip flexion as knee flexion is *initiated*. A short Rectus femoris, limiting knee flexion range of motion, will cause hip flexion as the knee flexion motion is *completed*. (Hip flexion in the prone position is seen as an anterior tilt of the pelvis with lumbar spine hyperextension.)

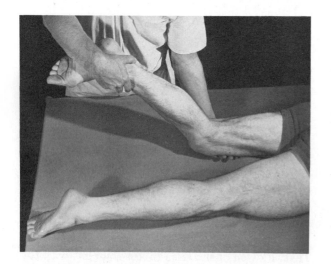

The action of the Gracilis as a knee flexor is illustrated by this figure. It is brought into action by the test position and pressure as used for the medial Hamstrings. The Gracilis has its origin on the pubis, and the medial Hamstrings arise from the ischium.

Popliteus

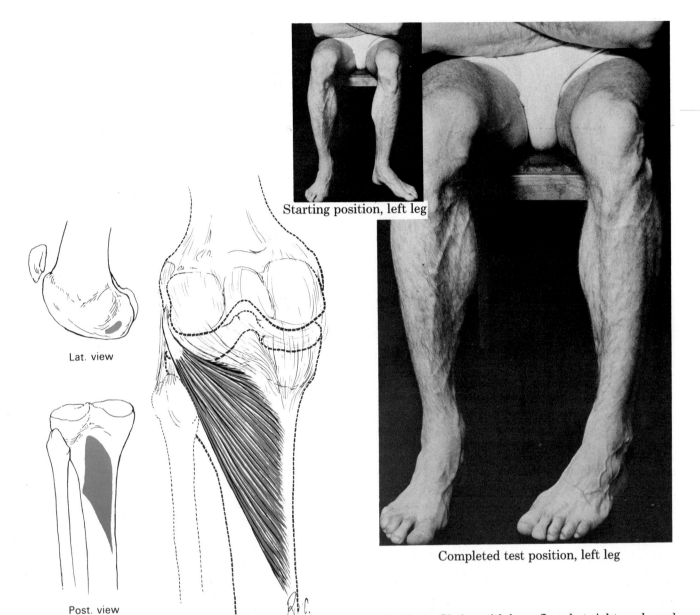

Starting position, left leg

Completed test position, left leg

Origin: Anterior part of groove on lateral condyle of femur, and oblique popliteal ligament of knee joint.

Insertion: Triangular area proximal to soleal line on posterior surface of tibia, and fascia covering the muscle.

Action: In nonweight-bearing (that is, *with the origin fixed*) the Popliteus medially rotates the tibia on the femur and flexes the knee joint. In weight-bearing (that is, *with the insertion fixed*), it laterally rotates the femur on the tibia and flexes the knee joint. This muscle helps to reinforce the posterior ligaments of the knee joint.

Nerve: Tibial L4, **5,** S1.

Patient: Sitting with knee flexed at right angle and with leg in lateral rotation of tibia on femur.

Fixation: None necessary.

Test: Medial rotation of the tibia on the femur.

Pressure: Seldom is resistance or pressure applied since the movement is not used as a test for the purpose of grading the Popliteus, but merely to indicate whether the muscle is active or not.

Weakness: May result in hyperextension of the knee and lateral rotation of the leg on the thigh. A Popliteus weakness is usually found in instances of imbalance between the lateral and medial Hamstrings in which the medial Hamstrings are weak and the lateral are strong.

Shortness: Results in slight flexion of the knee and medial rotation of the leg on the thigh.

157

Quadriceps Femoris

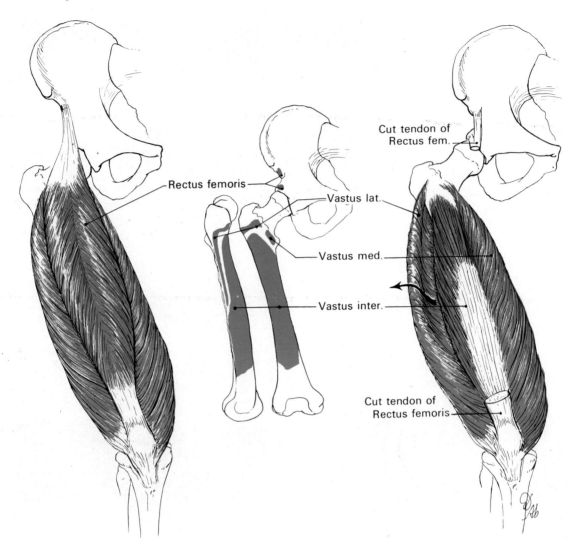

Origin of Rectus femoris: *Straight head* from anterior inferior iliac spine. *Reflected head* from groove above rim of acetabulum.

Origin of Vastus lateralis: Proximal part of intertrochanteric line, anterior and inferior borders of greater trochanter, lateral lip of gluteal tuberosity, proximal one-half of lateral lip of linea aspera, and lateral intermuscular septum.

Origin of Vastus intermedius: Anterior and lateral surfaces of proximal two-thirds of body of femur, distal one-half of linea aspera, and lateral intermuscular septum.

Origin of Vastus Medialis: Distal one-half of intertrochanteric line, medial lip of linea aspera, proximal part of medial supracondylar line, tendons of Adductor longus and Adductor magnus, and medial intermuscular septum.

Insertion: Proximal border of patella and through patellar ligament to tuberosity of tibia.

Action: The Quadriceps extends the knee joint, and the Rectus femoris portion flexes the hip joint.

Nerve: Femoral, L2, 3, 4.

The *Articularis genus* is a small muscle which may be blended with the Vastus intermedius, but usually is distinct from it. (Not shown in drawing.)

Origin: Anterior surface of distal part of body of femur.

Insertion: Proximal part of synovial membrane of knee joint.

Action: Draws articular capsule proximally.

Nerve: Branch of nerve to Vastus intermedius.

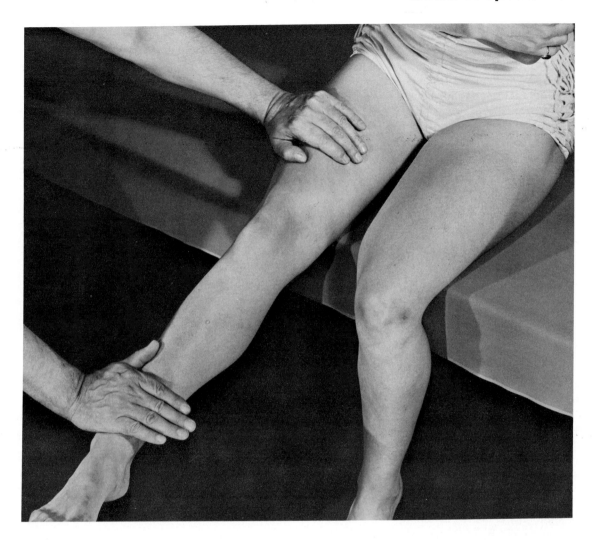

Patient: Sitting with knees bent over side of table. Arms may be folded across the chest or the subject may sit at the end of the table and hold on to either side of the table.

Fixation: The examiner may hold the thigh firmly down on the table; or, because the weight of the trunk is usually sufficient to stabilize the patient during this test, the examiner may put his hand under the distal end of the thigh to cushion that part against table pressure.

Test: Extension of the knee joint without rotation of the thigh.

Pressure: Against the leg above the ankle, in the direction of flexion.

Note: Inclining the body backward may be evidence of an attempt to release Hamstring tension when those muscles are contracted. When the Tensor fasciae latae is being substituted for the Quadriceps, it medially rotates the thigh and exerts a stronger pull if the hip is extended. If the Rectus femoris is the strongest part of the Quadriceps the patient will lean backward to extend the hip thereby obtaining maximum action of the Rectus femoris.

Weakness: Interferes with the function of stair climbing or walking up an incline, as well as getting up and down from a sitting position. The weakness results in knee hyperextension, not in the sense that such weakness permits a posterior knee position, but in the sense that walking with a weak Quadriceps requires that the patient lock the knee joint by slight hyperextension. Continuous thrust in the direction of hyperextension in growing children may result in a very marked degree of deformity.

Contracture: Knee extension.

Shortness: Restriction of knee flexion. A shortness of the Rectus femoris part of the Quadriceps results in restriction of knee flexion when the hip is extended, or restriction of hip extension when the knee is flexed.

Tests for Length of Hip Flexor Muscles

The Psoas major, Iliacus, Pectineus, Adductors longus and brevis, Rectus femoris, Tensor fasciae latae, and Sartorius compose the hip flexor group of muscles. The Iliacus, Pectineus, and Adductors longus and brevis are one-joint muscles. The Psoas major, along with the Iliacus, as the Iliopsoas, acts essentially as a one-joint muscle. The Rectus femoris, Tensor fasciae latae, and Sartorius are two-joint muscles, crossing the knee joint as well as the hip joint. While all three muscles flex the hip, the Rectus femoris and, to some extent, the Tensor extend the knee, while the Sartorius flexes the knee.

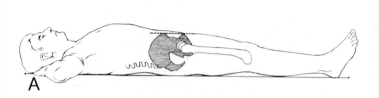

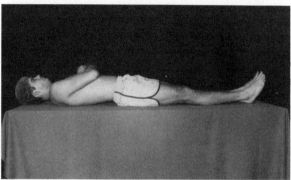

In fig. A, the pelvis is shown in neutral position, the low back in normal anterior curve, and the hip joint in zero position. Normal hip joint extension is considered to be approximately 10°. Normal length of hip flexors permits this range of motion in extension. The length may be demonstrated by moving the thigh in a posterior direction with the pelvis in a neutral position, or moving the pelvis in the direction of posterior tilt with the thigh in zero position.

In a subject with normal length of hip flexors, the low back will tend to flatten in the supine position. If the low back remains in a lordotic position as above, there is usually some hip flexor shortness.

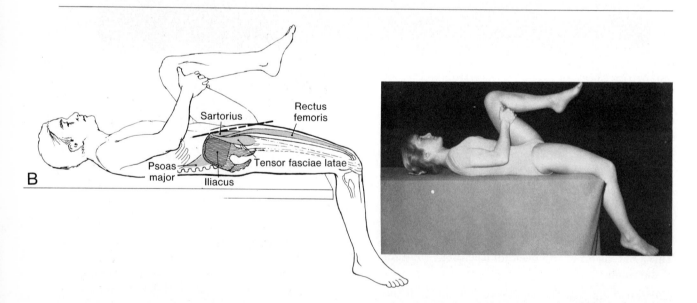

In fig. B, the pelvis is shown in 10° posterior tilt. This is equivalent to 10° hip joint extension, and, with the thigh touching the table, represents normal length of the one-joint hip flexors. In addition, the knee flexion (about 80°) indicates that the Rectus femoris and Tensor fasciae latae muscles are also normal in length. In order to maintain the pelvis in posterior tilt and the low back flat, as shown in fig. B, one thigh is held toward the chest during the length test of the hip flexors on the other leg.

This subject has normal length in the hip flexor muscles.

Tests for Length of Hip Flexor Muscles

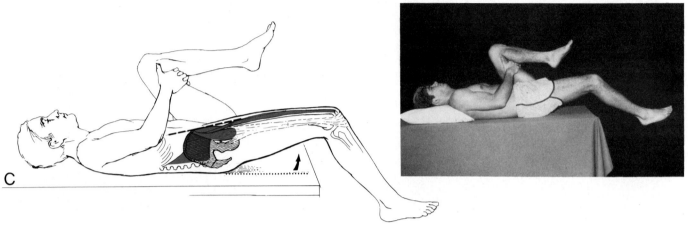

Fig. C shows shortness of both the one-joint and two-joint hip flexor muscles resulting in the inability to extend the hip joint.

This subject has shortness in both one-joint and two-joint hip flexors.

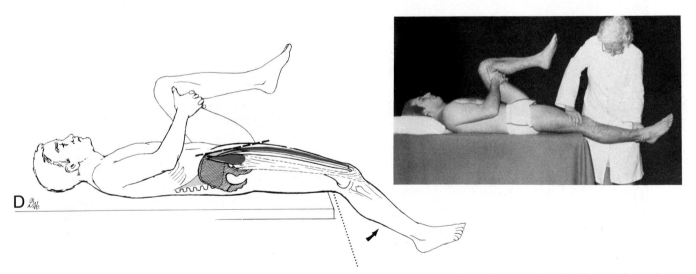

Fig. D shows a subject in whom the hip joint can be extended if the knee joint is allowed to extend. This means that the one-joint hip flexors are normal in length but the Rectus femoris and (probably) the Tensor fasciae latae are short.

This subject has shortness in the Rectus femoris and Tensor fasciae latae but not in the one-joint hip flexors. (See p. 162 for additional analysis of this subject.)

Tests for Length of Hip Flexor Muscles

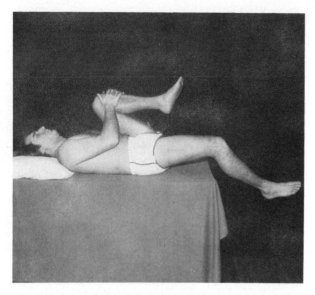

The test for hip flexor length shows limitation of hip extension and knee flexion.

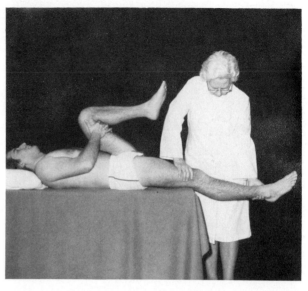

This test confirms that the shortness is not in the one-joint hip flexors but is in the Rectus femoris and Tensor fasciae latae.

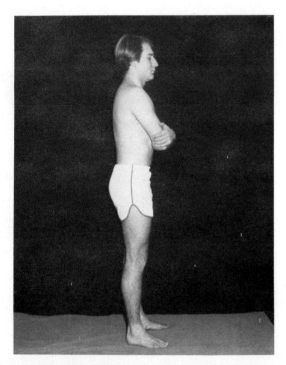

In standing, the subject does not have a lordosis. This fact would indicate that the shortness is *not* in the one-joint hip flexors.

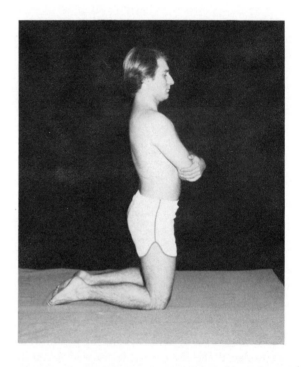

A kneeling position puts a stretch on the short Rectus femoris and Tensor fasciae latae over both the hip joints and knee joints, causing these muscles to pull the pelvis into anterior tilt and the back into a lordotic position.

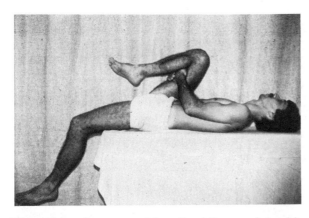

This subject has normal length of the one-joint hip flexors. The position of the knee indicates slight shortness in the Rectus femoris and Tensor fasciae latae.

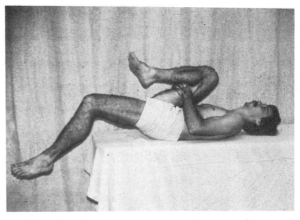

In this figure, the subject appears to have hip flexor shortness. The pelvis is in too much posterior tilt and low back in too much flexion. (See photograph of same subject below.) While it is important that the back be flat on the table, it is equally important that it not be flexed too much.

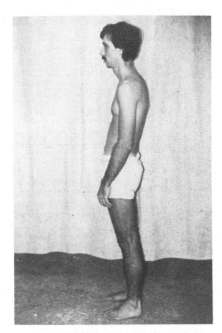

The position in standing is one in which the hip joints are in extension, the pelvis is displaced anteriorly and the upper trunk posteriorly. The one-joint hip flexors are elongated rather than shortened.

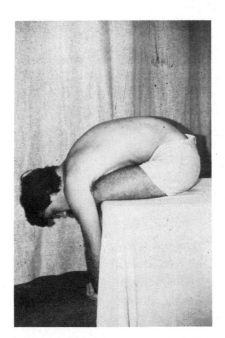

The excessive flexion in the low back permits the pelvis to go into too much posterior tilt and the low back into too much flexion during the hip flexor length test, as illustrated in top figure.

Iliopsoas

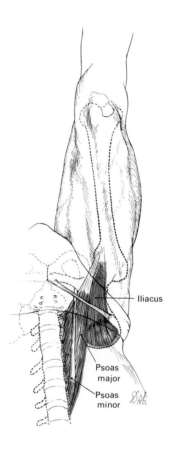

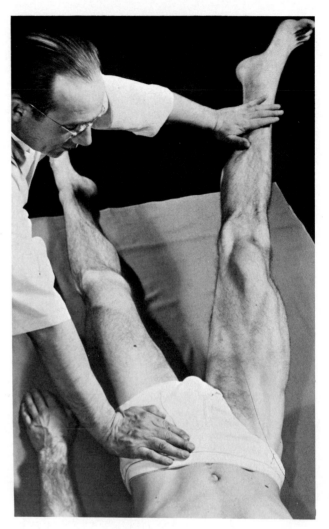

PSOAS MAJOR

Origin: Ventral surfaces of transverse processes of all lumbar vertebrae, sides of bodies and corresponding intervertebral discs of the last thoracic and all lumbar vertebrae and membranous arches which extend over the sides of the bodies of the lumbar vertebrae.

Insertion: Lesser trochanter of femur.

Nerve: Lumbar plexus, L1, **2, 3,** 4.

ILIACUS

Origin: Superior two-thirds of iliac fossa, internal lip of iliac crest, iliolumbar and ventral sacroiliac ligaments and ala of sacrum.

Insertion: Lateral side of tendon of Psoas major, and just distal to the lesser trochanter.

Nerve: Femoral, L(1), **2, 3,** 4.

ILIOPSOAS

Action: *With the origin fixed,* the Iliopsoas flexes the hip joint by flexing the femur on the trunk as in supine alternate leg raising, and may assist in lateral rotation and abduction of the hip joint. *With the insertion fixed* and acting bilaterally, the Iliopsoas flexes the hip joint by flexing the trunk on the femur as in the sit-up from supine position. The

Psoas major, acting bilaterally with the insertion fixed, will increase the lumbar lordosis; acting unilaterally, assists in lateral flexion of the trunk toward the same side.

ILIOPSOAS (with emphasis on psoas major)

Patient: Supine.

Fixation: The examiner gives fixation on the opposite iliac crest.

Test: Hip flexion in a position of slight abduction and slight lateral rotation. The muscle is not seen in the photograph because it lies deep to the Sartorius, the femoral nerve, and the blood vessels contained in the femoral sheath.

Pressure: Against the anteromedial aspect of the leg, in the direction of extension and slight abduction, directly opposite the line of pull of the Psoas major from the origin on the lumbar spine to the insertion on the lesser trochanter of the femur.

Weakness and contracture: See Hip flexors, p. 165.

(Psoas minor: see p. 166.)

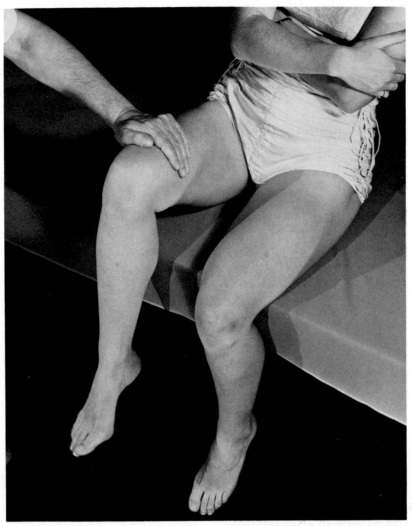

Patient: Sitting with legs over side of table. Arms may be folded across the chest or patient may hold on to the table.

Fixation: The weight of the trunk is usually sufficient to stabilize the patient during this test. If the trunk is weak, it is better to have the patient supine during the test.

Test: Hip flexion with the knee flexed.

Pressure: Against the anterior surface of the thigh in the direction of hip extension.

Note: Lateral rotation with abduction of the thigh as pressure is applied generally is evidence that the Sartorius is being substituted for direct hip flexor action, or that the Tensor fasciae latae is too weak to counteract the pull of the Sartorius. Medial rotation of the thigh shows the Tensor fasciae latae stronger than the Sartorius. If adductors are responsible for the flexion, the thigh will be adducted as it is flexed. If the anterior abdominals do not fix the pelvis to the trunk, the pelvis will flex on the thighs, and the hip flexors may hold against strong resistance but not at maximum height.

Weakness: Decreases the ability to flex the hip joint and results in marked disability in stair-climbing or walking up an incline, getting up from a reclining position, and bringing the trunk forward in the sitting position preliminary to rising from a chair. In marked weakness, walking is difficult because the leg must be brought forward by pelvic motion (produced by anterior or lateral abdominal muscle action) rather than by hip flexion. The effect of hip flexor weakness on posture is seen in figs. B and C, p. 219.

Contracture: Bilaterally, hip flexion deformity with increased lumbar lordosis. (See p. 283, fig. A.) Unilaterally, hip position of flexion, abduction and lateral rotation.

Shortness: In the standing position, shortness of the hip flexors is seen as a lumbar lordosis with an anterior pelvic tilt.

Tensor Fasciae Latae

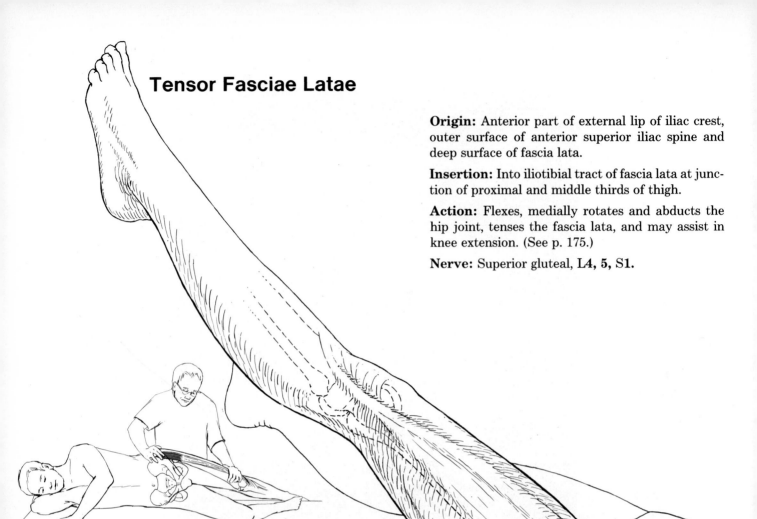

Origin: Anterior part of external lip of iliac crest, outer surface of anterior superior iliac spine and deep surface of fascia lata.

Insertion: Into iliotibial tract of fascia lata at junction of proximal and middle thirds of thigh.

Action: Flexes, medially rotates and abducts the hip joint, tenses the fascia lata, and may assist in knee extension. (See p. 175.)

Nerve: Superior gluteal, L4, 5, S1.

A

TEST FOR LENGTH OF TENSOR FASCIAE LATAE

The normal length of the Tensor fasciae latae permits the leg to drop in adduction toward the table as illustrated in fig. A. The thigh is neutral between medial and lateral rotation and is in very slight extension, the knee is extended, and the thigh drops in adduction toward the table allowing the toes to drop below the level of table-top.
The lateral trunk on the under side remains in contact with the table.

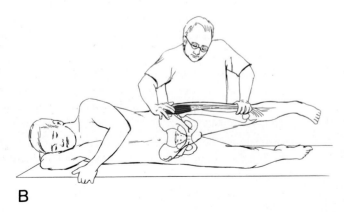

B

As seen in fig. B, if the leg fails to drop when the pelvis is fixed, it indicates a tightness of the Tensor fasciae latae and iliotibial tract.

PSOAS MINOR (con't. from p. 164)

This muscle is not a lower extremity muscle because it does not cross the hip joint. It is relatively unimportant, and only present in about 40% of the population.

Origin: Sides of bodies of twelfth thoracic and first lumbar vertebrae and from the intervertebral disc between them.

Insertion: Iliopectineal eminence, arcuate line of ilium, and iliac fascia.

Action: Flexion of pelvis on lumbar spine and vice versa.

Nerve: Lumbar plexus, L1, 2.

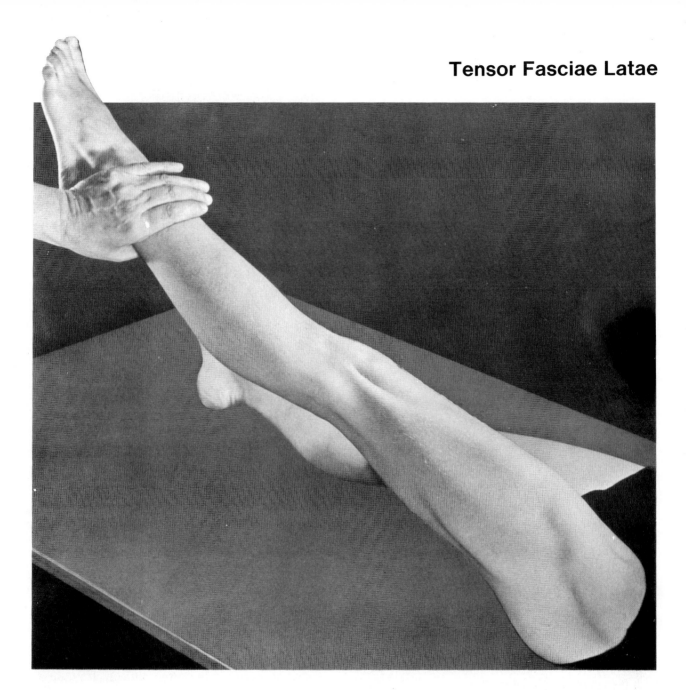

Patient: Supine.

Fixation: The patient may hold on to the table. Quadriceps action is necessary to hold the knee extended. Usually no fixation is necessary by the examiner, but if there is instability and the patient has difficulty in maintaining the pelvis firmly on the table, then one hand of the examiner should support the pelvis anteriorly on the opposite side.

Test: Abduction, flexion and medial rotation of the hip with the knee extended.

Pressure: Against the leg in the direction of extension and adduction. (Do not try to resist the rotation component.)

Weakness: In standing, there is a thrust in the direction of a bow-leg position, and the extremity tends to rotate laterally from the hip.

Shortness: The effect of shortness of the Tensor fasciae latae in standing depends upon whether the tightness is bilateral or unilateral. If bilateral, there is an anterior pelvic tilt, and sometimes bilateral knock-knee. If unilateral, the abductors of the hip and fascia lata are tight along with the tensor fasciae latae and there is an associated lateral pelvic tilt, low on the side of tightness. The knee on that side will tend toward a knock-knee position. If the Tensor fasciae latae and other hip flexor muscles are tight, there is an anterior pelvic tilt and a medial rotation of the femur, as indicated by the position of the patella.

Contracture: Hip flexion and knock-knee position.

Gluteus Minimus

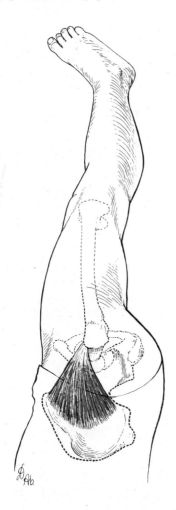

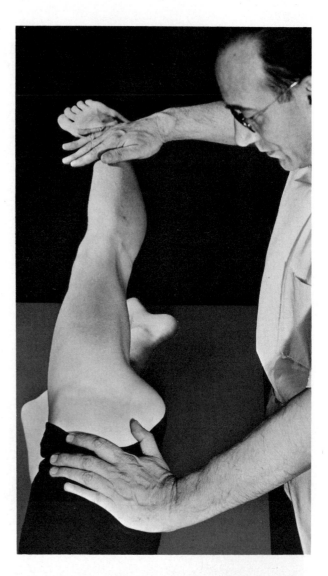

Origin: External surface of ilium between anterior and inferior gluteal lines, and margin of greater sciatic notch.

Insertion: Anterior border of greater trochanter of femur, and hip joint capsule.

Action: Abducts, medially rotates, and may assist in flexion of the hip joint.

Nerve: Superior gluteal, L4, **5,** S1.

Patient: Side lying.

Fixation: The examiner stabilizes the pelvis. (See note below.)

Test: Abduction of the hip in a position neutral between flexion and extension, and neutral in regard to rotation.

Pressure: Against the leg, in the direction of adduction and very slight extension.

Weakness: Lessens the strength of medial rotation and abduction of the hip joint.

Contracture and shortness: Abduction and medial rotation of the thigh. In standing, lateral pelvic tilt, low on the side of shortness, plus medial rotation of femur.

Note: In tests of the Gluteus minimus and medius, or the abductors as a group, the stabilization of the pelvis is necessary but often difficult. It requires a strong fixation by many trunk muscles, aided by stabilization on the part of the examiner. The slight flexion of the hip and knee of the underleg aids in stabilizing the pelvis against anterior or posterior tilt. The examiner's hand attempts to steady the pelvis to prevent the tendency to roll forward or backward, the tendency to tilt anteriorly or posteriorly and to prevent, if possible, any unnecessary hiking or dropping of the pelvis laterally. Any one of the six shifts in position of the pelvis may result primarily from trunk weakness, or such shifts may indicate an attempt to substitute anterior or posterior hip joint muscles or lateral abdominals in the movement of leg abduction. When the trunk muscles are strong, it is not too difficult to maintain good stabilization of the pelvis; but when trunk muscles are weak, the examiner may need the assistance of a second person to hold the pelvis steady.

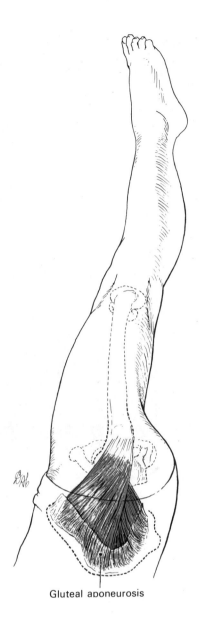

Gluteal aponeurosis

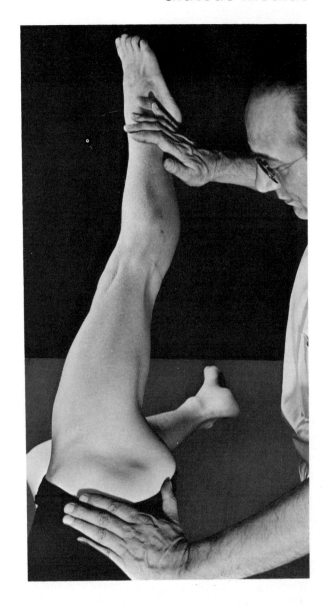

Origin: External surface of ilium between iliac crest and posterior gluteal line dorsally, and anterior gluteal line ventrally, gluteal aponeurosis.

Insertion: Oblique ridge on lateral surface of greater trochanter of femur.

Action: Abducts the hip joint. The anterior fibers medially rotate and may assist in flexion of the hip joint; the posterior fibers laterally rotate and may assist in extension.

Nerve: Superior gluteal, L4, **5,** S1.

Patient: Side lying.

Fixation: The muscles of the trunk and fixation by the examiner stabilize the pelvis. (See note under Gluteus minimus test, p. 168.)

Test, middle and posterior portions: Abduction of the hip with slight extension and slight external rotation. (Knee is maintained in extension.)

Pressure: Against the leg, in the direction of adduction and slight flexion. (Do not attempt to resist the rotation component.) The pressure is applied against the leg for the purpose of obtaining longer leverage. To determine normal strength, strong pressure is needed and can be obtained by the examiner only if the leverage is adequate. There is relatively little danger of injuring the fibular collateral ligament during the Gluteus medius test because the knee joint is reinforced by the strong iliotibial tract. (This tract is plainly visible in the photograph on page 167.)

Gluteus Medius

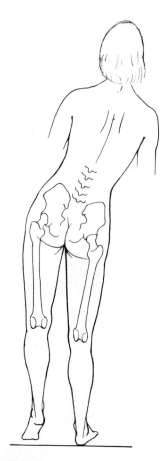

Marked weakness of right Gluteus medius

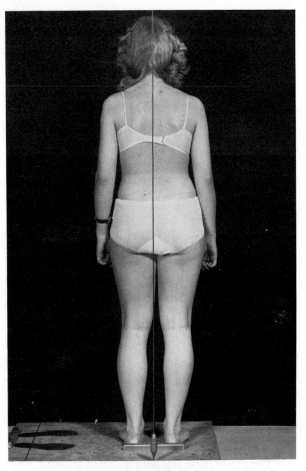

Slight weakness of right Gluteus medius

GLUTEUS MEDIUS

Weakness: The Gluteus medius muscle is the primary lateral stabilizer of the pelvis on the thigh when the opposite foot is raised from the floor as in standing on one leg or in walking. As such, a good deal of strength is necessary for normal function.

With *marked weakness* there will be a Gluteus medius limp in walking. This consists of displacement of the body weight laterally toward the side of the weak muscle in such a way that the hip joint is thrust in the position of hip abduction in relation to the pelvis (as illustrated). With *slight weakness* of the Gluteus medius, there will be a postural deviation in standing. (See photograph at right.)

Contracture: An abduction deformity which, in standing, may be seen as a lateral tilt of the pelvis low on the side of contracture, along with some actual abduction of the extremity.

Effect on posture: Slight hip abductor weakness unilaterally frequently is found in association with lateral pelvic tilt. The pelvis is high on the side of weakness, and the femur is adducted in relation to the pelvis. As a result of such unilateral pelvic tilt, the spine deviates with convexity toward the opposite (low) side. In other words, a slight weakness of the right Gluteus medius gives rise to a left C-curve. (See p. 291, left-hand column.) The accompanying illustration has been chosen because it represents the mild degree of postural deviation that is frequently seen.

The weakness in the abductors, as seen in faulty posture, usually is associated with handedness. Right-handed people generally show weakness of the right Gluteus medius and, although not as consistently, left-handed people generally show weakness of the left. Once the weakness is established, it further contributes to the faulty mechanics.

Medial Rotators of Hip Joint

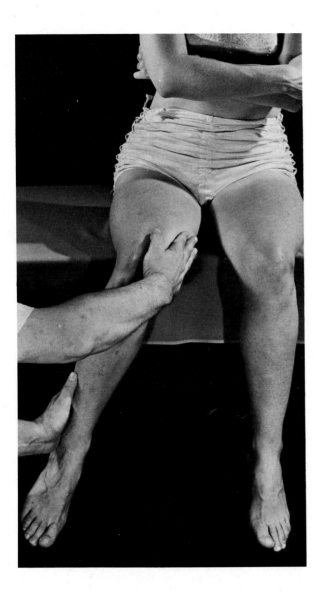

The medial rotators of the hip joint consist of the Tensor fascia latae, Gluteus minimus, and Gluteus medius (anterior fibers).

Patient: Sitting on table with knees bent over side of table.

Fixation: The weight of the trunk stabilizes the patient during this test. Stabilization is also given in the form of counterpressure as described below under Pressure.

Test: Medial rotation of the thigh, with the leg in position of completion of outward arc of motion.

Pressure: Counterpressure is applied by one hand of the examiner at the medial side of the lower end of the thigh. The other hand of the examiner applies pressure to the lateral side of the leg above the ankle, pushing the leg inward in an effort to rotate the thigh laterally.

Weakness: Results in lateral rotation of the lower extremity in standing and walking.

Contracture: Medial rotation of the hip, with a tendency toward knock-knee if the patient has been weight-bearing.

Shortness: Inability to laterally rotate the thigh through full range of motion. Inability to sit in a cross-legged position (tailor fashion).

Note: If the rotator test is done in a supine position, the pelvis will tend to tilt anteriorly if much pressure is applied, but this is not a substitution movement. Due to its attachments, the Tensor fasciae latae, when contracting to maximum, pulls forward on the pelvis as it medially rotates the thigh.

Lateral Rotators of Hip Joint

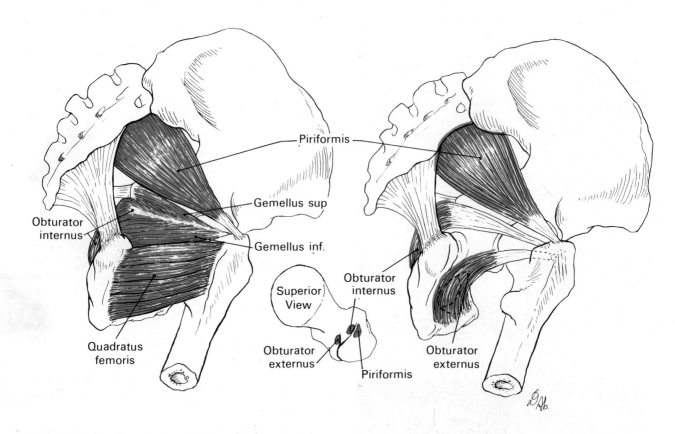

PIRIFORMIS

Origin: Pelvic surface of sacrum between, and lateral to, one, two, three, four pelvic sacral foramina, margin of greater sciatic foramen and pelvic surface of sacrotuberous ligament.

Insertion: Superior border of greater trochanter of femur.

Nerve: Sacral plexus, L(5), S1, 2.

QUADRATUS FEMORIS

Origin: Proximal part of lateral border of tuberosity of ischium.

Insertion: Proximal part of quadrate line extending distally from intertrochanteric crest.

Nerve: Sacral plexus, L4, 5, S1, (2).

OBTURATOR INTERNUS

Origin: Internal or pelvic surface of obturator membrane and margin of obturator foramen, and pelvic surface of ischium posterior and proximal to obturator foramen, and, to a slight extent, from the obturator fascia.

Insertion: Medial surface of greater trochanter of femur proximal to trochanteric fossa.

Nerve: Sacral plexus, L5, S1, 2.

OBTURATOR EXTERNUS

Origin: Rami of pubis and ischium, and external surface of obturator membrane.

Insertion: Trochanteric fossa of femur.

Nerve: Obturator, L3, 4.

GEMELLUS SUPERIOR

Origin: External surface of spine of ischium.

Insertion: With tendon of Obturator internus into medial surface of greater trochanter of femur.

Nerve: Sacral plexus, L5, S1, 2.

GEMELLUS INFERIOR

Origin: Proximal part of tuberosity of ischium.

Insertion: With tendon of Obturator internus into medial surface of greater trochanter of femur.

Nerve: Sacral plexus, L4, 5, S1, (2).

Action: All of the above muscles laterally rotate the hip joint. In addition, the Obturator externus may assist in adduction of the hip joint; and the Piriformis, Obturator internus and Gemelli may assist in abduction when the hip is flexed. The Piriformis may assist in extension.

Lateral Rotators of Hip Joint

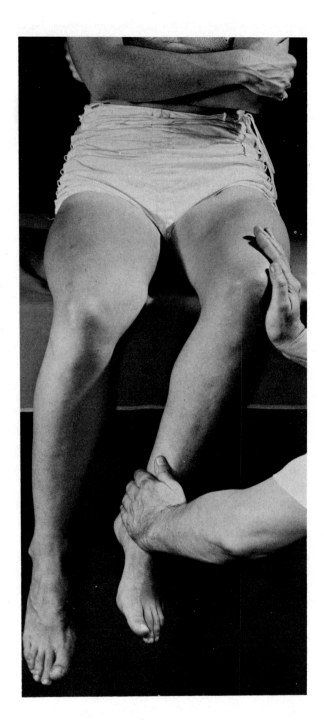

Patient: Sitting on table with knees bent over side of table.

Fixation: The weight of the trunk stabilizes the patient during this test. Stabilization is also given in the form of counterpressure as described below under Pressure.

Test: Lateral rotation of the thigh, with the leg in position of completion of the inward arc of motion.

Pressure: Counterpressure is applied by one hand of the examiner at the lateral side of the lower end of the thigh. The other hand of the examiner applies pressure to the medial side of the leg above the ankle, pushing the leg outward in an effort to rotate the thigh medially.

Weakness: Usually, medial rotation of the femur accompanied by pronation of the foot, and a tendency toward knock-knee position.

Contracture: Lateral rotation of the thigh, usually in an abducted position.

Shortness: The range of medial rotation of the hip will be limited. (Frequently there is excessive range of lateral rotation.) In the standing posture, there is a lateral rotation of the femur and out-toeing.

Gluteus Maximus

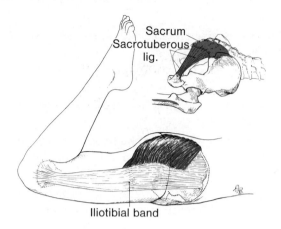

Sacrum
Sacrotuberous lig.

Iliotibial band

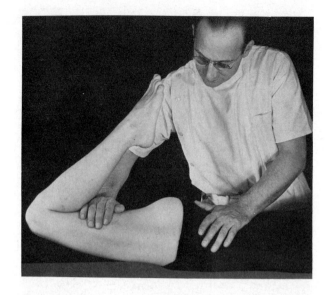

Origin: Posterior gluteal line of ilium and portion of bone superior and posterior to it, posterior surface of lower part of sacrum, side of coccyx, aponeurosis of erector spinae, sacrotuberous ligament and gluteal aponeurosis.

Insertion: Larger proximal portion and superficial fibers of distal portion of muscle into iliotibial tract of fascia lata. Deeper fibers of distal portion into gluteal tuberosity of femur.

Action: Extends, laterally rotates, and lower fibers assist in adduction of the hip joint. The upper fibers assist in abduction. Through its insertion into the iliotibial tract, helps to stabilize the knee in extension.

Nerve: Inferior gluteal L5, S1, 2.

Patient: Prone.

Fixation: Posteriorly, the back muscles, laterally, the lateral abdominal muscles, and, anteriorly, the *opposite* hip flexors fix the pelvis to the trunk.

Test: Hip extension with knee flexed 90° or more. The more the knee is flexed, the less the hip will extend, due to restricting tension of the Rectus femoris anteriorly.

Pressure: Against the lower part of the posterior thigh in the direction of hip flexion.

Weakness: Bilateral marked weakness of the Gluteus maximus makes walking extremely difficult, and necessitates the aid of crutches. The individual bears weight on the extremity in a position of posterolateral displacement of the trunk over the femur. Raising the trunk from a forward-bent position requires the action of the Gluteus maximus, and in cases of weakness patients must push themselves to an upright position by using their arms.

Gluteus Maximus

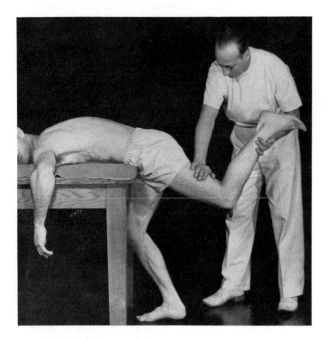

Fascia Lata

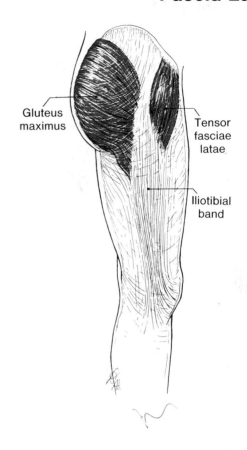

Gluteus maximus

Tensor fasciae latae

Iliotibial band

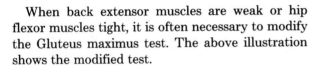

When back extensor muscles are weak or hip flexor muscles tight, it is often necessary to modify the Gluteus maximus test. The above illustration shows the modified test.

Patient: Trunk prone on table, legs hanging over the end of table.

Fixation: The patient usually needs to hold on to the table when pressure is applied.

Test: Extension of the hip: (1) with the knee passively flexed by the examiner, as illustrated, or (2) with the knee extended, permitting Hamstring assistance.

Pressure: This test presents a rather difficult problem in regard to application of pressure. If the Gluteus maximus is to be isolated as much as possible from the Hamstrings, it requires that knee flexion be maintained by the examiner otherwise the Hamstrings will unavoidably act in maintaining the antigravity knee flexion. Trying to maintain knee flexion passively and apply pressure to the thigh makes it difficult to obtain an accurate test.

If this test is used because of marked hip flexor tightness, it may be impractical to flex the knee, thereby increasing the Rectus femoris tension over the hip joint.

The extensive deep fascia which covers the gluteal region and the thigh like a sleeve is called the fascia lata. It is attached proximally to the external lip of the iliac crest, the sacrum and coccyx, the sacrotuberous ligament, the ischial tuberosity, the ischiopubic rami, and the inguinal ligament. Distally it is attached to the patella, the tibial condyles, and the head of the fibula. The fascia on the medial aspect of the thigh is thin while that on the lateral side is very dense, especially the portion between the tubercle of the iliac crest and the lateral condyle of the tibia, designated as the iliotibial tract. Upon reaching the borders of the Tensor fasciae latae and the Gluteus maximus, the fascia lata divides and invests both the superficial and deep surfaces of these muscles. In addition, both the Tensor fasciae latae and three-fourths of the Gluteus maximus insert into the iliotibial tract so that its distal extent serves as a conjoint tendon of these muscles. This structural arrangement permits both muscles to influence the stability of the extended knee joint.

Hip Adductors

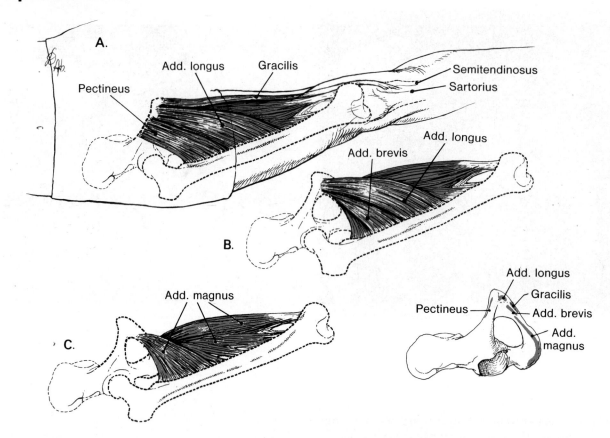

Stippled lines indicate muscle attachments located on posterior surface of femur

PECTINEUS

Origin: Surface of superior ramus of pubis ventral to pecten between iliopectineal eminence and pubic tubercle.

Insertion: Pectineal line of femur.

Nerve: Femoral and Obturator, L**2, 3,** 4.

ADDUCTOR MAGNUS

Origin: Inferior pubic ramus, ramus of ischium (anterior fibers), and ischial tuberosity (posterior fibers).

Insertion: Medial to gluteal tuberosity, middle of linea aspera, medial supracondylar line, and adductor tubercle of medial condyle of femur.

Nerve: Obturator, L**2, 3, 4,** and Sciatic, L**4,** 5, S1.

GRACILIS

Origin: Inferior half of symphysis pubis and medial margin of inferior ramus of the pubic bone.

Insertion: Medial surface of body of tibia, distal to condyle, proximal to insertion of Semitendinosus, and lateral to insertion of Sartorius.

Nerve: Obturator, L**2, 3, 4.**

ADDUCTOR BREVIS

Origin: Outer surface of inferior ramus of pubis.

Insertion: Distal two-thirds of pectineal line, and proximal half of medial lip of linea aspera.

Nerve: Obturator, L**2, 3, 4.**

ADDUCTOR LONGUS

Origin: Anterior surface of pubis at junction of crest and symphysis.

Insertion: Middle one-third of medial lip of linea aspera.

Nerve: Obturator L**2, 3, 4.**

Action: All the above adduct the hip joint. In addition, the Pectineus, Adductor brevis, and Adductor longus flex the hip joint. The anterior fibers of the Adductor magnus which arise from the rami of the pubis and ischium may assist in flexion, while the posterior fibers that arise from the ischial tuberosity may assist in extension. The Gracilis, in addition to adducting the hip joint, flexes and medially rotates the knee joint. (See p. 178 for discussion of rotation action on the hip joint.)

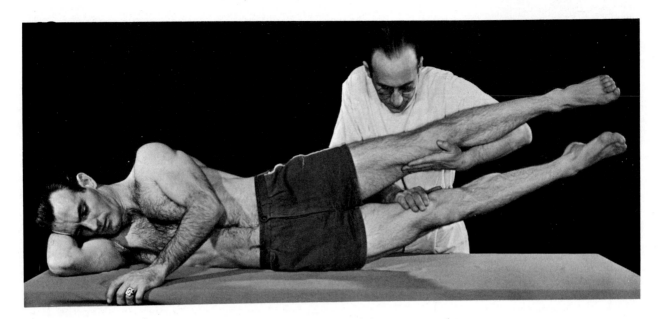

Patient: Lying on right side to test right, and vice versa, body in straight line, with lower extremities and lumbar spine straight.

Fixation: The examiner holds the upper leg in abduction. The patient should hold on to the table for stability.

Test: Adduction of the under extremity upward from the table without rotation, flexion, or extension of the hip, or tilting of the pelvis.

Pressure: Against the medial aspect of the distal end of the thigh in the direction of abduction (downward toward the table). Pressure is applied at a point above the knee to avoid strain of the tibial collateral ligament.

Note: Forward rotation of the pelvis with extension of the hip joint shows attempt to hold with lower fibers of Gluteus maximus. Anterior tilting of the pelvis, or flexion of the hip joint (with backward rotation of the pelvis on upper side), allows substitution by the hip flexors.

Adductor longus, Adductor brevis, and Pectineus aid in hip flexion. If the side-lying position is maintained and the hip tends to flex as the thigh is adducted during the test, it is not necessarily evidence of substitution but merely evidence that the adductors which flex the hip are doing more than the rest of the adductors that assist in this movement, or that hip extensors are not helping to maintain the thigh in neutral position.

Contracture: Hip adduction deformity. In standing, the position is one of lateral pelvic tilt, with the pelvis so high on the side of contracture, that it becomes necessary to plantar flex the foot on the same side, holding it in equinus in order for the toes to touch the floor. As an alternative, if the foot is placed flat on the floor, the opposite extremity must be either flexed at the hip and knee, or abducted to compensate for the apparent shortness on the adducted side.

Mechanical Axis of Femur and Rotation Action of Adductors

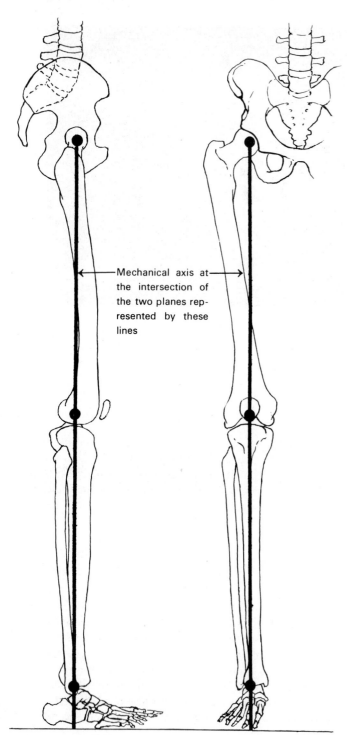

Mechanical axis at the intersection of the two planes represented by these lines

The following brief discussion about the rotator action of the adductors is not an attempt to solve the controversy that appears to exist but, rather, to present some of the reasons why a controversy exists.

On the accompanying illustration, it is important to note that in anatomical position, and from anterior view, the femur extends obliquely with the distal end more medial than the proximal. From lateral view, the shaft of the femur is convexly curved in an anterior direction. The *anatomical axis* of the femur extends longitudinally along the shaft. If rotation of the hip took place about this axis there would be no doubt that the adductors, attached as they are posteriorly along the linea aspera, would be lateral rotators.

However, rotation of the hip joint does not occur about the anatomical axis of the femur, but rather about the *mechanical axis* passing from the center of the hip joint to the center of the knee joint.

The muscles or major portions of muscles that insert on the part of the femur which is anterior to the mechanical axis will act as medial rotators of the femur. (See lateral view.) On the other hand, the muscles or major portions of the muscles that insert on the part of the femur posterior to the mechanical axis will act as lateral rotators.

When the position of the extremity in relation to the pelvis changes from that illustrated as the anatomical position, the actions of the muscles change. Thus, if the femur is medially rotated, a larger portion of the shaft comes to lie anterior to the mechanical axis with the result that more of the adductor insertions will be anterior to the axis and, therefore, will act as medial rotators. With increased lateral rotation more of the adductors will act as lateral rotators.

Besides the change that occurs with movement, there are normal variations of bone structure of the femur that tend to make variable the rotator action of the adductors.

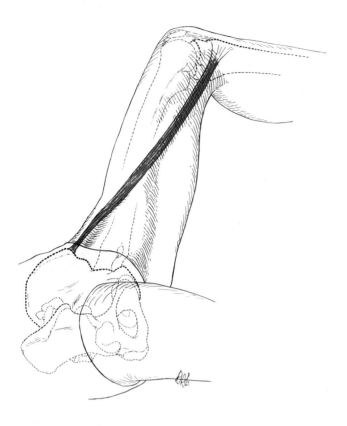

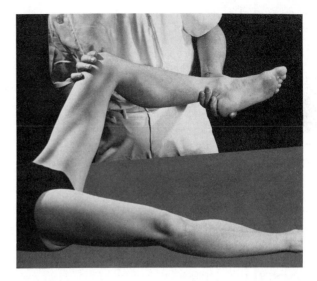

Origin: Anterior superior iliac spine and superior half of notch just distal to spine.

Insertion: Proximal part of medial surface of tibia near anterior border.

Action: Flexes, laterally rotates, and abducts the hip joint. Flexes and assists in medial rotation of the knee joint.

Nerve: Femoral, L2, 3, (4).

Patient: Supine.

Fixation: None necessary on the part of the examiner. The patient may hold on to the table.

Test: Lateral rotation, abduction, and flexion of the thigh, with flexion of the knee.

Pressure: Against the anterolateral surface of the lower thigh, in the direction of hip extension, adduction and medial rotation, and against the leg in the direction of knee extension. The examiner's hands are in position to resist the lateral rotation of the hip joint by pressure and counterpressure (in the same way as described under hip lateral rotator test, p. 173.) The examiner must resist the multiple action test movement by a combined resistance movement.

Weakness: Decreases strength of hip flexion, abduction, and lateral rotation. Contributes to anteromedial instability of the knee joint.

Contracture: Flexion, abduction, and lateral rotation deformity of the hip, with flexion of the knee.

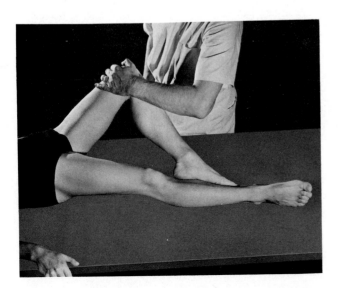

ERROR IN TESTING SARTORIUS

The position of the leg, as illustrated, resembles the Sartorius test position in its flexion, abduction, and lateral rotation. However, the ability to hold this position is essentially a function of the hip adductors (as seen in the accompanying photograph) and requires little assistance from the Sartorius.

LOWER EXTREMITY MUSCLES, Listed According to Spinal Segment Innervation and Grouped According to Joint Action

Spinal Segment

Lumb.					Sac.				HIP						KNEE		
1	2	3	4	5	1	2	3	**Muscle**	Flexion	Adduction	Med. Rotat.	Abduction	Lat. Rotat.	Extension	Extension	In flexion Lat. Rotat.	In flexion Med. Rotat.
1	2	3	4					Psoas major	Psoas maj.			Psoas maj.	Psoas maj.				
(1)	2	3	4					Iliacus	Iliacus			Iliacus	Iliacus				
	2	3	(4)					Sartorius	Sartorius			Sartorius	Sartorius				Sartorius
	2	3	4					Pectineus	Pectineus	Pectineus							
	2	3	4					Adductor long.	Add. long.	Add. long.	Add. long.						
	2	3	4					Adductor brev.	Add. brev.	Add. brev.	Add. brev.						
	2	3	4					Gracilis		Gracilis							Gracilis
	2	3	4					Quadriceps	Rect. fem.						Quadriceps		
	2	3	4					Add. mag. (ant.)	Add. m. (ant.)	Add. mag.							
		3	4					Obturator ext.		Obt. ext.			Obt. ext.				
			4	5	1			Add. mag. (post.)		Add. mag.				Ad. m. post.			
			4	5	1			Tibialis ant.									
			4	5	1			Ten. fas. lat.	Tensor f.l.		Tensor f.l.	Tensor f.l.			Tensor f.l.		
			4	5	1			Gluteus minimus	Glut. min.		Glut. min.	Glut. min.					
			4	5	1			Gluteus medius	G. med., ant.		G. med., ant.	Glut. med.	G. med., post.	G. med., post.			
			4	5	1			Popliteus									Popliteus
			4	5	1			Ext. dig. long.									
			4	5	1			Peroneus tertius									
			4	5	1			Ext. hall. long.									
			4	5	1			Ext. dig. brev.									
			4	5	1			Flex. dig. brev.									
			4	5	1			Flex. hall. brev.									
			4	5	1			Lumbricalis I									
			4	5	1			Abductor hall.									
			4	5	1			Peroneus longus									
			4	5	1			Peroneus brevis									
			(4)	5	1			Tibialis post.									
			4	5	1	(2)		Gemelli inferior				Gem. inf.	Gem. inf.				
			4	5	1	(2)		Quadratus fem.					Quadratus f.				
			4	5	1	(2)		Plantaris									
			4	5	1	2		Semimembranosus			Semimemb.			Semimemb.			Semimemb.
			4	5	1	2		Semitendinosus			Semitend.			Semitend.			Semitend.
			4	5	1	(2)		Flex. dig. long.									
				5	1	2		Gluteus maximus	G. max., low.			G. max., upp.	Glut. max.	Glut. max.			
				5	1	2		Biceps, short h.								Bic., s.h.	
				5	1	2		Flex. hall. long.									
				5	1	2		Soleus									
				(5)	1	2		Piriformis				Piriformis	Piriformis	Piriformis			
				5	1	2		Gemelli superior				Gem. sup.	Gem. sup.				
				5	1	2		Obturator int.				Obt. int.	Obt. int.				
				5	1	2	3	Biceps, long h.					Biceps l.h.	Biceps l.h.		Bic., l.h.	
			(4)	(5)	1	2		Lumb. II, III, IV									
					1	2		Gastrocnemius									
					1	2		Dorsal inteross.									
					1	2		Plantar inteross.									
					1	2		Abd. dig. min.									
					1	2		Adductor hall.									

LOWER EXTREMITY MUSCLES, Listed According to Spinal Segment Innervation and Grouped According to Joint Action (Continued)

KNEE	ANKLE		FOOT		METATARSOPHALANGEAL JOINT				Digs. 2-5 Prox. Interphal. Jts.		Digs. 1-5 Distal Interphal. Jts.	
Flexion	Dorsiflex.	Plant flex.	Eversion	Inversion	Extension	Flexion	Abduction	Adduction	Extension	Flexion	Extension	Flexion
Sartorius												
Gracilis												
	Tib. ant.			Tib. ant.								
Popliteus												
	Ext. d. long.		Ext. d. long.		(2-5 dig.) Ext. d. long				(2-5 dig.) Ext. d. long.		(2-5 dig.) Ext. d. long.	
	Peroneus t.		Peroneus t.									
	Ext. hall. l.			Ext. hall. l.	Ext. hall. l.				///	///	Ext. hall l.	
					(1-4 dig.) Ext. dig. br.				(1-4 dig.) Ext. dig. br.		(1-4 dig.) Ext. dig. br.	
						(2-5 dig.) Flex. dig. br.				(2-5 dig.) Flex. dig. br.		
						Flex. hall. br.			///	///		
						2nd dig. Lumb. I			(2nd dig.) Lumb. I		(2nd dig.) Lumb. I	
						Abd. hall.	Abd. hall.		///	///		
		Peroneus l.	Peroneus l.									
		Peroneus b.	Peroneus b.									
		Tib. post.		Tib. post.								
Plantaris		Plantaris										
Semimemb.												
Semitend.												
		Flex. dig. l.		Flex. dig. l.		(2-5 dig.) Flex. dig. l.				(2-5 dig.) Flex. dig. l.		(2-5 dig.) Flex. dig. l
Bic., s.h.												
		Flex. hall. l.		Flex. hall. l.		Flex. hall. l.			///	///		Flex. hall. l.
		Soleus										
Bic., l.h.												
						(3-5 dig.) Lumb. II-IV			(3-5 dig.) Lumb. II-IV		(3-5 dig.) Lumb. II-IV	
Gastroc.		Gastroc.										
						(2-4 dig.) Dor. int.	(2-4 dig.) Dor. int.		(2-4 dig.) Dor int.		(2-4 dig.) Dor int.	
						(3-5 dig.) Plant. int.		(3-5 dig.) Plant int.	(3-5 dig.) Plant. int.		(3-5 dig.) Plant int.	
							Abd. d. min.					
						Add. hall.		Add. hall.	///	///		

NECK, TRUNK, and LOWER EXTREMITY MUSCLE CHART

PATIENT'S NAME _____   CLINIC No. _____

LEFT   RIGHT

					EXAMINER DATE							
					Neck flexors							
					Neck extensors							
					Back extensors							
					Quadratus lumborum							
					Rectus abdominis							
					External oblique							
					Internal oblique							
					Lateral abdominals							
					Gluteus maximus							
					Gluteus medius							
					Hip abductors							
					Hip adductors							
					Hip medial rotators							
					Hip lateral rotators							
					Hip flexors							
					Tensor fasciae latae							
					Sartorius							
					Medial hamstrings							
					Lateral hamstrings							
					Quadriceps							
					Gastrocnemius							
					Soleus							
					Peroneus longus							
					Peroneus brevis							
					Peroneus tertius							
					Tibialis posterior							
					Tibialis anterior							
					Extensor hallucis longus							
					Flexor hallucis longus							
					Flexor hallucis brevis							
					1 Extensor digitorum longus 1							
					2 Extensor digitorum longus 2							
					3 Extensor digitorum longus 3							
					4 Extensor digitorum longus 4							
					1 Extensor digitorum brevis 1							
					2 Extensor digitorum brevis 2							
					3 Extensor digitorum brevis 3							
					4 Extensor digitorum brevis 4							
					1 Flexor digitorum longus 1							
					2 Flexor digitorum longus 2							
					3 Flexor digitorum longus 3							
					4 Flexor digitorum longus 4							
					1 Flexor digitorum brevis 1							
					2 Flexor digitorum brevis 2							
					3 Flexor digitorum brevis 3							
					4 Flexor digitorum brevis 4							
					1 Lumbricalis 1							
					2 Lumbricalis 2							
					3 Lumbricalis 3							
					4 Lumbricalis 4							
					Leg length							
					Thigh circumference							
					Calf circumference							

NOTES: _____

_____

_____

_____

_____

CHART FOR ANALYSIS OF MUSCLE IMBALANCE
LOWER EXTREMITY

Name:.. Date: 1st. Ex.-.................. 2nd. Ex.-...........................

Diagnosis:.. Onset:...............................Exam. of..............extremity

		2nd EX.	1st. EX.	1st. EX.	2nd. EX.		
	ILIOPSOAS / SARTORIUS / TENSOR FAS. LAT. / RECTUS FEMORIS } HIP FLEXORS					GLUTEUS MAXIMUS	
	HIP ADDUCTORS					GLUTEUS MEDIUS	
						GLUTEUS MINIMUS	
						TENSOR FASCIAE LATAE	
	HIP LATERAL ROTATORS					HIP MEDIAL ROTATORS	
	QUADRICEPS					MEDIAL HAMSTRINGS LATERAL	
	TIBIALIS ANTERIOR					SOLEUS	
						GASTROCNEMIUS & SOLEUS	
						PERONEUS LONGUS & BREVIS	
	TIBIALIS POSTERIOR					PERONEUS TERTIUS	
	FLEXOR DIGITORUM LONGUS — 1 2 3 4					1 2 3 4 DISTAL INTER-PHALANGEAL JOINT EXTENSORS	
	FLEXOR DIGITORUM BREVIS — 1 2 3 4					1 2 3 4 PROXIMAL INTER-PHALANGEAL JOINT EXTENSORS	
	LUMBRICALES & INTEROSSEI — 1 2 3 4					1 2 3 4 EXT. DIGITORUM LONGUS & BREVIS	
	FLEXOR HALLUCIS LONGUS					EXTENSOR HALLUCIS LONGUS & BREVIS	
	FLEXOR HALLUCIS BREVIS						
	ABDUCTOR HALLUCIS					ADDUCTOR HALLUCIS	

chapter 6

Trunk Muscles

Introduction
Abdominal muscles:
   Origins, insertions, actions, nerve supply
Definitions and descriptions of terms
Trunk raising:
   Analysis of movements and muscle actions,
   effect of holding feet down, test, grading,
   weakness, therapeutic exercises
Holding back flat during leg-lowering:
   Analysis of muscle actions, test, grading,
   weakness, therapeutic exercises

Marked abdominal weakness:
   Imbalance, testing, grading
Lateral trunk raising
Oblique trunk raising
Quadratus lumborum
Back extensors:
   Origins and insertions, test, back and hip
   extensors
Forward bending:
   Normal, variations, age-level differences

Trunk Muscles

Proper exercise is an important part of preventive medicine and the public has a right to know which exercises are beneficial and which are harmful, and to know the reasons why. It is the purpose of this chapter to provide accurate information of a technical nature in a manner that will help make the material useful to many people in the fields of health care and physical fitness.

The public is bombarded with fitness programs that include exercises that are supposed to strengthen abdominal muscles. Many of the exercises are inappropriate or ineffective, and some cause harm. Attention should be focused on the following concerns:

Subjects with marked weakness of abdominal muscles cannot perform trunk raising in the correct manner of curling the trunk, and, consequently, they do it incorrectly with the low back arching forward, subjecting the abdominal muscles to stretch and strain. (See p. 210.)

Subjects who do the curled-trunk sit-up correctly (with legs extended or flexed), but do it to excess in frequency and/or duration, may develop excessive flexibility of the back, and shortness of the hip flexors. These adverse effects may be more pronounced from doing the knee-bent sit-up than from doing the sit-up with legs extended. (See pp. 206 and 207.)

In many exercise programs, sit-ups are the only abdominal exercises included. The programs fail to include *proper pelvic tilt* exercises that strengthen the muscles most needed to help maintain good alignment of the trunk and pelvis in standing. (See p. 214.) Furthermore, when pelvic tilt exercises are done, they frequently are done without any action on the part of the abdominal muscles.

The actions of certain muscle groups of the trunk and hip joints are so closely allied in testing and exercise procedures that the action of one group cannot be defined without reference to the other. When there is imbalance between the muscles that should be working together, the normal patterns of movement are distorted and errors arise in tests and exercises. From a face-lying position, *hip extensors* act to stabilize the pelvis during *trunk extension.* (See p. 231.) From a sidelying position, *hip abductors* stabilize the pelvis during *lateral flexion.* (See p. 223.) Hip joint motion that accompanies trunk movements, in each of these instances, is limited to a few degrees. In both extension and lateral flexion, muscle imbalance between the trunk and hip muscles is seldom present in cases of faulty posture to the extent that it interferes with normal patterns of movement.

In contrast, muscle imbalance frequently exists between *abdominal and hip flexor* muscles in *trunk-raising forward* from a supine position, as in the sit-up.* Hip flexors usually are strong, not infrequently abdominals are weak, and have less endurance than hip flexors. Whenever there is muscle imbalance there is a tendency for stronger muscles to substitute for weaker ones in movements that ordinarily involve both groups. Because the hip joint moves through approximately 80° of flexion during the sit-up, irrespective of the position of the trunk or of the lower extremities, it is possible for the hip flexors to perform the trunk-raising when abdominals are weak.

Many people become aware of abdominal muscle weakness because of a painful low back, because of being unable to get up easily from a lying position, or simply because they are concerned about their appearance and posture. The traditional exercises of sit-ups and double-leg-raising have been offered as the panacea for strengthening these muscles. Unfortunately, they are not the "cure-all" that they are supposed to be. Much of the confusion has been caused by failure to distinguish the action of the abdominal muscles from that of the hip flexors during these exercises. (See pp. 203 and 213.) People with strong abdominal muscles can do sit-up or leg-raising exercises without harmful effects; those who have weakness are often affected adversely. When there is marked weakness, use of these two exercises should be *avoided* because they can further weaken and strain the abdominal muscles, instead of strengthening them. There is evolving a better understanding of the uses and abuses of double leg-raising and sit-ups but indiscriminate use of these exercises still persists.

---

* These muscles that act together in movements from the lying position oppose each other in the standing position.

Trunk muscles consist of back extensors that bend the trunk backward, lateral flexors that bend it sideways, and anterior abdominals that bend it forward. All of these muscles play a role in stabilizing the trunk, but the back extensor muscles are the most important. The loss of stability that accompanies paralysis or marked weakness of back muscles offers dramatic evidence of their importance. Fortunately, marked weakness of these muscles occurs very seldom. In persons who have faulty posture with roundness of the upper back, some weakness may exist in the extensors of the upper back, but the low back muscles are very seldom weak. The term "weak back" as frequently used in connection with low back pain mistakenly suggests that there is weakness of the low back muscles. The feeling of weakness that occurs along with the painful back is associated with the faulty alignment the body assumes and is often caused by weakness of abdominal muscles.

Despite the fact that the low back muscles are the most important, relatively little space will be devoted to them in this Chapter as compared to the detailed discussion of abdominal muscles. Testing back muscles is much less complicated than testing abdominal muscles, and, in the field of exercise, there are relatively few errors regarding back exercises, while there are many misconceptions and errors regarding proper abdominal exercises. Furthermore, in contrast to back muscles, weakness of abdominal muscles is prevalent. It is important to know how to test for strength and how to prescribe proper exercises for the abdominal muscles because of the affect that weakness of these muscles has on the overall posture and the relationship to painful postural problems.

Illustrations, definitions, and descriptions of basic concepts are used to help achieve this purpose. The illustrations of the abdominal muscles that follow on pages 189–193, and the accompanying text, provide information in detail about the origins, insertions, and actions of these muscles. This information is essential to understanding the functions of these important trunk muscles.

Rectus Abdominis

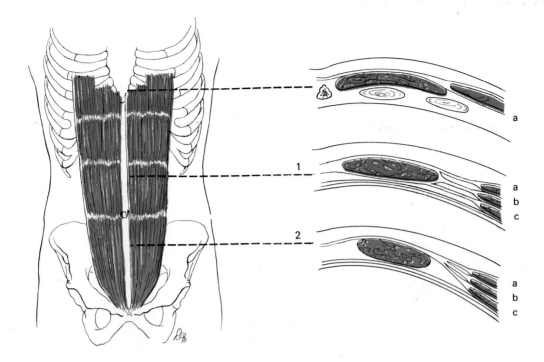

Origin: Pubic crest and symphysis.

Insertion: Costal cartilages of fifth, sixth, and seventh ribs, and xiphoid process of sternum.

Direction of fibers: Vertical.

Action: Flexes the vertebral column by approximating the thorax and pelvis anteriorly. With the pelvis fixed, the thorax will move toward the pelvis; with the thorax fixed, the pelvis will move toward the thorax.

Nerve: T5–12, ventral rami.

Weakness: A weakness of this muscle results in a decrease in the ability to flex the vertebral column. In the supine position, the ability to tilt the pelvis posteriorly or to approximate the thorax toward the pelvis is decreased, making it difficult to raise the head and upper trunk. In order for anterior neck flexors to raise the head from a supine position, it is essential that anterior abdominal muscles, particularly the Rectus abdominis, fix the thorax. With marked weakness of abdominal muscles an individual may not be able to raise the head even though neck flexors are strong. In the erect position, weakness of this muscle permits an anterior pelvic tilt and a lordotic posture (increased anterior convexity of the lumbar spine).

CROSS SECTIONS OF RECTUS ABDOMINIS AND ITS SHEATH

(1) Above the arcuate line the aponeurosis of the Internal oblique (b) divides. Its anterior lamina fuses with the aponeurosis of the External oblique (a) to form the ventral layer of the Rectus sheath. Its posterior lamina fuses with the aponeurosis of the Transversus abdominis (c) to form the dorsal layer of the Rectus sheath.

(2) Below the arcuate line the aponeuroses of all three muscles fuse to form the ventral layer of the Rectus sheath, and the transversalis fascia forms the dorsal layer. (See also p. 193.)

Obliquus Externus Abdominis

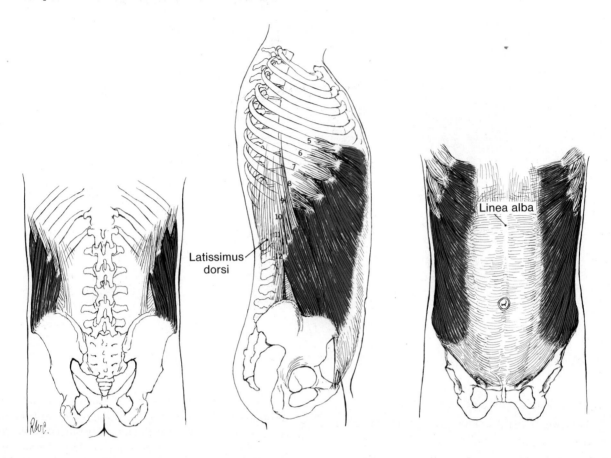

EXTERNAL OBLIQUE, ANTERIOR FIBERS

Origin: External surfaces of ribs five through eight, interdigitating with Serratus anterior.

Insertion: Into a broad, flat aponeurosis, terminating in the linea alba, a tendinous raphe which extends from the xiphoid.

Direction of fibers: The fibers extend obliquely downward and medialward with the uppermost fibers more medialward.

Action: Acting *bilaterally*, the anterior fibers flex the vertebral column approximating the thorax and pelvis anteriorly, support and compress the abdominal viscera, depress the thorax, and assist in respiration. Acting *unilaterally* in conjunction with the anterior fibers of the Internal oblique on the opposite side, the anterior fibers of the External oblique rotate the vertebral column, bringing the thorax forward (when the pelvis is fixed), or the pelvis backward (when the thorax is fixed). For example, with the pelvis fixed, the right External oblique rotates the thorax counterclockwise, and the left External oblique rotates the thorax clockwise.

Nerves to anterior and lateral fibers: T5–12.

EXTERNAL OBLIQUE, LATERAL FIBERS

Origin: External surface of ninth rib, interdigitating with Serratus anterior; and external surfaces of tenth, eleventh, and twelfth ribs, interdigitating with Latissimus dorsi.

Insertion: As the inguinal ligament, into anterior superior spine and pubic tubercle, and into external lip of anterior one-half of iliac crest.

Direction of fibers: Fibers extend obliquely downward and medialward, more downward than the anterior fibers.

Action: Acting *bilaterally*, the lateral fibers of the External oblique flex the vertebral column, with major influence on the lumbar spine, tilting the pelvis posteriorly. (See also action in relation to posture, p. 218.) Acting *unilaterally* in conjunction with the lateral fibers of the Internal oblique on the same side, the lateral fibers of the External oblique laterally flex the vertebral column, approximating the thorax and iliac crest laterally. These fibers also act to rotate the vertebral column as described above. This portion of the External oblique in its action on the thorax is comparable to that of the Sternocleidomastoid in its action on the head.

Obliquus Internus Abdominis

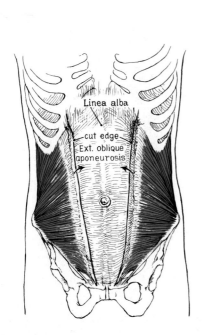
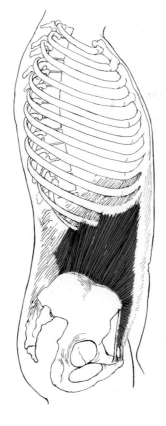
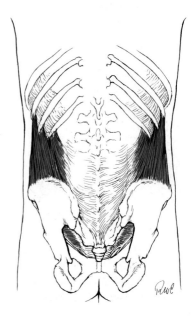

Linea alba

cut edge
Ext. oblique
aponeurosis

INTERNAL OBLIQUE, LOWER ANTERIOR FIBERS

Origin: Lateral two-thirds of inguinal ligament, and short attachment on iliac crest near anterior superior spine.

Insertion: With Transversus abdominis into crest of pubis, medial part of pectineal line, and into linea alba by means of an aponeurosis.

Direction of fibers: Fibers extend transversely across lower abdomen.

Action: The lower anterior fibers compress and support the lower abdominal viscera in conjunction with the Transversus abdominis.

INTERNAL OBLIQUE, UPPER ANTERIOR FIBERS

Origin: Anterior one-third of intermediate line of iliac crest.

Insertion: Linea alba by means of aponeurosis.

Direction of fibers: Fibers extend obliquely medialward and upward.

Action: Acting *bilaterally*, the upper anterior fibers flex the vertebral column, approximating the thorax and pelvis anteriorly, support and compress the abdominal viscera, depress the thorax, and assist in respiration. Acting *unilaterally*, in conjunction with the anterior fibers of the External oblique on the opposite side, the upper anterior fibers of the Internal oblique rotate the vertebral column, bringing the thorax backward (when the pelvis is fixed), or the pelvis forward (when the thorax is fixed). For example, the right Internal oblique rotates the thorax clockwise, and the left Internal oblique rotates the thorax counterclockwise on a fixed pelvis.

INTERNAL OBLIQUE, LATERAL FIBERS

Origin: Middle one-third of intermediate line of iliac crest, and thoracolumbar fascia.

Insertion: Inferior borders of tenth, eleventh, and twelfth ribs and linea alba by means of aponeurosis.

Direction of fibers: Fibers extend obliquely upward and medialward, more upward than the anterior fibers.

Action: Acting *bilaterally*, the lateral fibers flex the vertebral column, approximating the thorax and pelvis anteriorly, and depress the thorax. Acting *unilaterally* and in conjunction with the lateral fibers of the External oblique on the same side, the lateral fibers of the Internal oblique laterally flex the vertebral column, approximating the thorax and pelvis laterally. These fibers also act to rotate the vertebral column as described previously.

Nerves to anterior and lateral fibers: T7–12, Iliohypogastric and ilioinguinal, ventral rami.

191

Obliquus Externus and Obliquus Internus Abdominis

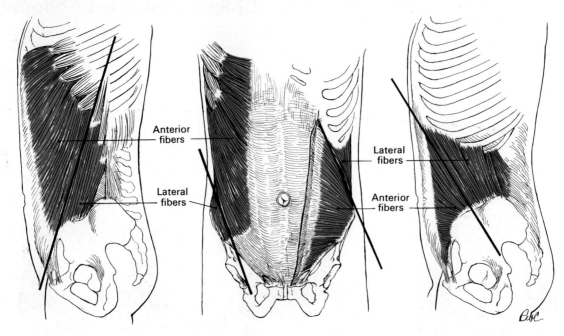

Obliquus externus abdominis   Obliquus internus abdominis

Weakness: Moderate or marked weakness of both External and Internal obliques decreases respiratory efficiency and decreases support of abdominal viscera.

Bilateral weakness of External obliques decreases the ability to flex the vertebral column and tilt the pelvis posteriorly. In standing, it results in either anterior pelvic tilt, or in an anterior deviation of the pelvis in relation to the thorax and lower extremities. (See p. 219, figs. B and C.)

Bilateral weakness of Internal obliques decreases the ability to flex the vertebral column.

Cross-sectional weakness of External oblique on one side, and Internal oblique on the other allows separation of costal margin from opposite iliac crest resulting in rotation and lateral deviation of the vertebral column. With weakness of the right External oblique and left Internal oblique (as seen in a right thoracic, left lumbar scoliosis) there is a separation of the right costal margin from the left iliac crest. The thorax deviates toward the right and rotates posteriorly on the right. With weakness of the left External and right Internal obliques the reverse occurs.

Unilateral weakness of lateral fibers of External oblique and Internal oblique on the same side allows separation of the thorax and iliac crest laterally, resulting in a C-curve convex toward the side of weakness. Weakness of the lateral fibers of the

left External and Internal obliques gives rise to a left C-curve.

Shortness: *Bilateral shortness of anterior fibers of External and Internal oblique* muscles causes the thorax to be depressed anteriorly contributing to flexion of the vertebral column. In standing, this will be seen as a tendency toward kyphosis and depressed chest. In a kyphosis-lordosis posture, the lateral portions of the Internal oblique may shorten, but the lateral portions of the External oblique may be elongated. This same tendency may occur in a posture showing anterior deviation of the pelvis and posterior deviation of the thorax.

Cross-sectional shortness of External oblique on one side and Internal oblique on the other causes rotation and lateral deviation of the vertebral column. Shortness of left External oblique and right Internal oblique, as seen in advanced cases of right thoracic, left lumbar scoliosis, causes rotation of the thorax forward on the left.

Unilateral shortness of lateral fibers of External oblique and Internal oblique on same side causes approximation of the iliac crest and thorax laterally resulting in a C-curve convex toward the opposite side. Shortness of the lateral fibers of the right Internal and External obliques may be seen in a left C-curve.

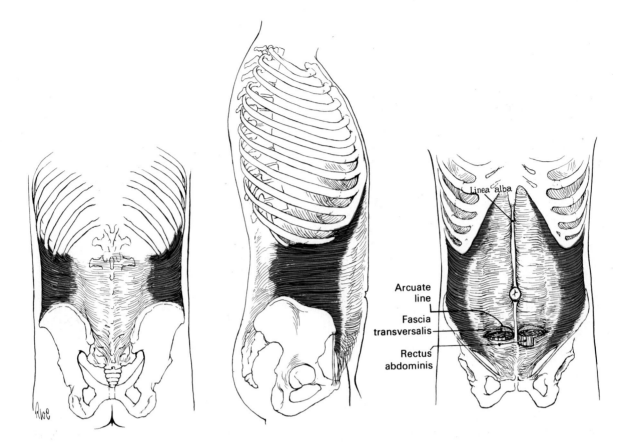

Origin: Inner surfaces of cartilages of lower six ribs, interdigitating with the diaphragm; thoracolumbar fascia; anterior three-fourths of internal lip of iliac crest; and lateral one-third of inguinal ligament.

Insertion: Linea alba by means of a broad aponeurosis, pubic crest and pecten of pubis.

Direction of fibers: Transverse (horizontal).

Action: Acts like a girdle to flatten the abdominal wall and compress the abdominal viscera; upper portion helps to decrease the infrasternal angle of the ribs as in expiration. This muscle has no action in lateral trunk flexion except that it acts to compress the viscera and stabilize the linea alba thereby permitting better action by anterolateral trunk muscles.

Nerve: T7–12, Iliohypogastric, ilioinguinal, ventral divisions.

Weakness: Permits a bulging of the anterior abdominal wall thereby, indirectly, tending to affect an increase in lordosis. (See accompanying photograph.) During flexion in the supine position, and hyperextension of the trunk in the prone position, there tends to be a bulging laterally if the Transversus abdominis is weak.

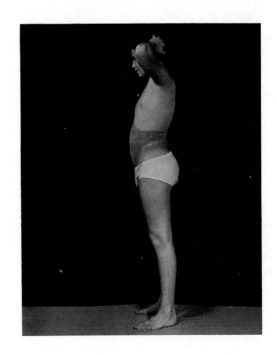

Definitions and Descriptions of Terms

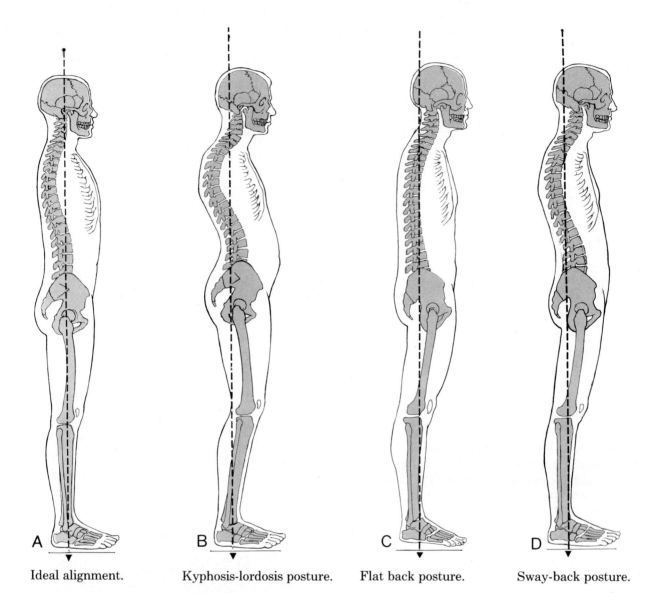

A | Ideal alignment.

B | Kyphosis-lordosis posture.

C | Flat back posture.

D | Sway-back posture.

The *normal curves of the spine* consist of a curve convex forward in the neck (cervical region), convex backward in the upper back (thoracic region), and convex forward in the low back (lumbar region). These may be described as slight extension of the neck, slight flexion of the upper back, and slight extension of the low back. Along with the normal curve in the low back the *pelvis is in a neutral position*. In fig. A, the two bony prominences at the front of the pelvis are in the same vertical plane indicating that it is in a neutral position.

In faulty postural positions, the pelvis may be tilted forward, backward, or sideways. Any tilting of the pelvis involves simultaneous movements of the low back and hip joints. In an *anterior pelvic tilt*, fig. B, the pelvis tilts forward and the low back arches forward creating an increased forward curve (lordosis) in the low back. The hip joints flex slightly by the pelvis tilting forward toward the front of the thighs. In *posterior pelvic tilt* (figs. C and D), the pelvis tilts backward, and the hip joints extend by the pelvis tilting posteriorly toward the back of the thigh. In *lateral pelvic tilt*, one hip is higher than the other and the spine curves with convexity toward the low side. (For lateral pelvic tilt, see p. 291.)

DEFINITIONS AND DESCRIPTIONS OF TERMS

The following definitions intentionally are not in alphabetical order, but, hopefully, in a meaningful sequence. Included are definitions relating to the trunk and hip joints that are considered essential to understanding the functions of the trunk muscles.

The *trunk*, or torso, is the body excluding the head, neck, and limbs. The *thorax* (rib cage), the *abdomen* (belly), the *pelvis* (hip bones), and the *low back* are all parts of the trunk. The term *"trunk raising"* may be used to describe raising the trunk against gravity from various positions: From face-lying (prone), trunk raising backward; from side-lying, trunk raising sideways; from back-lying (supine), trunk raising forward. The term may apply also to raising the trunk in standing from positions of forward bending, side bending, or back bending.

The thorax is *elevated* (chest lifted up and forward) by straightening the upper back, bringing the rib cage out of a slumped position. The thorax is *depressed* when sitting or standing in a slumped position or it may be pulled downward and inward by action of certain abdominal muscles.

The trunk is joined to the thighs at the hip joints. The movement of *hip flexion* means bending (forward) at the hip joint. It may be done by bringing the front of the thigh closer to the pelvis, as in leg-raising, or by tilting the pelvis forward toward the thigh as in the sit-up movement. Positions of the pelvis in good and faulty postural alignment are illustrated on the facing page.

In addition to understanding pelvic tilts in relation to the upright posture, it is necessary to understand how the pelvis tilts during such exercises as sit-ups and double leg raising. *During a "curled-trunk sit-up"* with legs† extended, the pelvis first tilts posteriorly, accompanied by flattening of the low back and extension of the hip joints. (See p. 200, fig. C.) After the trunk-curl phase is completed, the pelvis tilts anteriorly (forward) toward the thigh in hip flexion (fig. D) but still remains in posterior tilt in relation to the trunk, maintaining the flat back position. *During a "sit-up with low back arched,"* on the other hand, the pelvis tilts anteriorly toward the thigh as the sit-up begins and remains tilted anteriorly. (See p. 210.) *During "double-leg-raising",* if abdominal muscles are weak, the pelvis tilts anteriorly, causing the low back to arch, as the legs are lifted. When abdominal muscles are

† In this chapter, the term "*leg*" applies to the whole lower extremity. In both common usage and in medical literature, when referring to "straight-leg-raising tests for hamstring

strong, the pelvis can be held in posterior tilt with the low back flat as the legs are lifted. (See p. 213 for more details.)

Extension of the spine moves the head and trunk in a backward direction. In the neck and low back, extension results in an increase in the normal curves as the spinal column bends backward; in the upper back it results in a decrease of the normal curve and straightening the spine since this part does not bend backward. When done against gravity, as lifting the head and shoulders from a face-lying position, the movement is performed by back extensor muscles with fixation of the pelvis by hip extensors.

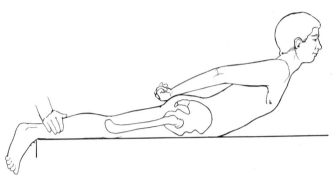

Hyperextension of the spine is movement beyond the "normal" range of motion or may refer to a position or movement greater than the normal anterior curve. Hyperextension may vary from slight to extreme.

Flexion of the spine moves the head and trunk in a forward direction. In the upper back, the spine normally curves convexly backward. Flexion increases this curve resulting in a rounding of the upper back. In the neck, the spine normally curves convexly forward. Flexion, by tilting the head forward, usually results only in straightening the cervical spine. Seldom does flexion continue to the point that the neck curves convexly backward. Similarly, in the low back, the spine normally curves convexly forward. When a person bends the trunk

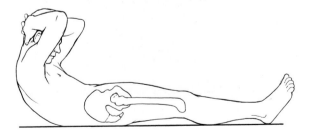

length," and to "double-leg-raising exercises," the term "leg" is used in this manner. [The anatomical term "leg" refers only to the portion between the knee and the ankle.]

Definitions and Descriptions of Terms

forward in flexion, the forward curve in the low back straightens and the low back appears flat. In the neck and low back, straightening the spine may be considered normal flexion. This concept results in some confusion because one tends to think of a straight spine being extended. It is, therefore, easier to understand if one speaks in terms of the motion in low back ranging from a lordotic to a flat back position. When done against gravity, as in curling the trunk from a back-lying position, the movement of trunk flexion is done by abdominal muscles.

Overemphasis on curling the trunk in many exercise programs for children and adults is conducive to increasing flexion and quite a few people have more back flexion than "normal." Both *excessive flexion of the low back* (spine convex backward) and *excessive extension* (lordosis) can result in problems of low back pain.

"*Trunk curl*" refers to flexion of the spine only (i.e., upper back curved convexly backward, and the neck and low back straightened). When abdominal muscles are strong and hip flexor muscles are very weak, only the trunk curl can be completed when attempting to do a sit-up. The subject illustrated below has strong abdominal muscles but the hip flexors are paralyzed. (Leg braces were left on so legs could be "propped" in the knee-bent position for two of the photographs.)

"*Sitting position*" is one in which the trunk is upright and the hips are flexed. To "*sit down*" means to move from an upright to a sitting position by flexing at the hip joints but may not require hip flexor muscle action. To "*sit up*" means to move from a reclining to a sitting position by flexing at the hip joints and, when done unassisted, can be performed only by hip flexor muscles. *Alone or in combination, the word sit should be used only in connection with movement that involves hip joint flexion.*

The *sit-up exercise*, therefore, is the movement of coming from a supine to a sitting position by flexing the hip joints, and is performed by hip flexors. It may be combined with various trunk and leg positions as illustrated on the facing page.

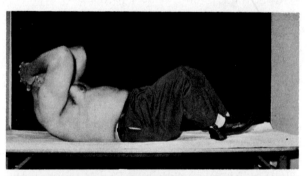

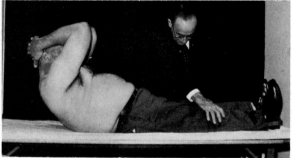

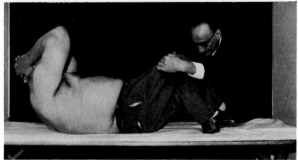

A subject with *strong abdominal muscles* and *paralyzed hip flexor muscles* can perform only the trunk curl. He is unable to flex the trunk toward the thighs (i.e., hip joint flexion) in an effort to perform the sit-up. Regardless of whether legs are extended or flexed, and whether or not the legs are held down, no flexion can occur at the hip joints in the absence of hip flexor muscle action. The abdominal muscles complete their range of motion by curling the trunk only.

It may be noted that the subject does not raise the trunk as high from the table with legs flexed as with legs extended. The pelvis moves more freely in posterior tilt with legs flexed and, as the abdominal muscles shorten, both the pelvis and thorax move with the result that the thorax is not raised as high from the table as would occur if the pelvis were stabilized by the legs being in extension.

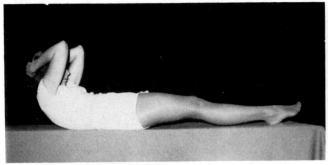 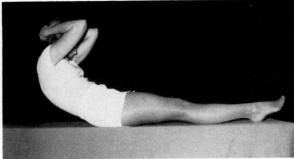

(1) *"Curled-trunk sit-up with legs extended"* consists of flexion of the spine (trunk curl), performed by abdominal muscles, followed by flexion of the hip joints (sit-up), performed by hip flexors.

(2) *"Curled-trunk sit-up with hips and knees flexed"* (knee-bent sit-up), starts from a position of hip flexion (flexion of thigh toward pelvis) and consists of flexion of the spine (trunk curl) performed by abdominal muscles, followed by further flexion of the hip joints (by flexion of pelvis toward thigh) performed by hip flexors.

 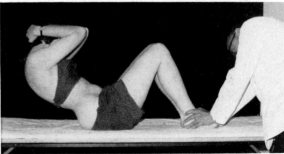

(3) *"Sit-up with low back arched"* (with legs extended or flexed) occurs when the abdominal muscles are very weak and consists of flexion of the hip joints by action of the hip flexors, accompanied by hyperextension of the low back (lordosis). With strong hip flexors, the entire trunk-raising movement can be performed. (Compare with photographs on p. 196 in which no hip joint flexion occurs in absence of hip flexors.)

Abdominal Muscle Action During Trunk Raising

Because the oblique abdominal muscles are essentially fan-shaped, a part of one muscle may function in a somewhat different role than another part of the same muscle. A consideration of the attachments and the line of pull of the fibers, coupled with clinical observations of patients with marked or spotty weakness as well as those with normal musculature, leads one to certain conclusions regarding the action of the various muscles or segments of the abdominal muscles.

As trunk flexion is *slowly* initiated by raising the head and shoulders from a supine position, one can observe, in a subject with normal abdominal muscles, that the rib cage is depressed anteriorly, the ribs flare outward, and the infrasternal angle is increased. The pelvis tilts posteriorly, simultaneously with the head and shoulder raising. As the trunk is raised in flexion on the thighs, the ribs tend to be compressed laterally and the infrasternal angle decreases.

The Rectus abdominis acts to depress the ribs anteriorly and tilt the pelvis posteriorly. The flaring outward of the rib cage and the accompanying increase in the infrasternal angle is compatible with action of the Rectus abdominis and the Internal oblique. (See fig. A.) As the hip flexors come into action and exert the strong force to tilt the pelvis anteriorly, the External oblique helps maintain the lumbar spine in flexion by pulling the pelvis in the direction of posterior tilt. (See fig. B.) As the External oblique contracts, the ribs are compressed laterally and the infrasternal angle decreases. The action of the External oblique may be observed also in cases of scoliosis in which muscle imbalance exists between the right and left External oblique muscles. It it not uncommon to observe that flexion of the spine may begin with a rather symmetrical pull and as the effort is made to raise the trunk in flexion on the thighs there will be a forward rotation of the thorax on the side of the stronger External oblique.

No test movement can cause an approximation of parts to which the lower transverse fibers of the Internal oblique are attached since these fibers extend across the lower abdomen from ilium to ilium as do the lower fibers of the Transversus abdominis. In posterior pelvic tilt and trunk-raising movements, however, this part will become firm, if strong.

Electromyographic studies may either confirm or modify the conclusions drawn from clinical observations. To date the information is rather scarce with reference to abdominal muscles, but Crowe et al.,[23] in a report on an electromyographic study of the sit-ups, report—"activity appears first in the upper rectus abdominis, followed about 0.2 to 0.3 seconds later by the lower rectus and Internal oblique abdominals." Also, " . . . subjects tested were asked to curl the thorax until the scapulae were just off the floor and then to hold this position for 2–3 seconds. The resulting electromyograms recorded intense activity in all the abdominal muscles . . .".

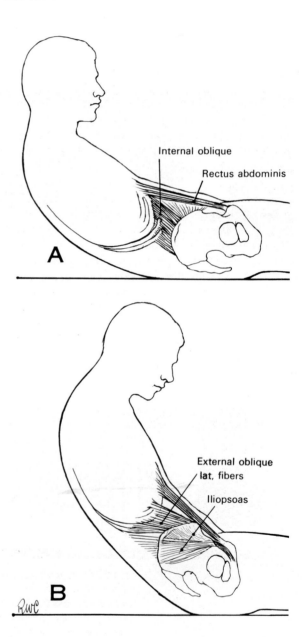

AN ANALYSIS OF MOVEMENTS AND MUSCLE ACTIONS DURING A CURLED-TRUNK SIT-UP

The illustrations on pp. 200 and 201, show the various stages of movement of the spine and hip joints that occur during a curled-trunk sit-up. On pages 202–204, the illustrations are repeated and the accompanying text describes the associated muscle actions.

Outlines of the basic figures have been made from photographs. To these have been added drawings of the femur and pelvis, and a dotted line representing part of the vertebral column. The broken line from the anterior-superior spine to the symphysis pubis is the line of reference for the pelvis and is the same as that representing the vertical plane in the anatomical position. A solid line parallel to the broken line has been drawn through the pelvis to the hip joint and continues as a reference line through the femur to indicate the position of the hip joint at the various stages of movement.

Specific degrees, based on the average normal ranges of motion presented here and in Chapter 2, are used to help explain the movements that occur. Because of individual variations with respect to ranges of motion of the spine and hip joints, there will be variations in the manner in which subjects perform these movements.

For this particular analysis, it is assumed that the abdominal and erector spinae muscles, as well as the hip flexor and extensor muscles, are normal in length and strength and that the spine and hip joints permit normal range of motion.

Normal hip joint extension is given as 10°. From the standpoint of stability in standing, it is desirable to have a few degrees of extension; it is not desirable to have more than a few degrees. In the upright or supine position with hips and knees extended, a posterior pelvic tilt of 10° results in 10° of hip joint extension. It occurs because the pelvis is tilted posteriorly toward the back of the thigh instead of the thigh being moved posteriorly toward the pelvis. Flattening the lumbar spine accompanies the posterior pelvic tilt. Flexion to the point of straightening or flattening the low back is considered normal flexion on the basis that it is an acceptable and desirable range of motion.

With the knee flexed, the hip joint can flex approximately 125° from the zero position to an acute angle of approximately 55° between the femur and the pelvis. With the knee extended (as in the straight-leg-raising test for hamstring length) the leg can be raised approximately 80° from the table. The equivalent of this is a trunk raising movement, with legs extended, in which the pelvis is flexed toward the thighs through a range of approximately 80° from the table.

For convenience in measuring joint motion, the trend is to use the anatomical position as zero. Thus, the straight position of the hip joint is considered to be zero position. However, it is necessary to adhere to geometric terms when describing angles and the number of degrees in angles.

On pp. 200 and 201, the right-hand column is headed "Angle of Flexion." This refers to the hip joint angle anteriorly between the reference line through the pelvis and the line through the femur, and is expressed in geometric terms. Changes in the angle of flexion represent corresponding changes in overall length of the hip flexors.

The second column from the right is headed "Hip Joint" and listed here are the degrees through which the hip joint has moved, first in extension, then in flexion, from the zero position.

Movements of Spine and Hip Joints During a Curled-Trunk Sit-up with Legs Extended

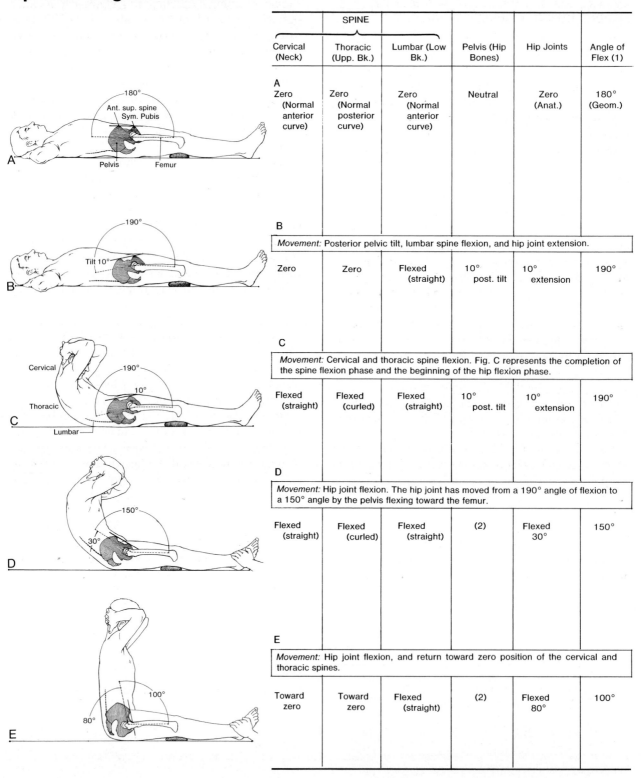

	SPINE				
Cervical (Neck)	Thoracic (Upp. Bk.)	Lumbar (Low Bk.)	Pelvis (Hip Bones)	Hip Joints	Angle of Flex (1)
A Zero (Normal anterior curve)	Zero (Normal posterior curve)	Zero (Normal anterior curve)	Neutral	Zero (Anat.)	180° (Geom.)
B *Movement:* Posterior pelvic tilt, lumbar spine flexion, and hip joint extension.					
Zero	Zero	Flexed (straight)	10° post. tilt	10° extension	190°
C *Movement:* Cervical and thoracic spine flexion. Fig. C represents the completion of the spine flexion phase and the beginning of the hip flexion phase.					
Flexed (straight)	Flexed (curled)	Flexed (straight)	10° post. tilt	10° extension	190°
D *Movement:* Hip joint flexion. The hip joint has moved from a 190° angle of flexion to a 150° angle by the pelvis flexing toward the femur.					
Flexed (straight)	Flexed (curled)	Flexed (straight)	(2)	Flexed 30°	150°
E *Movement:* Hip joint flexion, and return toward zero position of the cervical and thoracic spines.					
Toward zero	Toward zero	Flexed (straight)	(2)	Flexed 80°	100°

Note: The small roll under the knees places the thigh and hip joints in position similar to the anatomical position.

(1) Angle of flexion: Angle between pelvis and femur, anteriorly.
(2) Pelvis remains in posterior tilt in relation to trunk but moves in the direction of anterior tilt toward the thigh.

Movements of Spine and Hip Joints During a Curled-Trunk Sit-Up with Hips and Knees Flexed

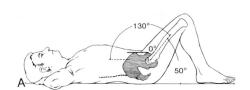

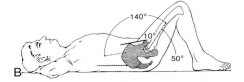

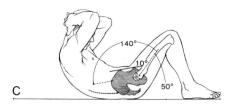

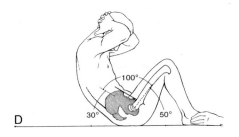

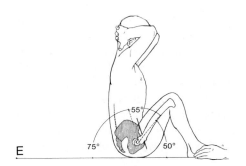

| | SPINE | | | | | |
	Cervical (Neck)	Thoracic (Upp. Bk.)	Lumbar (Low Bk.)	Pelvis (Hip Bones)	Hip Joints	Angle of Flex (1)
A	Zero (Normal anterior curve)	Zero (Normal posterior curve)	Zero (Normal anterior curve)	Neutral	(Anat.) 50°	(Geom.) 130°
B	*Movement:* Lumbar spine flexion and 10° decrease in hip joint flexion by virtue of posterior pelvic tilt.					
	Zero	Zero	Flexed (straight)	10° post. tilt	50° flexion of thigh	140°
C	*Movement:* Cervical and thoracic spine flexion. Fig. C represents completion of the spine flexion and the beginning of flexion of the pelvis toward the flexed thigh.					
	Flexed (straight)	Flexed (curled)	Flexed (straight)	10° post. tilt	50° flexion of thigh	140°
D	*Movement:* Hip joint flexion. The hip joint has moved from a 140° angle of flexion to a 100° angle by the pelvis flexing toward the femur.					
	Flexed (straight)	Flexed (curled)	Flexed (straight)	(2)	80° (50 thigh + 30 pelvis)	100°
E	*Movement:* Hip joint flexion, and a return toward zero position of the cervical and thoracic spines. On the basis of 125° being complete flexion, the hip joint has reached the position of complete flexion.					
	Toward zero	Toward zero	Flexed (straight)	(2)	125° (50 thigh + 75 pelvis)	55°

(1) Angle of flexion: Angle between pelvis and femur, anteriorly.

(2) Pelvis remains in posterior tilt in relation to trunk but moves in the direction of anterior tilt toward the thigh.

Actions of Abdominal and Hip Flexor Muscles During a Curled-Trunk Sit-Up

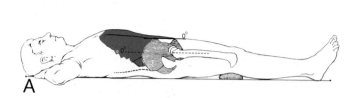

ZERO POSITION OF SPINE, PELVIS, AND HIP JOINTS

Figs. A and A¹ may be regarded as hypothetical starting positions. In reality, especially with knees bent, the low back tends to flatten (i.e., the lumbar spine flexes) when a normally flexible individual assumes the supine position.

In fig. A, the length of the hip flexors corresponds with the zero anatomical position of the hip joints.

ZERO POSITION OF SPINE AND PELVIS, FLEXION OF HIP JOINTS

In fig. A¹, because of the flexed position of the hips, the one-joint hip flexors are shorter in length than in fig. A. In relation to its overall length, the Iliacus is at about 40% of its range of motion which is within the middle third of the overall range.

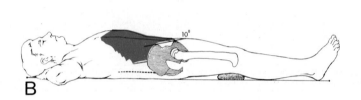

POSTERIOR PELVIC TILT, LUMBAR SPINE FLEXION, AND HIP JOINT EXTENSION

Figs. B and B¹ represent a stage of movement in which the pelvis is tilted posteriorly prior to beginning the trunk raising. (Note the 10° posterior pelvic tilt.) In testing, this movement often is performed as a separate stage to ensure lumbar spine flexion.

In fig. B, the hip flexors have lengthened, and the one-joint hip flexors (chiefly the Iliacus) have reached the limit of length permitted by the hip joint extension. At this length they help stabilize the pelvis by restraining further posterior pelvic tilt.

In fig. B¹, the hip flexor length is slightly more than in fig. A¹ because the pelvis has tilted posteriorly 10° away from the femur.

POSTERIOR PELVIC TILT, LUMBAR SPINE FLEXION, AND HIP JOINT FLEXION

Posterior pelvic tilt exercises frequently are used with the intention of strengthening the abdominal muscles. Too often the tilt is done without any benefit to the abdominals. The subject performs the movement by contracting the buttocks muscles (hip extensors) and, in the case of the knee-bent position, by pushing with the feet to help "rock" the pelvis back into posterior tilt.

To ensure that the pelvic tilt is performed by the abdominal muscles, there must be an upward-inward pull by these muscles with the front and sides of the lower abdomen becoming very firm. (See p. 221.)

Often it is necessary to discourage use of the buttocks muscle in order to force action by the abdominals in performing the tilt.

Actions of Abdominal and Hip Flexor Muscles During a Curled-Trunk Sit-Up

SPINE FLEXION PHASE (TRUNK CURL) COMPLETED

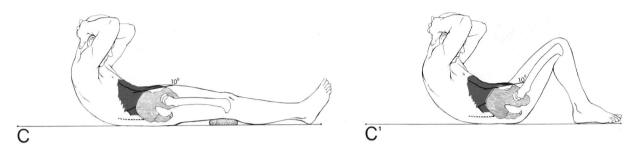

In figs. C and C[1], the neck (cervical spine), the upper back (thoracic spine), and the low back (lumbar spine) are flexed. The low back remained in the same degree of flexion as in figs. B and B[1] where it reached maximum flexion for this subject.

When the posterior tilt is not done as a separate movement, as in figs. B and B[1], it occurs simultaneously with the beginning phase of trunk raising (i.e., the trunk-curl phase) unless the abdominal muscles are extremely weak or, with legs extended, the hip flexors are so short they prevent posterior tilt.

In figs. C and C[1], the abdominal muscles have shortened to their fullest extent with the completion of spine flexion. In fig. C, the hip flexors have remained lengthened to the same extent as in fig. B.

In fig. C[1], the one-joint hip flexors have not reached the limit of their length and, therefore, do not act passively to restrain posterior tilt. The hip flexors contract to stabilize the pelvis, and, on palpation of the superficial hip flexors, there is evidence of firm contraction as the subject begins to lift the head and shoulders from the table.

HIP FLEXION PHASE (SIT-UP) INITIATED

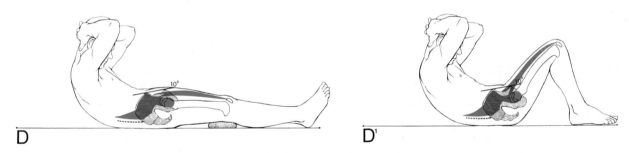

With flexion of the spine complete (as in figs. C, C[1], D, and D[1]), no further movement in the direction of coming to a sitting position can occur except by flexion of the hip joints.

Since abdominal muscles do not cross the hip joint, these muscles cannot assist in the movement of hip flexion.

From a supine position, hip flexion can be performed only by the hip flexors acting to bring the pelvis in flexion toward the thighs.

Figs. D and D[1] represent the beginning of the sit-up phase as well as the end of the trunk-curl phase.

Actions of Abdominal and Hip Flexor Muscles During a Curled-Trunk Sit-Up

HIP FLEXION PHASE (SIT-UP) CONTINUED

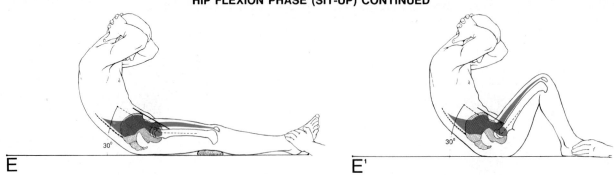

E

E¹

Figs. E and E¹ show a point in the arc of movement between the completed trunk curl (as seen in C, C¹, D, and D¹) and the full sit-up. The abdominal muscles maintain the trunk in flexion, and the hip flexors have lifted the flexed trunk upward toward the sitting position through an arc of about 30° from the table.

When it is necessary, the feet *may* be held down at the initiation of, and during, the hip flexion phase. (See p. 205.) Prior to the hip flexion phase, the feet must not be held down.

HIP FLEXION PHASE (SIT-UP) COMPLETED

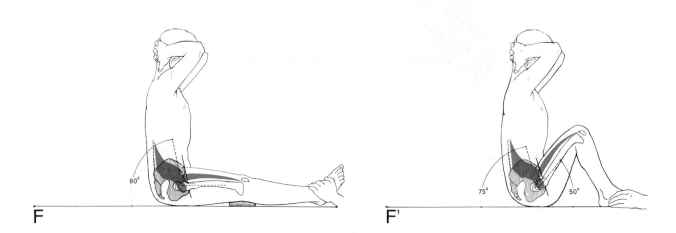

F

F¹

In figs. F and F¹, as the subjects reach the sitting position, the cervical and thoracic spines are no longer fully flexed, and the abdominal muscles relax to some extent.

In fig. F, the hip flexors have moved the pelvis in flexion toward the thigh, completing an arc of about 80° from the table. In this position, with knees extended, and lumbar spine flexed, the hip joint is as fully flexed as the range of normal Hamstring length permits. The lumbar spine remains flexed because moving from the flexed position of the low back to the zero position (normal anterior curve) would require that the pelvis be tilted ten more degrees in flexion toward the thigh and the Hamstring length does not permit it.

In fig. F¹, the hip flexors have moved the pelvis in flexion toward the thigh through an arc of about 75° from the table. The lumbar spine remains in flexion because the hip joint has already reached the 125° of full flexion. Further flexion of the hip joints by tilting the pelvis forward (and bringing the low back into normal anterior curve) could be done only if the flexion of the thigh were decreased by moving the heels farther from the buttocks in this sitting position.

204

Effect of Holding Feet Down During Trunk Raising

EFFECTS OF HOLDING FEET DOWN DURING TRUNK RAISING

The center of gravity of the body is generally given as being at approximately the level of the first sacral segment, and this point is above the hip joint. If half the body weight is above the center of gravity, then more than half the body weight is above the hip joint. (Basmajian[22] states that the lower extremities constitute about one-third of the body weight.) For most people, this means that the force exerted by the trunk in supine position is greater than that exerted by both lower extremities. *Usually*, double-leg-raising (with knees straight) can be initiated without over-balancing the weight of the trunk in the supine position. *Seldom* can the straight trunk or hyperextended trunk (see p. 210) be raised from supine toward a sitting position without some outside force being applied (such as pressure downward on the feet) in addition to that exerted by the extended extremities.

On the other hand, if the trunk curls sufficiently as the trunk raising is started, the center of gravity of the body moves downward toward, or below, the hip joints. As this occurs, the curled trunk can be raised in flexion toward the thighs without having the feet held down. Most adolescents (especially those in whom the legs are long in relation to the trunk) and most adult females can perform the sit-up with legs extended and without the feet being held down. In contrast, many men may need to have some added force applied (usually very little) at the point where the trunk curl is completed and the hip flexion phase begins.

For the curled-trunk sit-up to be used as a test for strength of abdominal muscles, it must be made certain that the ability to curl the trunk is being measured. The trunk curl must precede the hip flexion phase in the trunk raising movement. When the feet are not held down, the pelvis tilts posteriorly as the head and shoulders are raised in initiating the trunk curl. With feet held down, the hip flexors are given fixation and the trunk raising can immediately become a sit-up movement with flexion at the hip joints. Hence, to help ensure that the test determines the ability to curl the trunk before the hip flexion phase starts, the feet must not be held down during the trunk flexion phase.

The question is asked frequently whether holding the feet down causes any problem if abdominal strength is normal. The answer is that it might not if one is performing only a few sit-ups, but it can make a great deal of difference if many repetitions are performed. One or two curled-trunk sit-ups, properly done, determines normal strength. It does not determine endurance. An individual may grade normal, and perform several sit-ups properly. With repeated sit-ups, whether legs are extended or flexed, the abdominal muscles may fatigue and the person may "slip into" doing a sit-up with the low back arched. This situation arises frequently because abdominal muscles apparently do not have the endurance exhibited by hip flexors.

On the assumption that it would make no difference if the feet were held down from the beginning of the exercise, the transition to an arched-back sit-up could and would go undetected. If the feet were not held down during the initial spine flexion phase, the inability to curl the trunk would become obvious as fatigue sets in. One might find that an individual could do fifty or one hundred sit-ups with the feet held down, yet no more than five if the feet were not held down. This would indicate a hip-flexor take-over at initiation of the trunk-raising after the first five.

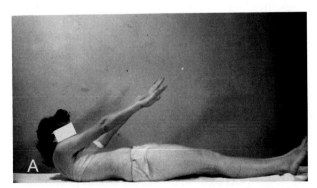

The above photograph shows an individual with marked abdominal muscle weakness who, with arms in a relatively easy (60%) test position, is unable to flex the lumbar spine and complete the sit-up when feet are not held down.

The above photograph shows the same individual as in fig. A who, with arms in 100% test position is able to perform the sit-up by hip flexor action because the feet are held down. As a test this latter measures only hip flexor strength.

Sit-Up Exercises: Indications and Contraindications

The subject in the above photograph with arms in 100% test position, and with knees flexed, can flex the vertebral column but cannot raise the trunk any higher from the table than illustrated.

With feet held down, the subject immediately begins the hip flexion phase and can continue to a full sitting position, as seen in the series of photographs of this same individual on p. 210.

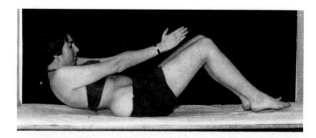

In this illustration, the subject is making an effort to sit up with the arms in an easy test position, feet not held down. It is obvious that she goes immediately into the hip flexion phase. Legs tend to extend in an effort to move the center of gravity of the lower extremities more distally and off-set the force exerted by the trunk. The same problems exist with respect to the stabilization of the feet whether knees are extended or flexed.

The ability to do a curled-trunk sit-up should be considered a normal accomplishment. People should be able to get up easily from a supine position without having to roll over on the side or push themselves up with their arms. When there is weakness in either or both of the muscle groups involved in the curled-trunk sit-up (namely, abdominal and hip flexor muscles) efforts should be made to correct the weakness and restore the ability to perform the movement correctly. While hip flexors may exhibit some weakness associated with postural problems, it is rarely to the degree that it interferes with performing the sit-up (hip flexion) movement. The problem in performing the trunk curl is due to weakness of abdominal muscles. Using the sit-up exercise to try to correct the abdominal weakness is a mistake because, when marked weakness exists, the hip flexors initiate and perform the movement with the low back hyperextended.

For many years, sit-ups were done most frequently with legs extended. In more recent years, emphasis has been placed on doing the exercise in the knee-bent position which automatically flexes the hips in the supine position. *The sit-up, whether legs are straight or bent, is a strong hip flexor exercise*, the difference being in the *arc* of hip-joint motion through which the hip flexors act. With legs extended, they act through an arc from zero to about 80°; with hips and knees flexed, from about 50° (the starting position) to completion of range at 125° flexion for a total of about 75°.

Ironically, the knee-bent sit-up has been advocated as a means of minimizing the action of the hip flexors. The idea has persisted for many years, both among professional and lay people, that having the hips and knees bent in the back-lying position would put the hip flexors "on a slack"‡ and would "rule out" or "eliminate" the action of the hip flexors while doing the sit-up; and that in this position the sit-up would be performed by abdominal muscles. *These ideas are not based on facts; they are false and misleading.* The abdominal muscles can only curl the trunk and cannot perform the hip flexion part (which is the major part) of the trunk-raising movement. Furthermore, the Iliacus is a one-joint muscle which is expected to complete the movement of hip flexion and, as such, is not put on a slack. The two-joint Rectus femoris is not put on a slack because it is lengthened over the knee joint while it is shortened over the hip joint. Except for the Sartorius to some extent, other hip flexors are not put on a slack in the knee-bent position.

‡ For discussion of the term "slack", see p. 5.

Sit-Up Exercises: Indications and Contraindications

If hip flexors are not short, an individual, when starting the trunk raising with legs extended, will curl the trunk and the low back will flatten before starting the hip flexion phase. The danger of hyperextension will occur only if the abdominals are too weak to maintain the curl—a reason not to continue into the sit-up.

The real problem in doing sit-ups with legs extended compared to the apparent advantage of flexing the hips and knees stems from dealing with many subjects who have short hip flexors. In the supine position, the person with short hip flexors will lie with the low back hyperextended (arched forward). The hazard of doing the sit-up from this position is that the hip flexors will further hyperextend the low back, putting a strain on the low back while doing the exercise, and will increase the tendency toward a lordotic posture in standing. The knee-bent position, on the other hand, releases the downward pull by the short hip flexors, allowing the pelvis to tilt posteriorly and the low back to flatten, thereby relieving strain on the low back.

Instead of recognizing and treating the problem of the short hip flexors, the "solution" has been to "give in" to them by flexing the hips and knees. But there are problems that arise from this solution. The same hazard of coming up with the low back hyperextended can occur with knees bent, and does occur when the abdominal muscles are too weak to curl the trunk. (See p. 210.) In trying to come up, the subject requires more pressure than usual to hold the feet down, or more extension of the legs, or is aided by performing the movement quickly with added momentum. Sometimes it is advocated (inadvisedly) that the arms be placed overhead and brought quickly forward to help in performing the sit-up. This added momentum enables the subject to sit up with the low back hyperextended—a movement which can cause severe strain on the low back.

Since the exercise with knees bent is equally as strong a hip flexor exercise as with the knees extended, the effect of increasing hip flexor strength is present in this exercise also. The possibility exists that the knee-bent sit-up, in which the hip flexors shorten to a greater extent than with legs extended may be even more conducive to developing shortness in the muscles and increasing the lordosis.

While normal flexibility of the back is a desirable feature, excessive flexibility is not. Hazards of doing the knee-bent sit-up also relate to the danger of hyperflexion of the trunk (spine curving convexly backward). The first sacral segment is approximately the level of the center of gravity in the body in the anatomical position or supine with legs extended. With hips and knees bent, the center of gravity moves cranially (toward the head). The lower extremities exert less force in counter-balancing the trunk during the sit-up with hips and knees bent than with legs extended. There are two alternatives in accomplishing the sit-up from this knee-bent position: Outside pressure must be exerted to hold the feet down (more than is required for those few who need it with the legs extended), or the trunk must curl excessively to move the center of gravity downward. This excessive flexion is portrayed as an exaggerated thoracic curve (marked rounding of the upper back) or as abnormal flexion involving the thoraco-lumbar area (roundness extending into the low back area). The latter is accentuated when the knee-bent sit-up is done without the feet being held down and with heels close to the buttocks.

The people most in danger of being adversely affected by repeated sit-ups with the knees bent are children and youths because they start with more flexibility than adults. Those adults who have low back pain that is associated with excessive low back flexibility also may be affected adversely by this exercise. An interesting phenomenon that occurs in some subjects who have done a great number of knee-bent sit-ups is that they show excessive flexion in sitting or in forward bending, but a lordosis in standing.

It is unfortunate that the ability to do a certain number of sit-ups, regardless of how they are performed, is used as a measure of physical fitness. Coupled with push-ups, these two exercises probably are stressed more than any others in fitness programs. Done to excess, these exercises increase the tendency toward round upper back and forward shoulders—problems all too common in the general population.

When, how, and to what extent the knee-bent position should be used is discussed on page 212.

Abdominal Muscle Test I: Trunk Raising Forward

Preliminary to doing the curled-trunk sit-up test, the strength of the neck flexors and hip flexors should be determined. Also, the length of back muscles, hip flexors, and Hamstrings should be determined in order that restriction of motion not be confused with muscle weakness.

Valid testing for the strength of the abdominal muscles requires that the curled-trunk sit-up be done *slowly*, starting with hips and knees extended, and with heels remaining in contact with the table, but *feet not held* down by the examiner during the *initial* spine flexion phase.

Patient: Supine, legs extended with small roll under knees. If the hip flexor muscles are short, thereby limiting posterior tilting of the pelvis and flexion of the lumber spine, knees should be allowed to flex sufficiently to allow the back to flatten. A small pillow should be put under the knees so the position is maintained *passively, not actively.*

Fixation: None necessary during the *initial phase* of the test in which the spine is flexed and the thorax and pelvis approximated anteriorly. Holding the feet down must be *avoided* during this phase because stabilizing the feet would allow hip flexors to come immediately into action as trunk raising is initiated. This is to be avoided since the test must ascertain the ability to curl the trunk *before* hip flexors begin flexion of the pelvis toward the thigh. (See photographs on pp. 197 and 205.)

During the *second phase* when the trunk (with the spine held in flexion) is flexed toward the thighs, the hip flexors should be afforded adequate fixation. At this point the feet may be held down if the force exerted by the extended lower extremities does not counterbalance the force exerted by the flexed trunk. Adolescents and adult females seldom need the feet held down; adult males often need the stabilization of the feet.

Test movement: Have the patient *slowly* complete a curled-trunk sit-up by first tilting the pelvis posteriorly and flexing the spine, raising in sequence the head, shoulders, and thorax, followed by flexion of the trunk toward the thighs (hip joint flexion).

Although the abdominal muscles only flex the spine, there is a reason for continuing into the sit-up phase. The fact that the second phase of the movement adds the very strong hip flexor action required to lift the entire trunk introduces, at the moment this is initiated, a second force opposing the abdominal muscles. Up to that point, the weight of the head, upper extremities, and thorax has been the force which the abdominal muscles must over-come in order to approximate the thorax toward the pelvis. At the moment the hip flexors are required to flex the pelvis on the thighs, they exert a force which tends to extend the lumbar spine and pull the pelvis anteriorly in the direction of separating the pelvis from the thorax, directly resisting the pull of the abdominal muscles.

The crucial point in the test for abdominal muscle strength is at the moment the hip flexors come into strong action. The abdominal muscles, at this point, must be able to oppose the force of the hip flexors in addition to maintaining the trunk curl.

While it is not necessary to complete the sit-up beyond the beginning of the hip flexion phase, it is useful and practical to do so for purposes of testing. There are variations in the flexibility of the spine and some individuals raise the head and shoulders higher than others in the trunk-curl part of the test movement, so it is not possible to stipulate a certain height in inches that the shoulders should be raised, nor a specific angle (i.e., 45°) at which the trunk should be held in order to be certain that the movement has proceeded into the hip flexion phase. It would be impractical to "cut off" the action at a given height or angle when the subject is supposed to continue smoothly from the trunk-curl phase into the hip flexion phase. Consequently, it is expedient, albeit not necessary, to allow the subject to complete the sit-up for purposes of testing.

At the point in the sit-up where the strong hip flexor action occurs, subjects with weak abdominal muscles will quickly substitute with the hip flexors, tilting the pelvis anteriorly and arching the low back. When this occurs, a *test position is used instead of the test movement* in order to determine the extent of weakness and to grade the strength.

Test position: The examiner raises the subject to the point of that individual's maximum spine flexion. (Flexibility of the back is highly variable.) Then a second person holds the legs firmly down to give fixation for the hip flexors. The subject is asked to hold this test position. If weakness is present in the abdominal muscles, very quickly and almost simultaneously the thorax will drop back, the low back will arch, and the pelvis will tilt anteriorly. Some subjects will be able to keep the trunk, as a whole, from dropping down to the table because of the quick "take over" by the hip flexors, but the sudden shift in position is quite obvious.

Pressure: None by the examiner. Changing the arm position from hands behind the head, to forearms across the chest, to arms forward (as illustrated on p. 209) decreases the resistance offered by the head, trunk, and upper extremities as the center of gravity of the body shifts caudally.

Trunk Raising Forward: Grading

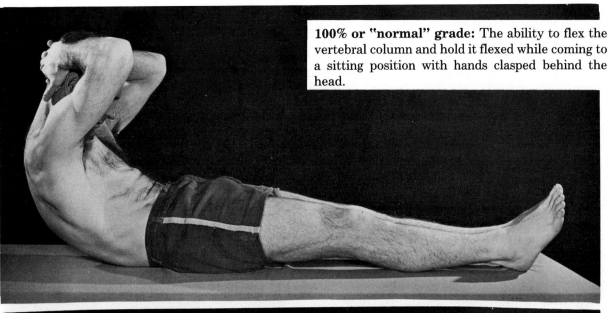

100% or "normal" grade: The ability to flex the vertebral column and hold it flexed while coming to a sitting position with hands clasped behind the head.

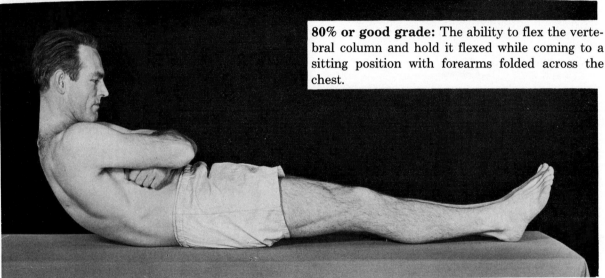

80% or good grade: The ability to flex the vertebral column and hold it flexed while coming to a sitting position with forearms folded across the chest.

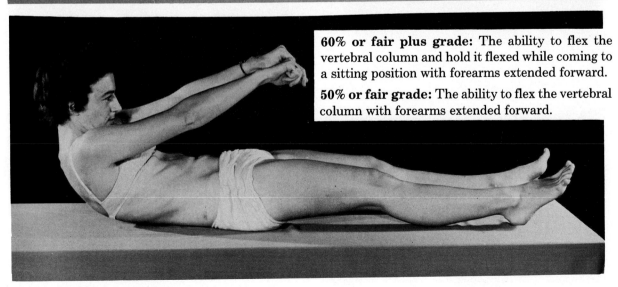

60% or fair plus grade: The ability to flex the vertebral column and hold it flexed while coming to a sitting position with forearms extended forward.

50% or fair grade: The ability to flex the vertebral column with forearms extended forward.

Abdominal Muscle Weakness: Sit-Up with Low Back Hyperextended

When abdominal muscles are too weak to curl the trunk, the hip flexors tilt the pelvis forward and hyperextend the low back as they raise the trunk to a sitting position. Some people cannot do the sit-up unless the feet are held down from the start. Usually these are the people who have marked weakness of the abdominal muscles. They should practice the trunk curl only and *avoid* doing the sit-up in the manner illustrated here.

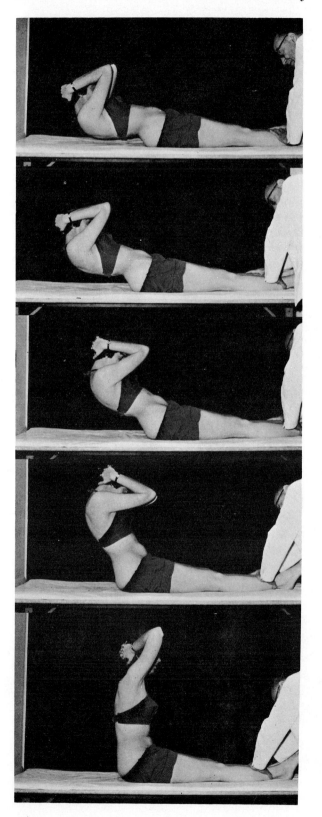

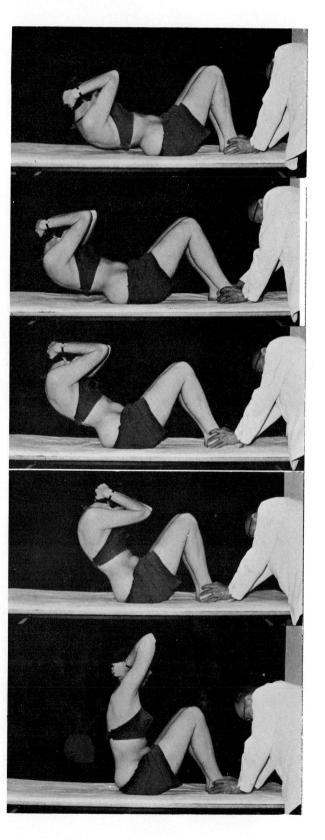

Therapeutic Exercises: Trunk curl

For purposes of strengthening those abdominal muscles that show weakness on the trunk-raising test, it is desirable, in most instances, to have the subject perform only the trunk-curl part of the movement. Doing so provides the advantage of exercising the abdominal muscles without strong hip flexor exercise. In addition, according to Nachemson and Elfstron[24], there is less intradiscal pressure when doing only the trunk curl as compared to completing the sit-up.

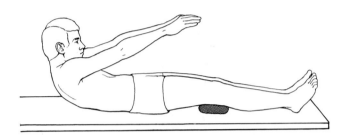

Abdominal Exercise: Trunk curl. In the back-lying position, place a small roll under the knees. Tilt the pelvis to flatten the low back on the table by pulling up and in with muscles of the lower abdomen. With arms extended forward, raise the head and shoulders upward from the table. Raise the upper trunk as high as the back will bend, but *do not try to come to a sitting position.*

Abdominal Exercise: Assisted trunk curl. When the abdominal muscles are very weak and the subject cannot lift the shoulders upward from the table, modify the above exercise by placing a wedge-shaped pillow (or the equivalent) under the head and shoulders.

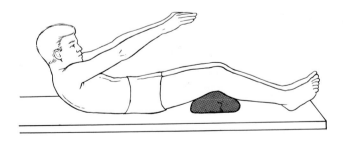

Abdominal Exercise: Modified when hip flexors are short. When hip flexor muscles are short and restrict the posterior pelvic tilt movement, modify the above trunk-curl exercise by temporarily placing a pillow under the knees to passively flex the hips. (See explanation p. 212.)

Temporary Use of Knee-bent position

Temporary use of knee-bent position. When one-joint hip flexors are short, they hold the pelvis in anterior tilt and the low back in hyperextension when standing or when supine with legs extended. From this position, it is difficult, if not impossible, to do posterior-pelvic-tilt exercises to strengthen abdominal muscles. Since the head and shoulder raising movement involves a simultaneous posterior pelvic tilt, there is interference with this exercise also.

As an effort is made to tilt the pelvis, the short hip flexors become taut and prevent the movement. To release this restraint and make it easier to tilt the pelvis, the knee-bent position has been widely advocated.

This position, obviously, gives in to the short, tight hip flexors. It also makes it relatively easy to perform the tilt, oftentimes merely by pressing the feet against the table in order to "rock the pelvis back." With shortness of hip flexors, the hips and knees should be bent, but *only as much as is needed*, to allow the pelvis to tilt back, and this position should be maintained *passively* by using a large enough roll or pillow under the knees. From this position the pelvic tilt and trunk-curl exercises may be done to strengthen abdominal muscles.

While bending the hips and knees initially is needed and justified, the *position should not be continued indefinitely*. The extent and duration of modifying the exercise become important. The goals set should be based on the desired end result and exercises should be directed toward attaining it. A desired end result in standing is the ability to maintain good alignment of the pelvis with the legs straight, i.e., the hip joints and knee joints in good alignment. Working toward this goal in exercise is accomplished by minimizing and gradually decreasing the amount of hip flexion permitted by the knee-bent position.

Tilting the pelvis posteriorly with the legs extended as much as possible moves the pelvis in the direction of elongating hip flexors while strengthening abdominals. Although this movement is not sufficient to stretch the hip flexors, it helps establish the pattern of muscle action necessary when attempting to correct a faulty lordotic posture in standing. Concurrently with doing proper abdominal exercise, the hip flexors should be stretched so that in time the individual will be capable of doing the posterior tilt with legs extended. (See pp. 160 and 289.)

Action of Hip Flexor and Abdominal Muscles During Leg-Raising and Lowering

Double-leg-raising from a supine position is flexion of the hips with knees extended. With the knee extensors holding the knees straight, the hip flexors raise the legs upward. No abdominal muscles cross the hip joints, so these muscles cannot assist directly in the leg-raising movement. The role of the hip flexors is made very clear by observing the loss of function when hip flexors are paralyzed.

In order to perform the double-leg-raising movement from a supine position, the pelvis must be stabilized in some manner. Although the abdominal muscles cannot enter directly into the leg-raising movement, the strength or weakness of these muscles directly affects the trunk position and the way in which the pelvis is stabilized. Leg-raising, through hip flexor action, exerts a strong pull downward on the pelvis in the direction of tilting it anteriorly. The abdominal muscles pull upward on the pelvis in the direction of tilting it posteriorly.

Attempted Double-Leg-Raising When Hip Flexors are Paralyzed

A subject with *strong abdominal muscles and very weak or paralyzed hip flexors* cannot begin to lift the legs upward from a supine position. The only active movement that occurs in attempting to raise the legs is that the pelvis is drawn forcefully into posterior tilt. Passively, the thighs may be raised slightly from the table secondary to the tilting of the pelvis, as illustrated above, or they may remain flat on the table if anterior hip joint structures are relaxed.

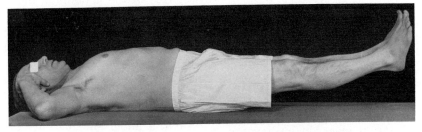

If the subject has strong abdominal muscles, the back can be held flat on the table by the abdominals holding the pelvis in posterior tilt during the leg-raising movement.

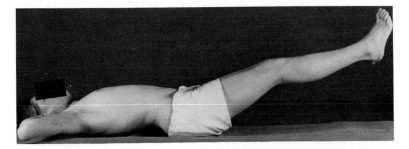

If the abdominal muscles are weak, the pelvis tilts anteriorly as the legs are lifted. As this tilt occurs, the back hyperextends, often causing pain, and the weak abdominal muscles are put on a stretch and vulnerable to strain.

Abdominal Muscle Test II: Holding Low Back Flat During Leg-Lowering

Two separate tests are described in this text for testing the strength of abdominal muscles. The first one consists of a curled-trunk sit-up. (See p. 208) In it, the curling of the trunk and lifting the weight of the head and shoulders requires action by the anterior neck and abdominal muscles, followed by hip flexor action throughout the sit-up phase.

In the second test, the abdominal muscles contract to tilt the pelvis posteriorly, flatten the low back on the table, and hold it flat, as the legs are lowered. Lowering the legs, by the lengthening contraction (gradual release) of the hip flexors increases the leverage and furnishes a gradually increasing resistance against the abdominal muscles.

Emphasis needs to be placed on this second test because it is the more important of the two from the standpoint of ascertaining the strength of those parts of the abdominal musculature that most affect the alignment of the body in standing. Strong abdominal muscles play a vital role in maintaining good posture and in preventing faulty postural positions that may lead to low back pain. Unless evaluation procedures include the leg-lowering test, weakness of important segments of the abdominal musculature may not be detected and there would be no concerted effort to apply the proper exercises to correct existing weakness. It is not uncommon to find that the sit-up test grades normal and the leg-lowering test grades very weak. Repeated sit-up exercises do not correct the problem exhibited by the leg-lowering test.

In this test, the strength of the abdominal muscles is determined by their ability to maintain the pelvis in a position of posterior tilt and the lumbar spine flat on the table while resisting the strong downward force exerted by the hip flexors as the legs are lowered. Because this test involves hip flexor action in the leg-lowering and Quadriceps action in holding the knees straight, the strength of these muscles should be determined before doing this abdominal muscle test.

The resistance encountered by the abdominal muscles is least when the legs are near the 90° (vertical) position and gradually increases as the legs are lowered. The lowest angle to which the legs can be lowered with the back remaining flat determines the grade (See facing page.)

Patient: Supine on a firm surface. (The test must not be done on a table with a soft pad.) Forearms are folded across the chest with fingers touching opposite shoulder to ensure that elbows are not down on the table. For a more difficult test, hands may be placed up beside the head, but in testing patients with any painful condition of the back the arms should be folded across the chest.

Fixation: None should be applied to the trunk because the test is actually one to determine whether the abdominal muscles can fix the pelvis in approximation to the thorax during the leg-lowering. Giving support to the trunk would be giving assistance. Allowing the patient to hold on to the table, or rest the hands or elbows on the table, would interfere with the test, also, by assisting the abdominals, and should not be permitted.

Test: The examiner assists the patient in raising the legs to a right angle or has the patient raise them *one at a time* to that position. Have the patient tilt the pelvis posteriorly to flatten the back on the table by contracting the abdominal muscles, and *hold it flat as he slowly lowers the legs. Attention must be focused on the position of the low back and pelvis during the leg lowering movement.* The subject should not be permitted to raise the head and shoulders during the test. When grading the strength of the abdominal muscles, the angle between the extended legs and the table is noted at the moment the pelvis starts to tilt anteriorly and the low back starts to arch from the table. To help detect the moment this occurs the examiner is aided by placing one hand at the low back and the other at the anterior-superior spine of the ilium. However, when testing a patient with a painful back condition, it is important that one hand of the examiner be free to support the legs the moment the abdominals cease to hold and the pelvis starts to tilt. In such an instance, the other hand of the examiner should be on the anterior-superior spine to detect the beginning tilt, rather than at the low back.

Pressure: None by the examiner, but lowering the legs provides increasing resistance against the abdominal muscles which are attempting to maintain the lumbar spine in flexion and the pelvis in posterior tilt.

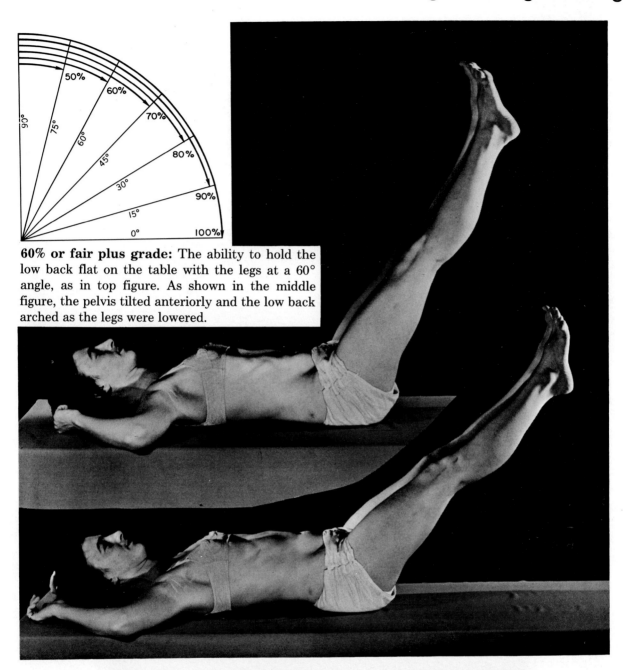

60% or fair plus grade: The ability to hold the low back flat on the table with the legs at a 60° angle, as in top figure. As shown in the middle figure, the pelvis tilted anteriorly and the low back arched as the legs were lowered.

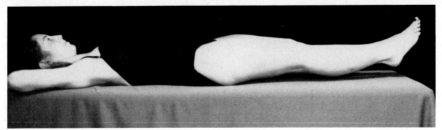

80% or good grade: The ability to hold the low back flat on the table with the legs at a 30° angle. In the bottom figure on p. 216 the legs are at a 40° angle and the grade at that point would be 70%.

100% or normal grade: The ability to hold the low back flat on the table as the legs are raised from, or lowered to, the fully extended position. Of necessity, the legs are shown elevated a few degrees for the photograph.

Abdominal Muscle Action During Leg-Raising or Lowering

When discussing the actions of the abdominal muscles, it should be recognized that various segments of the abdominal musculature are closely allied and interdependent. However, the oblique muscles are essentially fan-shaped and different segments may have different actions. From a mechanical standpoint, the pelvis can be tilted posteriorly by an upward pull on the pubis, an oblique pull in an upward and posterior direction on the anterior iliac crest, or a downward pull posteriorly on the ischium. The muscles or parts of muscles which are aligned in these directions of pull are the

Rectus abdominis, the lateral fibers of the External oblique, and the hip extensors. These muscles may act to tilt the pelvis posteriorly whether the subject is standing erect or lying supine. However, in the supine position, during double-leg-lowering, the hip extensors are not in a position to assist in maintaining the flexion of the lumbar spine and the posterior pelvic tilt. Consequently, the Rectus abdominis and External oblique muscles assume the major role in the effort to maintain the position of the low back and pelvis during the leg-lowering movement.

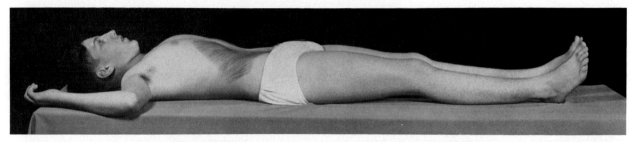

The lateral fibers of the External oblique act to tilt the pelvis posteriorly and may do so with little or no assistance from the Rectus abdominis.

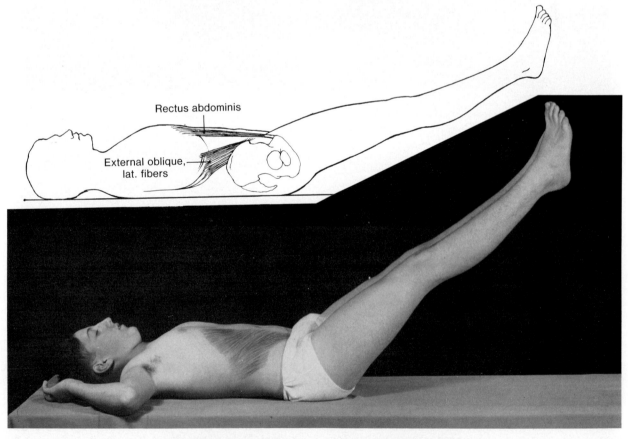

Rectus abdominis

External oblique, lat. fibers

To maintain the pelvis in a position of posterior tilt and the low back flat on the table as legs are raised or lowered requires action by the Rectus abdominis and the External oblique.

Abdominal Muscle Weakness: Leg-Lowering with Low Back Hyperextended

A subject with *marked weakness of abdominal muscles* and *strong hip flexors* can hold the extended extremities in flexion on the pelvis and lower them slowly but the low back arches, increasingly, as the legs approach the horizontal. The force exerted by the weight of the extremities and by the hip flexors holding the extremities in flexion on the pelvis, pulls the pelvis in anterior tilt overcoming the force of the abdominal muscles which are attempting to pull in the direction of posterior tilt.

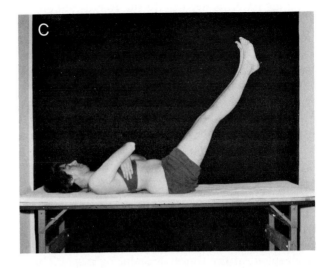

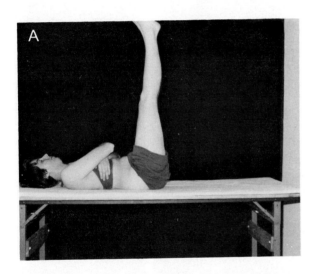

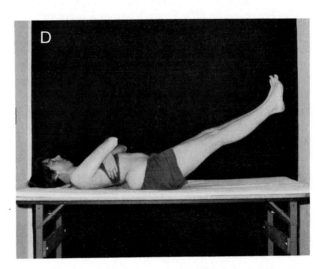

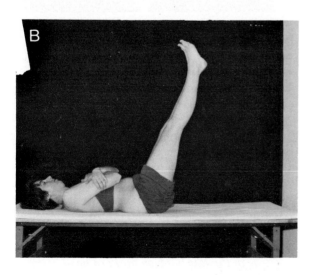

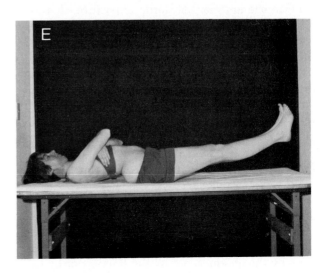

External Oblique in Relation to Posture

The muscles that hold the pelvis in posterior tilt during leg lowering are chiefly the Rectus abdominis and External oblique. In *many* instances, abdominal strength is normal on the trunk raising test, but the muscles grade very weak on the leg-lowering test. Since the Rectus must be strong in order to do the trunk curl, the inability to keep the low back flat during the leg-lowering cannot be attributed to that muscle. It is logical to attribute the lack of strength to the External oblique not to the Rectus. Furthermore, the postural deviations that exist in persons who show weakness on the leg-lowering test are associated with elongation of the External oblique.

There are two types of posture which exhibit this weakness: (1) Anterior tilt, lordotic posture, and (2) anterior displacement of the pelvis with posterior displacement of the thorax (See facing page.) The lateral fibers of the External oblique extend diagonally from posterolateral rib cage to anterolateral pelvis. By this line of pull, they are in a position to help maintain good alignment of the thorax in relation to the pelvis, or to restore the alignment when there is displacement. (See accompanying photographs.)

The difference in grades between the trunk raising test and the leg-lowering test is often very marked. Examination frequently reveals that the leg lowering grades only 50% to 60% in persons who can perform 100 or more sit-ups. It becomes very clear in such situations that the trunk raising exercise does not improve the ability to hold the low back flat during leg-lowering. Indeed, it appears that repeated and persistent trunk flexion exercises may contribute to continued weakness of the lateral fibers of the External oblique.

The type of postural deviation that occurs depends to a great extent on associated muscle weakness. In the anterior tilt, lordotic posture, there is often hip flexor tightness along with the abdominal weakness; in the anterior-posterior displacement, there is often hip flexor weakness.

The type of exercise indicated for strengthening the obliques depends upon what other muscles are involved and what postural problems are associated with the weakness. The manner in which movements are combined in exercises determines whether they will be therapeutic for the individual. For example, alternate leg-raising along with pelvic tilt exercises would be contraindicated in cases of hip flexor shortness, but would be indicated in cases of hip flexor weakness.

To correct anterior pelvic tilt, posterior pelvic tilt exercises are indicated. The movement should be done by the obliques, not by the Rectus nor by the hip extensors. The effort must be made to pull upward and inward with the abdominal muscles, making them very firm, particularly in the area of the lateral External oblique fibers. (See p. 221.)

To exercise the obliques when there is anterior displacement of the pelvis and posterior displacement of the thorax, the same effort should be made to pull upward and inward with the lower abdominal muscles but the pelvic tilt is not emphasized. In this type of faulty posture, there is a posterior pelvic tilt along with the hip flexor weakness. Contracting the lateral fibers of the External oblique in standing must be accompanied by *straightening, not flexing*, the upper back as the muscles act to shift the thorax forward and the pelvis back by the diagonal line of pull. Properly done, this movement brings the chest up and forward and restores the normal anterior curve in the low back (See below.).

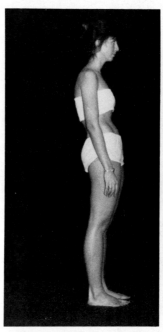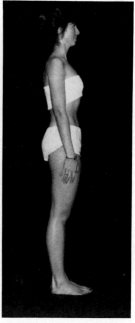

When properly done, the "wall-sitting" exercise and the "wall-standing" exercise (p. 301) stress the muscles of the lower abdomen and lateral fibers of the External oblique.

Such expressions as "make the lower abdomen cave in", or "hide the tummy under the chest", or, in the vernacular of the military, "suck in your gut", are all used to try to encourage the subject to exert strong effort in the exercise.

Proper exercise of abdominal muscles should be a part of preventive medicine and physical fitness programs. Good strength in these muscles is essential to the maintenance of good posture, but caution should be taken to avoid overdoing both the trunk curl and the pelvic tilt exercises. *The normal anterior curve in the low back should not be obliterated in the standing posture.*

External Oblique in Relation to Posture

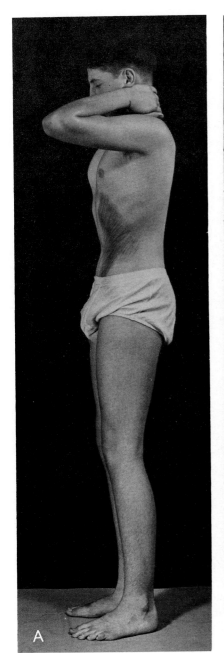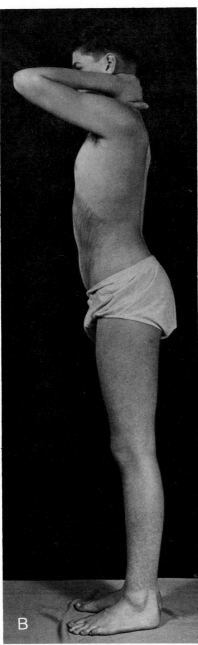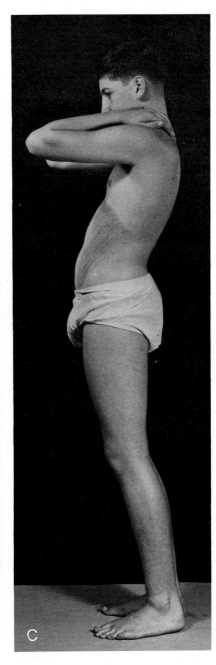

In standing, the lateral fibers of the External oblique, acting bilaterally, assist in tilting the pelvis posteriorly. By their diagonal line of pull, these fibers act to maintain good postural alignment of the pelvis and thorax as seen in fig. A. When weakness exists, the effect on posture may be that the pelvis tilts anteriorly as in fig. B, or that the pelvis is displaced anteriorly and the thorax posteriorly as in fig. C. External oblique muscle weakness, with the latter type of postural deviation, is seen frequently in conjunction with Iliopsoas weakness (see facing page) because of the accompanying hyperextension of the hip joints. Not illustrated, but rather obvious should be the fact that the lateral fibers of the Internal oblique are in a shortened position in fig. C. (See drawing of lateral view of Internal oblique p. 191.)

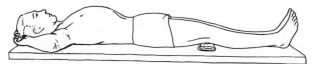

Posterior Pelvic Tilt. In the back-lying position, place a small roll under the knees. With hands up beside the head, tilt the pelvis to flatten the low back on the table by pulling up and in with the muscles in the lower abdomen. Hold the low back flat and breathe in and out easily, relaxing the upper abdominal muscles. There should be good chest expansion during inspiration but the low back should not arch upward from the table. Hold the position several seconds, then relax. Repeat several times. (See photographs p. 221.)

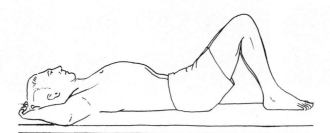

Posterior Pelvic Tilt and Leg Sliding. In the back-lying position, bend the knees and place the feet flat on the table. With hands up beside the head, tilt the pelvis to flatten the low back on the table by pulling up and in with the muscles in the lower abdomen. Hold the back flat and slide the heels along the table. Straighten the legs as far as possible with the back held flat. Keeping the back flat, return the knees to bent position, *sliding one back at a time.* Try to breathe in and out about three times while the feet slide down and return to bent position. (Hip flexor exercise is held to a minimum by not lifting the legs or feet from the table.) Repeat several times. (See photographs p. 221.)

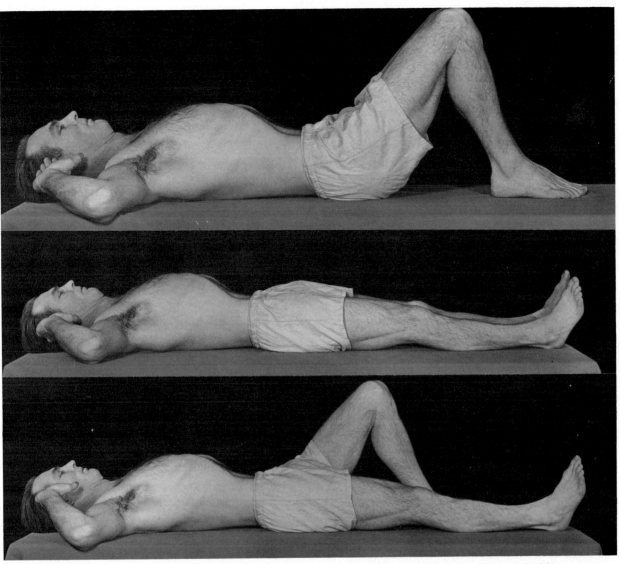

Posterior pelvic tilt and leg sliding exercise done correctly to exercise the External oblique.

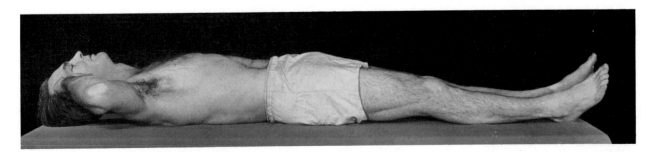

Pelvic tilt may be done with the Rectus abdominis but should not be done in this manner when emphasis is on strengthening the External oblique.

Lateral Trunk Raising

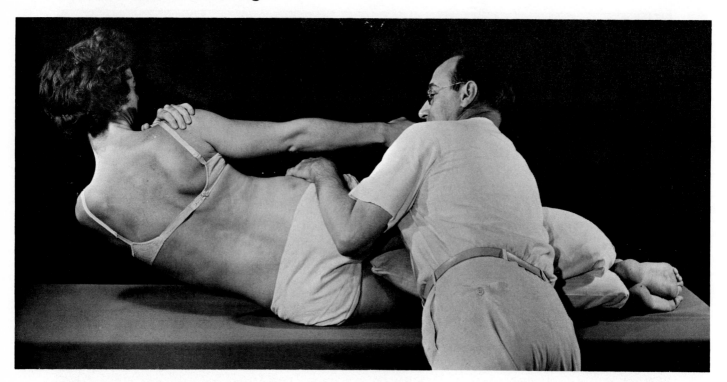

Trunk raising sideways is a combination of lateral trunk flexion, and hip abduction (the latter produced by downward tilting of the pelvis on the thigh). The lateral trunk muscles entering into the movement are the lateral fibers of the External and Internal obliques, the Quadratus lumborum, the Latissimus dorsi, and the Rectus abdominis on the side being tested.

Before doing the test for lateral trunk muscles, one should test the strength of hip abductors, adductors, and lateral neck flexors; and test for the range of motion in lateral flexion.

Patient: Side-lying with pillow between thighs and legs, and with head, upper trunk, pelvis, and lower extremities in a straight line. The top arm is extended down along the side, and fingers are closed so patient will not hold on to the thigh and attempt to pull himself up with his hand. The under arm is forward across the chest with the hand holding the upper shoulder to rule out assistance by pushing up with the elbow.

Fixation: Hip abductors must fix the pelvis to the thigh. The opposite adductors also help stabilize the pelvis. The legs must be held down by the examiner to counterbalance the weight of the trunk, but must not be held so firmly as to prevent the upper leg from moving slightly downward to accommodate for the displacement downward of the pelvis on that side. If the pelvis is pushed upward or not allowed to tilt downward, the subject will be unable to raise the trunk sideways even though the lateral abdominal muscles are strong.

Test: Trunk raising directly sideways.

Pressure: None necessary. The body weight offers sufficient resistance.

Grading: The ability to raise the trunk laterally from a side-lying position to a point of maximum lateral flexion, 100% or normal; to raise the under shoulder about four inches from the table, 80% or good; to raise the under shoulder slightly (one or two inches), 50% or fair.

Note: Test of the lateral trunk muscles may reveal an imbalance in the oblique muscles. In trunk raising sideways, if the legs and the pelvis are held steady, that is, not permitted to twist forward or backward from the direct side-lying position, the thorax may be rotated forward or backward as the trunk is laterally flexed. A forward twist of the thorax denotes a stronger pull by the External oblique, while a backward twist denotes a stronger pull by the Internal oblique. If the back hyperextends as the patient raises himself, the Quadratus lumborum and the Latissimus dorsi show a stronger pull, indicating that anterior abdominal muscles cannot counterbalance this pull to keep the trunk in straight line with the pelvis.

Lateral Trunk Flexors and Hip Abductors

Strong lateral trunk muscles and strong hip abductor muscles.

Subject can raise the trunk in lateral flexion through subject's full range of motion.

Strong lateral trunk muscles and paralyzed hip abductor muscles.

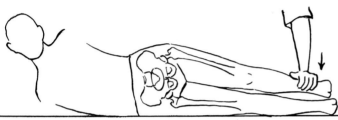

Subject can laterally flex the trunk but the under shoulder will scarcely be raised from the table. The pelvis will be drawn upward as the head is raised laterally and the iliac crest and costal margin are approximated.

Weak lateral trunk muscles and strong hip abductor muscles.

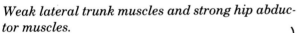

Subject cannot raise the trunk in true lateral flexion. Under certain circumstances the patient may be able to raise the trunk from the table laterally even though lateral trunk muscles are quite weak. If the trunk can be held rigid, the hip abductor muscles may raise the trunk in abduction on the thigh. The rib cage and iliac crest will not be approximated laterally as they are when the lateral trunk muscles are strong. By decreasing the pressure with which he is providing fixation for the hip abductors, the examiner can make it necessary for the lateral abdominals to attempt initiation of the movement.

Strong lateral trunk muscles and strong hip abductor muscles.

Subject can abduct the extremity through subject's full range of motion.

Strong lateral trunk muscles and paralyzed hip abductor muscles.

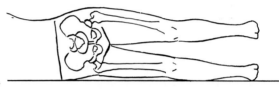

In attempting to raise the extremity in abduction, the movement that occurs is elevation of the pelvis by the lateral trunk muscles. The extremity may be drawn upward into the position as illustrated, but the hip joint is not abducted. Actually the thigh has dropped into a position of adduction and is held there by the joint structures rather than by hip muscle action.

Weak lateral trunk muscles and strong hip abductor muscles.

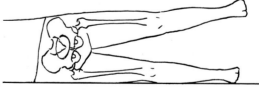

The extremity can be lifted in hip abduction, but without the fixation by the lateral abdominal muscles, it cannot be raised high from the table. Due to the weakness of the lateral trunk muscles, the weight of the extremity tilts the pelvis downward.

Abdominal Muscle Imbalance and Deviations of Umbilicus

When marked weakness and imbalance exist in abdominal muscles it is possible to determine, to some extent, the imbalance by observing the deviations of the umbilicus. The umbilicus will deviate toward a strong or away from a weak segment. If, for example, three segments, the left External and both left and right Internal obliques are equally strong and there is marked weakness of the right External, the umbilicus will deviate decidedly toward the left Internal. This happens not because the left Internal is the strongest, but because it has no opposition in the right External. This shows deviation away from a weak segment.

On the other hand, deviation may mean that there is one strong segment and that the other three are weak, and the deviation will be toward the strongest segment. The relative strengths must then be determined by palpation, and by the extent to which the umbilicus deviates during the performance of localized test movements.

There are times when the umbilicus deviates, not because of active muscle contraction, but because of a stretching of the muscle. The examiner must be sure the muscles he is testing are actively contracting before he uses the deviations of the umbilicus as being indicative of strength or weakness.

To obtain true deviations, the abdominal muscles should first be in a relaxed position. The knees may be bent sufficiently to relax the back flat on the table. Then the patient may be asked to attempt head raising or tilting the pelvis posteriorly (even though the back is already flat). If resistive arm and leg movements are used in testing, they should be started from this relaxed position also. Movements should be such that they produce actual shortening of the muscle. Movements should not be so strenuous that they cause the muscle to stretch. When weakness is very apparent, the initial test should be a mild active movement, with resistance gradually applied. It should be noted, first, to what extent the muscle can approximate its origin and insertion; and second, how much pressure can be added before the pull "breaks" and the muscle starts to stretch.

An individual unfamiliar with the examination of abdominal muscles may find it very difficult to be sure of the deviations of the umbilicus. If a tape or cord is held transversely, then diagonally, over the umbilicus as the test movements are performed, the direction of the deviation can be readily determined. The umbilicus may deviate up or down from the transverse tape showing uneven pull of upper and lower rectus. If it also shows a deviation from the tape held diagonally over the umbilicus, it will exhibit an imbalance between the obliques.

Lines made with ink or skin pencil on the anterior iliac crests, the costal margins, just above the pubis, and below the sternum may be an aid to the examiner. As the test movement is done, the tape is held from umbilicus to the various marks. Actual shortening or stretch of the segments can be detected as a movement is attempted.

ARM MOVEMENTS USED IN TESTING ABDOMINAL MUSCLES

Arm movements are performed against resistance or held against pressure when used in abdominal muscle testing because unresisted arm movements do not demand appreciable action of trunk muscles for fixation.

Normally an upward movement of the arms in the forward plane requires fixation by back muscles; a downward movement in the forward plane requires fixation by abdominal muscles. When abdominal weakness exists, however, fixation for the downward pull or push of the arm may be provided by the back muscles. For example, if a patient is placed in a supine position and given resistance to a downward pull of both arms, normal abdominal muscles will contract to fix the thorax firmly toward the pelvis. However, if extensive abdominal weakness is present, the back will arch from the table, and the thorax will pull away from the pelvis until it is firmly fixed by extension of the thoracic spine. The arching of the back stretches the abdominal muscles, and they may appear firm under tension. The examiner must be careful not to mistake this tautness for firmness due to actual contraction of the muscles.

In cross-sectional or diagonal arm movements, if the abdominal muscles are normal, the External oblique on the same side as the arm and the Internal oblique on the opposite side contract to fix the thorax to the pelvis. If cross-sectional weakness exists in that line of pull, the opposite oblique muscles may act to give fixation. The examiner should understand these substitute actions in order to do an accurate examination.

Testing and Grading Marked Abdominal Weakness

Objective grading of the antero-lateral abdominal muscles is not difficult when strength is 50% or fair, and above. Below 50% or fair, it is more difficult to grade accurately. The tests and grades described here furnish guidelines for grading the weak muscles.

When marked imbalance exists in the abdominal muscles, one must observe the deviations of the umbilicus (see facing page) and rely on palpation for grading.

Preliminary to doing the tests listed below, it is necessary to test the strength of the anterior neck muscles.

ANTERIOR ABDOMINAL MUSCLES, MAINLY RECTUS ABDOMINIS

40% or fair minus: In a supine position with knees slightly flexed (rolled towel under knees), the patient is able to tilt the pelvis posteriorly and maintain the pelvis and thorax approximated as the head is raised from the table.

20% or poor: In the same position as above the patient is able to tilt the pelvis posteriorly, but as the head is raised the abdominal muscles cannot hold against that resistance anteriorly, and the thorax moves away from the pelvis.

5% or trace: In the supine position, when attempting to depress the chest or tilt the pelvis posteriorly, the anterior abdominal muscles can be felt to contract but there is no approximation of the pelvis and thorax.

OBLIQUE ABDOMINAL MUSCLES

40% or fair minus: With the examiner giving moderate resistance against a diagonally downward pull of the arm, the cross-sectional pull of the oblique abdominal muscles will be very firm on palpation, and will pull the costal margin toward the opposite iliac crest. If the arm is weak, pushing the shoulder upward from the table and holding against pressure may be substituted for the arm movement.

With the straight leg held in approximately 60° hip flexion, the examiner applies moderate pressure against the thigh in a downward-outward direction. The strength of oblique muscles should be sufficient to pull the iliac crest toward the opposite costal margin. (This test can be used only if hip flexor strength is good.)

20% or poor: The patient is able to approximate the iliac crest toward the opposite costal margin.

5% or trace: The oblique muscles can be felt to contract when an effort is made to pull the costal margin toward the opposite iliac crest but there will be no approximation of these parts.

LATERAL TRUNK MUSCLES

40% or fair minus: In side-lying, there will be firm fixation and approximation of the rib cage and iliac crest laterally during active leg abduction, and arm adduction against resistance.

20% or poor: In the supine position, the patient is able to approximate the iliac crest and rib cage laterally as an effort is made to elevate the pelvis laterally or adduct the arm against resistance.

5% or trace: In the supine position, the lateral abdominal muscles can be felt to contract as an effort is made to elevate the pelvis laterally or adduct the arm against resistance, but there is no approximation of thorax and lateral iliac crest.

RECORDING GRADES OF ABDOMINAL MUSCLE STRENGTH

Abdominal muscle grades are recorded in two different ways, depending on the amount of strength.

When strength is 50% or better in the trunk raising and leg-lowering tests, it is usually sufficient to grade and record on the basis of these tests, fig. A. There seldom is intrinsic imbalance between parts of the Rectus or the obliques to necessitate grading parts separately if these tests show 50% or better.

When marked weakness or imbalance exists, it is necessary to indicate findings in relation to specific muscles (fig. B).

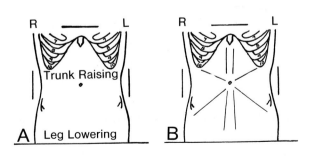

Oblique Trunk Raising

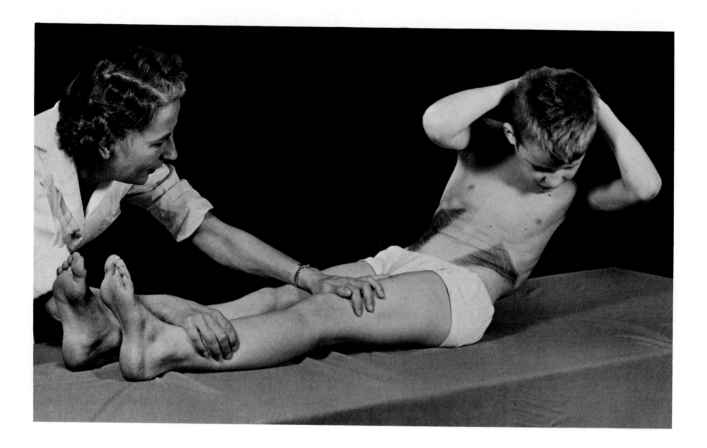

To raise the trunk obliquely forward combines trunk flexion and rotation. It is accomplished by the action of the External oblique on one side combined with that of the Internal oblique on the opposite side, and the Rectus abdominis.

The oblique trunk raising test is usually performed after the anterior trunk raising and leg-lowering tests have been done. The tests reveal to the examiner much about the relative strength of abdominal and hip flexor muscles.

Patient: Supine. (For arm position, see below under Grading.)

Fixation: Assistant supports the legs after examiner has placed the patient in test position.

Test: Patient clasps hands behind head. Examiner places patient in position of trunk flexion and rotation, and asks patient to hold that position. If the

muscles are weak, the trunk will derotate or extend, and there may be flexion of the pelvis on the thighs in an effort to hold the hyperextended trunk up from the table.

Resistance: None in addition to the weight of the trunk. Resistance is varied by position of the arms.

Grading: The ability to hold the test position with hands clasped behind the head, 100% or normal; with forearms folded across the chest, 80% or good; with arms forward, 60% or fair plus. (See illustrations of arm positions, p. 209.) A grade of 50% or fair is given if the patient, with forearms extended forward, can hold a position of flexion and rotation of the trunk with the scapular region of the lower shoulder off the table. (See p. 225 regarding tests and grading in cases of marked abdominal weakness.)

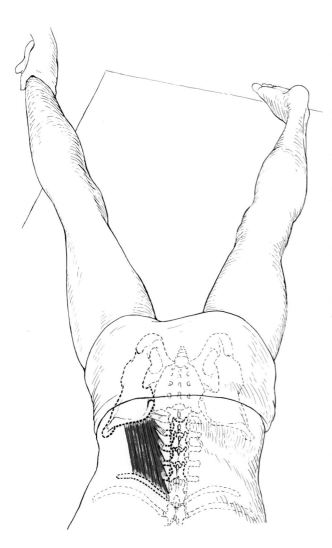
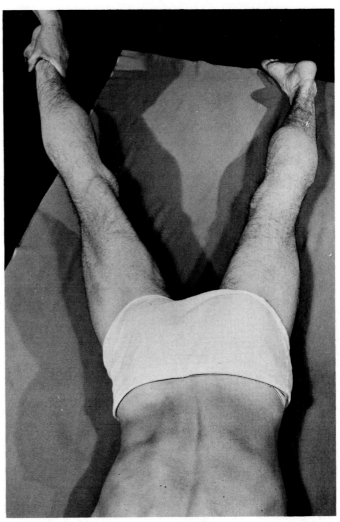

Origin: Iliolumbar ligament, iliac crest. Occasionally from upper borders of transverse processes of lower three or four lumbar vertebrae.

Insertion: Inferior border of last rib and transverse processes of upper four lumbar vertebrae.

Action: Laterally flexes the lumbar vertebral column and depresses the last rib. Bilaterally, fix the last two ribs, acting together with the diaphragm, in respiration.

Nerve: Lumbar plexus, T12, **L1, 2, 3.**

Patient: Prone.

Fixation: By muscles which hold the femur firmly in the acetabulum.

Test: Elevation of the pelvis laterally. The extremity is placed in slight extension and in the degree of abduction which corresponds with the line of fibers of the Quadratus lumborum.

Resistance: Given in the form of traction on the extremity, directly opposing the line of pull of the Quadratus lumborum. If hip muscles are weak, pressure may be given against the posterolateral iliac crest opposite the line of pull of the muscle.

The Quadratus lumborum acts along with other muscles in lateral trunk flexion. It is difficult to palpate this muscle because it lies deep to the Erector spinae. Although the Quadratus lumborum enters into the motion of elevation of the pelvis in the standing position or in walking, the standing position does not offer a satisfactory position for testing. Elevation of the right side of the pelvis in standing, for example, depends as much, if not more, on the downward pull by the abductors of the left hip joint as it does on the upward pull of right lateral abdominals.

The test illustrated should not be considered as limited to Quadratus lumborum action, but as giving the most satisfactory differentiation that can be obtained. This text does not recommend attempting to grade numerically the strength of this muscle but merely to record whether it appears weak or strong.

Origins and Insertions of Neck and Back Extensors

	Origin	Insertion
Erector spinae (Superficial) Iliocostalis: lumborum	Common origin from anterior surface of broad tendon attached to medial crest of sacrum, spinous processes of lumbar and 11th and 12th thoracic vertebrae, posterior part of medial lip of iliac crest, supraspinous ligament, and lateral crests of sacrum.	By tendons into inferior borders of angles of lower 6 or 7 ribs.
thoracis	By tendons from upper borders of angles of lower 6 ribs.	Cranial borders of angles of upper 6 ribs, and dorsum of transverse process of 7th cervical vertebra.
cervicis	Angles of 3rd, 4th, 5th, and 6th ribs.	Posterior tubercles of transverse processes of 4th, 5th, and 6th cervical vertebrae.
Longissimus: thoracis	In lumbar region it is blended with Iliocostalis lumborum, posterior surfaces of transverse and accessory processes of lumbar vertebrae, and anterior layer of thoracolumbar fascia.	By tendons into tips of transverse processes of all thoracic vertebrae, and by fleshy digitations into lower 9 or 10 ribs between tubercles and angles.
cervicis	By tendons from transverse processes of upper 4 or 5 thoracic vertebrae.	By tendons into posterior tubercles of transverse processes of 2nd through 6th cervical vertebrae.
capitis	By tendons from transverse processes of upper 4 or 5 thoracic vertebrae, and articular processes of lower 3 or 4 cervical vertebrae.	Posterior margin of mastoid process.
Spinalis: thoracis	By tendons from spinous processes of first 2 lumbar and last 2 thoracic vertebrae.	Spinous processes of upper 4–8 (variable) thoracic vertebrae.
cervicis	Ligamentum nuchae, lower part; spinous process of 7th cervical.	Spinous process of axis and, occasionally, into spinous processes of C3 and C4.
capitis	Inseparably connected with Semispinalis capitis.	See below.
Transversospinalis (Deep) Semispinalis: (1st layer) thoracis	Transverse processes of lower thoracic vertebrae.	Spinous processes of upper thoracic 4–8 (variable), and lower 2 cervical vertebrae.
cervicis	Transverse processes of upper 5 or 6 thoracic vertebrae.	Cervical spinous processes, 2nd through 5th.
capitis	Tips of transverse processes of upper 6 or 7 thoracic and 7th cervical vertebrae, and articular processes of cervical 4th, 5th, and 6th.	Between superior and inferior nuchal lines of occipital bone.
Multifidi (2nd layer)	*Sacral region:* Posterior surface of sacrum, medial surface of posterior superior iliac spine, and posterior sacroiliac ligaments. *Lumbar region:* *Thoracic region:* } Transverse processes of L5 through C4. *Cervical region:*	Spanning 2 to 4 vertebrae, inserted into spinous process of a vertebra above.
Rotatores (11) (3rd layer)	Transverse processes of vertebrae.	Lamina of the vertebra above.
Interspinales	Placed in pairs between spinous processes of contiguous vertebrae. *Cervical:* 6 pairs *Thoracic:* 2 or 3 pairs; between 1st and 2nd, (2nd and 3rd), and 11th and 12th. *Lumbar:* 4 pairs	
Intertransversarii	Small muscles placed between transverse processes of contiguous vertebrae in cervical, thoracic, and lumbar regions.	
Splenius cervicis	Spinous processes of 3rd through 6th thoracic vertebrae.	Posterior tubercles of transverse processes of first 2 or 3 cervical vertebrae.
capitis	Caudal 1/2 of ligamentum nuchae; spinous process of 7th cervical vertebra; spinous processes of first 3 or 4 thoracic vertebrae.	Mastoid process of temporal bone, and on occipital bone inferior to lateral 1/3 of superior nuchal line.

Head, Neck, and Back Extensors

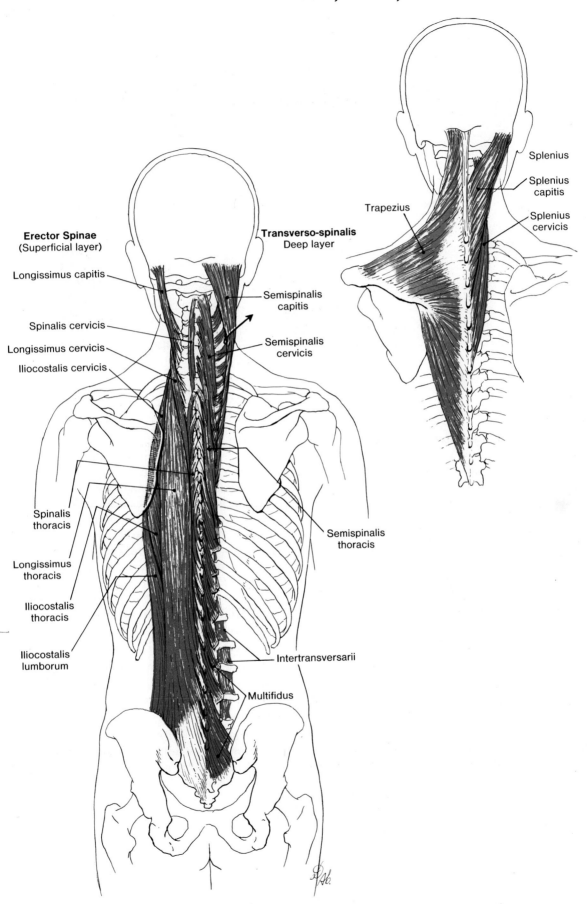

Erector Spinae
(Superficial layer)

Longissimus capitis

Spinalis cervicis

Longissimus cervicis

Iliocostalis cervicis

Spinalis thoracis

Longissimus thoracis

Iliocostalis thoracis

Iliocostalis lumborum

Transverso-spinalis
Deep layer

Semispinalis capitis

Semispinalis cervicis

Semispinalis thoracis

Intertransversarii

Multifidus

Splenius

Splenius capitis

Splenius cervicis

Trapezius

Back Extensors

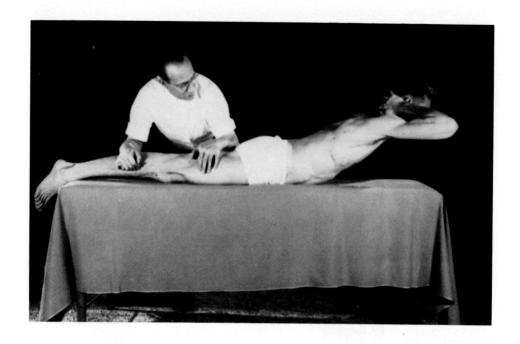

In the trunk extension test, back extensors are assisted by the Latissimus dorsi, Quadratus lumborum, and Trapezius.

The neck extensor muscles and the hip extensor muscles should be tested before doing the back extensor test.

Patient: Prone. (For arm positions, see below under Grading.)

Fixation: Hip extensors must give fixation of the pelvis to the thighs, and the examiner must stabilize the legs firmly on the table.

Test: Trunk extension.

Pressure: (See below under Grading.)

Grading: The ability to complete spine extension with hands clasped behind head, 100% or normal; the ability to complete spine extension with hands clasped behind back, 80% or good; with hands clasped behind the back, the ability to lift the thorax so that the xiphoid process of the sternum is raised slightly from the table, 50% to 60% or fair to fair plus.

When marked weakness is present, usually such weakness extends throughout the entire back. However, the cervical extensors may be able to lift the head, and a head raising movement can furnish slight resistance against other back extensors. Palpation and observation of the muscle actions in the upper and lower back must constitute the basis for grading when marked weakness exists.

When the low back is strong and the upper back is weak, an attempt to raise the thorax will result in the back extensors extending the low back, anteriorly tilting the pelvis, but the thorax will not be lifted from the table.

Weakness: *Bilateral* weakness of the back extensor muscles results in a lumbar kyphosis and an increased thoracic kyphosis. *Unilateral* weakness results in a lateral curvature with convexity toward the weak side.

Contracture: *Bilateral* contracture of the low back muscles results in a lordosis. *Unilateral* contracture results in a scoliosis with convexity toward the opposite side.

Back Extensors and Hip Extensors

For back extensors to raise the trunk from a prone position, the hip extensors must fix the pelvis in extension on the thigh (fig. A). For hip extensors to raise the extremity backward from a prone position through the few degrees of true hip joint extension (approximately 10°) requires fixation by back extensors to stabilize the pelvis (fig. B). Raising the extremity higher is accomplished by hyperextension of the lumbar spine and anterior tilting of the pelvis (fig. C). In this latter movement the back extensors are assisted by the hip flexors on the opposite side, helping to tilt the pelvis anteriorly.

If slight tightness exists in hip flexors, there is no range of extension in the hip joint, and all the movement in the direction of leg-raising backward is accomplished by lumbar spine hyperextension and pelvic tilt.

Normally, extension of the hip joints and extension of the lumbar spine are initiated simultaneously and are not two separate movements.

The illustrations on this page show the variations that occur depending upon the strength of the two primary muscle groups.

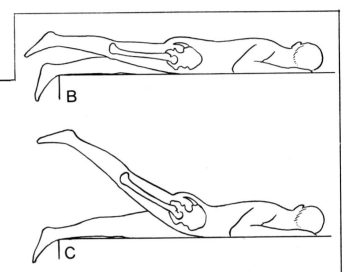

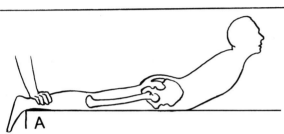

A subject with *strong back extensor muscles* and *strong hip extensor muscles* can raise the trunk in extension, fig. A, the hip in extension, fig. B, and the hip and trunk in extension, fig. C.

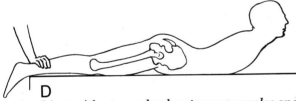

A subject with *strong back extensor muscles* and *markedly weak or paralyzed hip extensor muscles* can hyperextend the lumbar spine, but the trunk cannot be lifted high from the table, fig. D. In an effort to lift the extremity, the back contracts to fix the pelvis on the trunk, but, with little or no strength in the hip extensors, the thigh cannot be extended on the pelvis. The unopposed pull of the back muscles results in hyperextension of the back, and the hip joint is passively drawn into flexion despite the effort to extend it, fig. E.

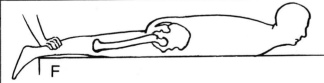

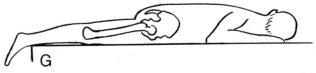

A subject with *weak or paralyzed back extensor muscles* and *strong hip extensor muscles* cannot raise the trunk in extension. The hip extensors, in their action to fix the pelvis, are unopposed, and the pelvis tilts posteriorly and the lumbar spine flexes, fig. F. In an effort to lift the extremity, the hip extensors contract but the extremity cannot be lifted because the back muscles are unable to stabilize the pelvis. The pelvis tilts posteriorly by the pull of the hip extensors and the weight of the extremity, fig. G, instead of tilting anteriorly as it would if the back extensors were normal.

Forward Bending Test for Length of Posterior Muscles

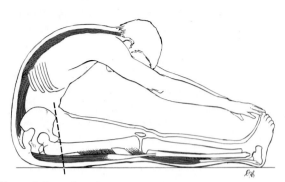

Normal length of back, Hamstring, and Gastroc-soleus muscles.

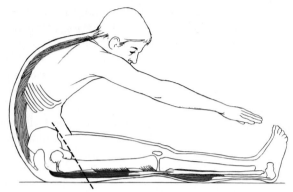

Excessive length of back muscles, short Hamstrings, normal length of Gastroc-soleus.

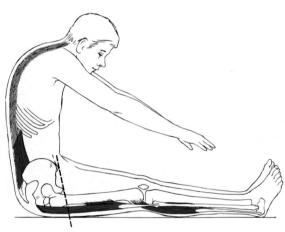

Excessive length of upper back muscles, slight shortness of muscles in mid and lower back, and of Hamstrings and Gastroc-soleus.

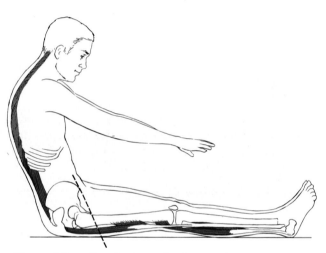

Normal length of upper back muscles, short low back, Hamstring, and Gastroc-soleus muscles.

Flexion slightly limited in lower thoracic area, excessive length of Hamstrings, normal Gastroc-soleus.

Normal length of upper back muscles, contracture of low back muscles with paralysis of extremity muscles.

The ability to touch fingertips to toes is a desirable accomplishment for most adults. This subject shows Hamstring length and back flexibility within normal limits.

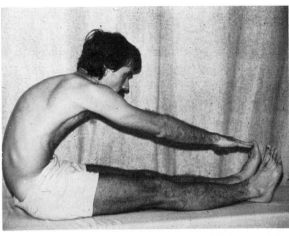

The excessive flexibility of the back overcompensates for the shortness of the Hamstring muscles. (See p. 153 for Hamstring length test on this subject.)

The subject is unable to touch his toes because of shortness in the Hamstrings and Gastroc-soleus, and slight limitation of flexibility in the mid-back area. The upper back shows some excessive flexion.

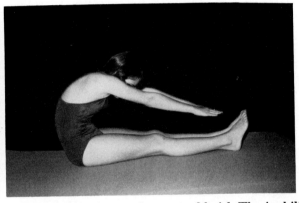

This subject is a twelve-year-old girl. The inability to touch the toes is typical of this age.[25] (See also p. 234.) Sometimes the leg length is the determining factor; sometimes, as in this case, there is slight shortness of the Hamstrings at this age.

Although the lower thoracic area lacks some flexibility, the subject is able to reach beyond the toes because of the excessive Hamstring length.

This six-year-old girl touches her toes easily. There is good contour of the back and normal Hamstring length.

Normal Flexibility According to Age Level[25]

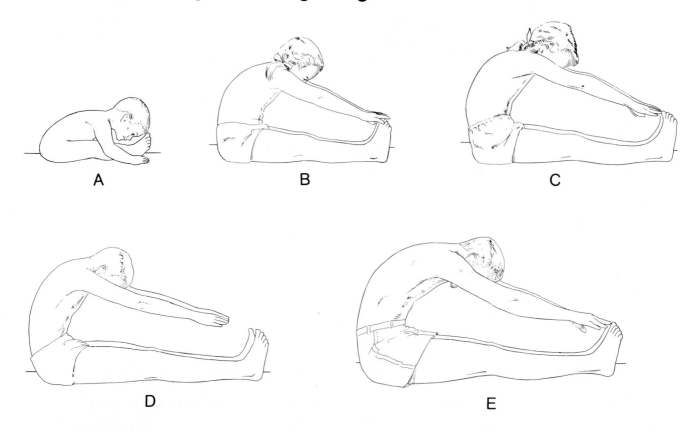

A   B   C

D   E

The ability to touch the toes with the finger tips may be considered normal for young children and adults. However, between the ages of eleven and fourteen many individuals who show no signs of muscle or joint tightness are unable to complete this movement. The reason seems to be that the proportionate length of the trunk and lower extremities is different in individuals of this age group from that of younger and older age groups.

The five drawings above are representative of the majority of individuals in each of the following age groups: fig. A, one to three years; fig. B, four to seven years; fig. C, eight to ten years; fig. D, eleven to fourteen years, and fig. E, fifteen years and over.

The change from the apparently extreme flexibility of the youngest child to the apparently limited flexibility of the child in fig. D occurs gradually over a period of years as the legs become proportionately longer in relation to the trunk. Standards of performance for children, which involve forward bending, should take into consideration the normal variations in the ability to complete the range of this movement.

chapter 7

Facial, Eye, and Neck Muscles; Muscles of Deglutition; Respiratory Muscles;

Cranial nerves and deep facial muscles (illustrated)
Facial muscles (illustrated)
Facial muscles: Origin, insertion, test
Cranial nerve and muscle chart
Muscle tests recorded on nerve-muscle charts
Anterior and lateral neck muscles (illustrated)
Suprahyoid and infrahyoid muscles (illustrated)
Head and neck muscles: Origins and insertions

Head and neck muscles: Actions
Anterior neck flexors
Anterolateral neck flexors
Posterolateral neck extensors
Muscles of deglutition
Respiratory muscle chart
Diaphragm
Muscles of respiration

Cranial Nerves and Deep Facial Muscles

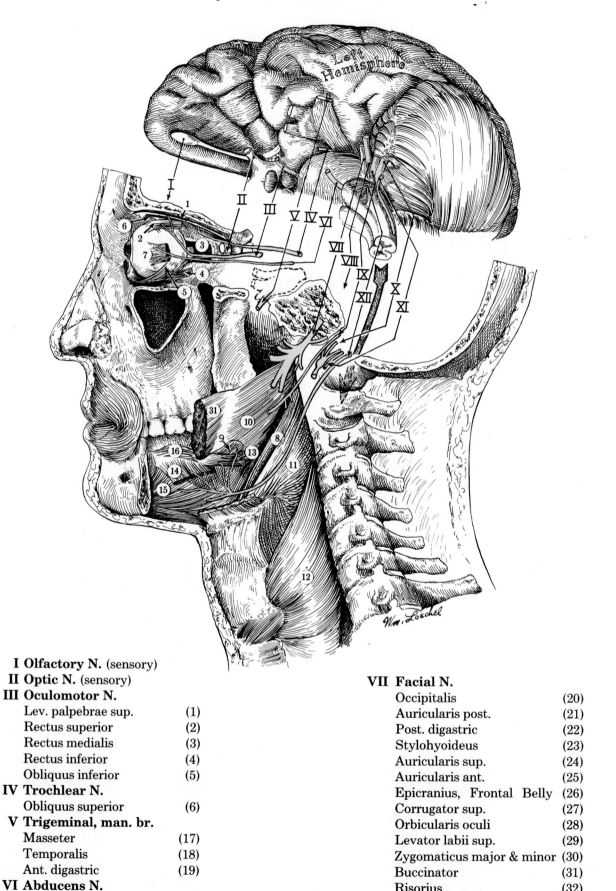

I Olfactory N. (sensory)

II Optic N. (sensory)

III Oculomotor N.

Lev. palpebrae sup.	(1)
Rectus superior	(2)
Rectus medialis	(3)
Rectus inferior	(4)
Obliquus inferior	(5)

IV Trochlear N.

| Obliquus superior | (6) |

V Trigeminal, man. br.

Masseter	(17)
Temporalis	(18)
Ant. digastric	(19)

VI Abducens N.

| Rectus Lateralis | (7) |

VII Facial N.

Occipitalis	(20)
Auricularis post.	(21)
Post. digastric	(22)
Stylohyoideus	(23)
Auricularis sup.	(24)
Auricularis ant.	(25)
Epicranius, Frontal Belly	(26)
Corrugator sup.	(27)
Orbicularis oculi	(28)
Levator labii sup.	(29)
Zygomaticus major & minor	(30)
Buccinator	(31)
Risorius	(32)
Orbicularis oris	(33)

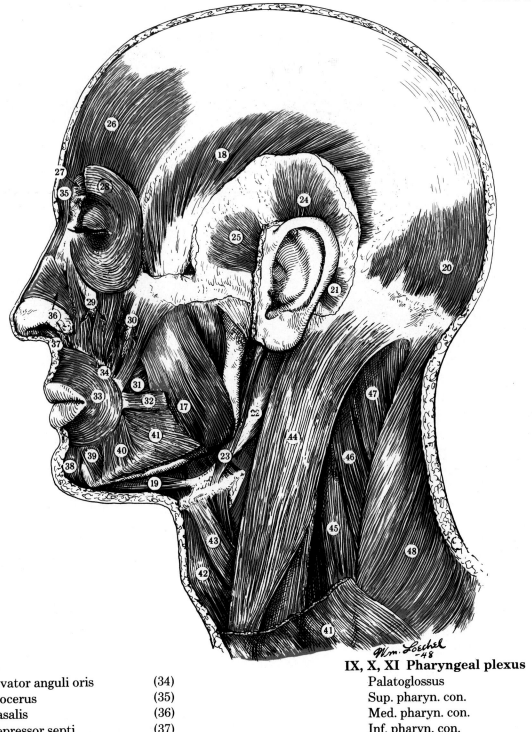

Levator anguli oris	(34)	**IX, X, XI Pharyngeal plexus**	
Procerus	(35)	Palatoglossus	(9)
Nasalis	(36)	Sup. pharyn. con.	(10)
Depressor septi	(37)	Med. pharyn. con.	(11)
Mentalis	(38)	Inf. pharyn. con.	(12)
Depressor labii inf.	(39)	**XII Hypoglossal N.**	
Depressor anguli oris	(40)	Styloglossus	(13)
Platysma	(41)	Hyoglossus	(14)
VIII Vestibulococh. N. (sen)		Genioglossus	(15)
IX Glossopharyngeal N.		Tongue intrinsics	(16)
Stylopharyngeus	(8)	**Misc. from cerv. nerves**	
X Vagus N.		Sternohyoid	(42)
XI Accessory N. (spinal portion)		Omohyoid	(43)
Sternocleidomastoid	(44)	Scalenus medius	(45)
Trapezius	(48)	Levator scapulae	(46)
		Splenius capitus	(47)

2

Epicranius, Frontal Belly

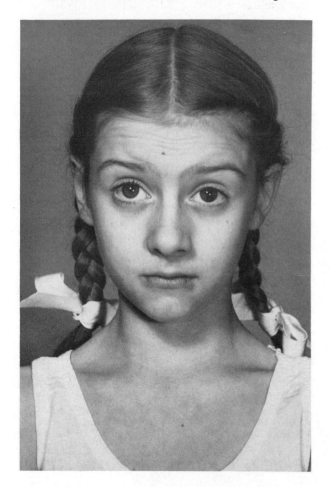

Origin of medial fibers: Continuous with those of the Procerus.

Origin of intermediate fibers: Blended with the Corrugator and Orbicularis oculi.

Origin of lateral fibers: Blended with Orbicularis oculi over zygomatic process of frontal bone.

Insertion: Fibers join galea aponeurotica below coronal suture.

Test: Have the patient raise the eyebrows, wrinkling the forehead as in surprise or fright.

Corrugator Supercilii

Origin: Medial end of superciliary arch.

Insertion: Deep surface of skin above middle of orbital arch.

Test: Have the patient draw the eyebrows together as in frowning.

Nasalis, Alar Portion

Depressor Septi and Transverse Portion Nasalis

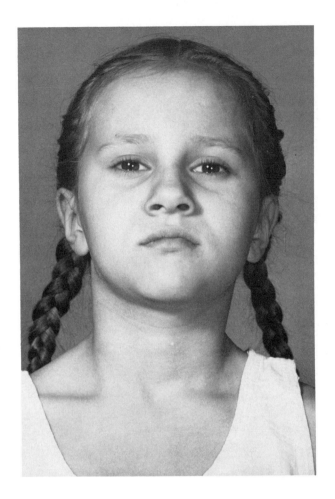

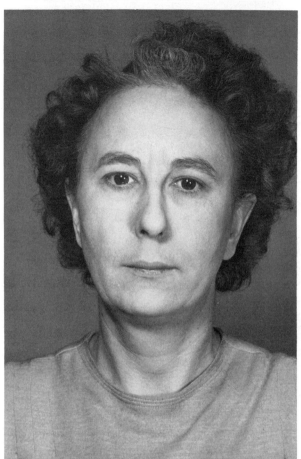

Origin: Greater alar cartilage.

Insertion: Integument at point of nose.

Test: Have the patient widen the apertures of the nostrils, as in forced or difficult breathing.

DEPRESSOR SEPTI

Origin: Incisive fossa of maxilla.

Insertion: Into septum and back part of ala of nose.

NASALIS, TRANSVERSE PORTION

Origin: Above and lateral to incisive fossa of maxilla.

Insertion: With aponeurosis of Procerus.

Test: Have the patient draw the point of the nose downward, narrowing the nostrils.

Procerus

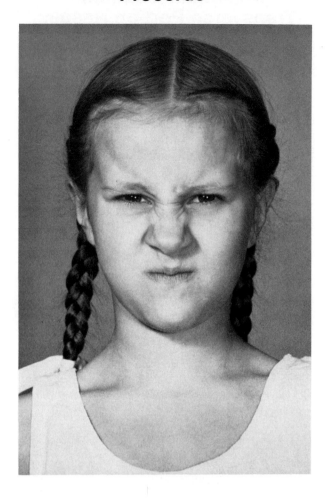

Levator Anguli Oris

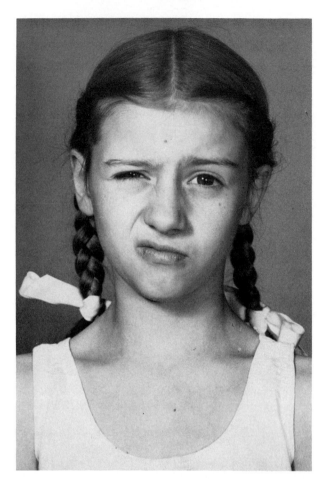

Origin: Fascia covering lower part of nasal bone and upper part of lateral nasal cartilage.

Insertion: Into skin over lower part of forehead between eyebrows.

Test: Have the patient pull the skin of the nose upward, forming transverse wrinkles over the bridge of the nose.

Origin: Canine fossa.

Insertion: Angle of the mouth, blending with adjacent muscles.

Test: Have the patient draw the angle of the mouth straight upward, deepening the furrow from the side of the nose to the side of the mouth as in sneering. Suggest that the patient try to show his "eye" (canine) tooth on one side, then the other.

Risorius

Zygomaticus Major

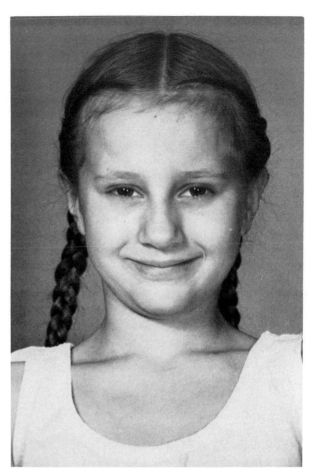

Origin: Fascia over Masseter.

Insertion: Into skin at angle of mouth.

Test: Have the patient draw the angle of the mouth backward.

Origin: Zygomatic bone.

Insertion: Angle of mouth, blending with adjacent muscles.

Test: Have the patient draw the angle of the mouth upward and outward as in smiling.

Levator Labii Superioris, etc.

Depressor Labii Inferioris and Platysma

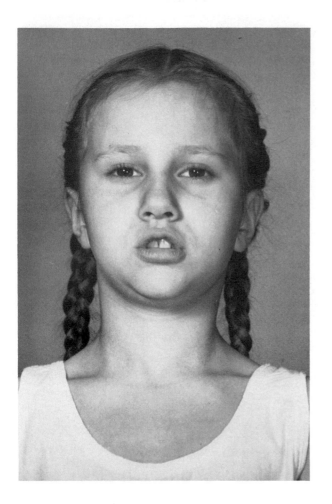

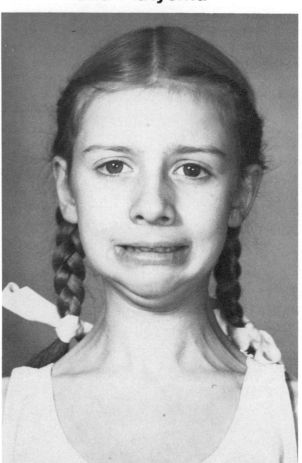

LEVATOR LABII SUPERIORIS

Origin: Lower margin of orbit.

Insertion: Muscular substance of upper lip.

ZYGOMATICUS MINOR

Origin: Malar surface of zygomatic bone.

Insertion: Upper lip.

LEVATOR LABII SUPERIORIS ALAEQUE NASI

Origin: Upper part of frontal process of maxilla.

Insertion: Greater alar cartilage and skin of nose, and lateral part of upper lip.

Test: Have the patient raise and protrude the upper lip as if to show his upper gums.

DEPRESSOR LABII INFERIORIS

Origin: Oblique line of mandible.

Insertion: Integument of lower lip, blending with Orbicularis oris.

PLATYSMA

Origin: Fascia covering superior portion of Pectoralis major and Deltoid.

Insertion: Lower border of mandible, posterior fibers blending with muscles about angle and lower part of mouth.

Test: Have the patient draw the lower lip and angle of the mouth downward and outward, tensing the skin over the neck.

Orbicularis Oris

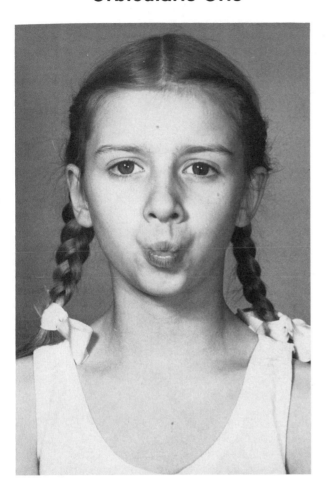

Origin: Numerous strata of muscular fibers surrounding the orifice of the mouth, derived in part from other facial muscles.

Insertion: For most part, into external skin and mucous membrane.

Test: Have the patient close the lips and protrude them forward as in whistling.

Buccinator

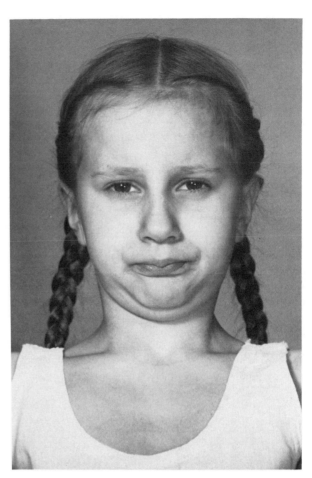

Origin: Outer surface of alveolar processes of maxilla and mandible, and anterior border of pterygomandibular tendinous band.

Insertion: Angle of mouth, blending with Orbicularis oris.

Test: Have the patient press the cheeks firmly against the side teeth, pulling back the angle of the mouth as in blowing a trumpet. (Drawing the chin backward, as seen in this illustration, is not part of the Buccinator action.)

Mentalis

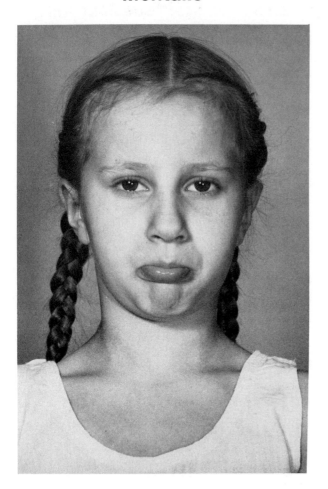

Depressor Anguli Oris

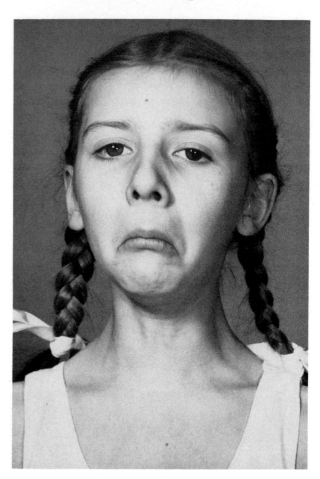

Origin: Incisive fossa of mandible.

Insertion: Integument of chin.

Test: Have the patient raise the skin of the chin. Secondarily, the lower lip will protrude somewhat as in pouting.

Origin: Oblique line of mandible.

Insertion: Angle of mouth, blending with adjacent muscles.

Test: Have the patient draw down the angles of the mouth.

Pterygoideus Medialis and Lateralis

Temporalis, Masseter, and Pterygoideus Medialis

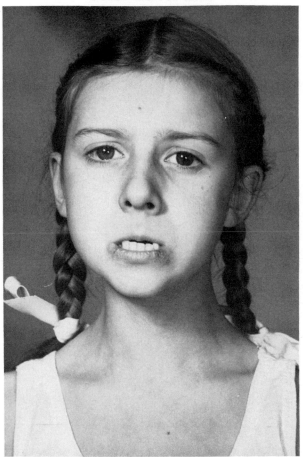

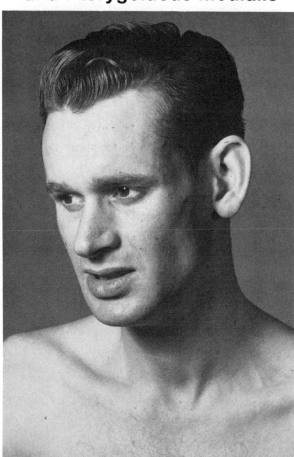

PTERYGOIDEUS MEDIALIS

Origin: Medial surface of lateral pterygoid plate, pyramidal process of palatine bone, and tuberosity of maxilla.

Insertion: Inferior and posterior part of medial surface of ramus and angle of manibular foramen.

PTERYGOIDEUS LATERALIS

Origin of superior head: Lateral surface of great wing of sphenoid, and infratemporal crest.

Origin of inferior head: Lateral surface of lateral pterygoid plate.

Insertion: Depression, anterior part of condyle of mandible, anterior margin of articular disk of temperomandibular articulation.

Test: Have the patient protrude the lower jaw.

MASSETER

Origin of superficial portion: Zygomatic process of maxilla, and anterior two-thirds of zygomatic arch.

Insertion: Inferior one-half of lateral surface of ramus of mandible.

Origin of profundus portion: Posterior one-third of inferior border and medial surface of zygomatic arch.

Insertion: Superior one-half of ramus and lateral surface of coronoid process of mandible.

TEMPORALIS

Origin: Temporal fossa and fascia.

Insertion: Coronoid process and anterior border of ramus of mandible.

TEMPORALIS, MASSETER, AND PTERYGOIDEUS MEDIALIS

Test: Have the patient bite firmly. (Mouth is slightly open only to show that teeth are being clenched.)

Levator Palpebrae Superioris, etc.

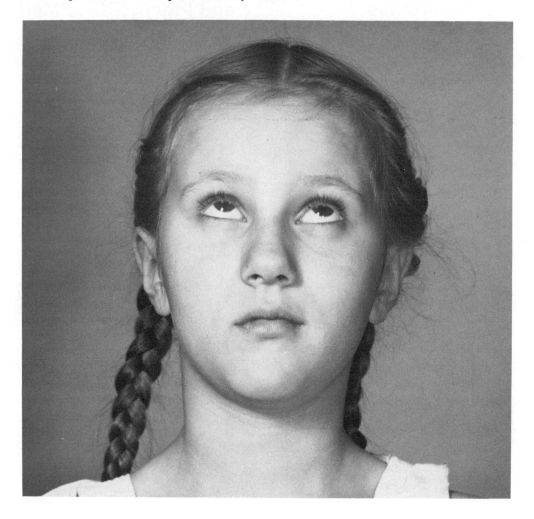

LEVATOR PALPEBRAE SUPERIORIS

Origin: Inferior surface of small wing of sphenoid.

Insertion of lamina superficialis: Palpebral part of Orbicularis oculi, and deep surface of skin of upper eyelid.

Insertion of lamina profunda: Superior margin of superior tarsus. In addition, a third set of fibers blend with the expansion from the sheath of the Rectus superior and insert into the fornix of the conjunctiva.

Test: Have the patient raise the upper lid.

RECTUS SUPERIOR AND INFERIOR

See facing page for origin and insertion.

OBLIQUUS INFERIOR

Origin: Orbital surface of maxilla.

Insertion: Into external part of sclera, between Rectus superior and Rectus lateralis, posterior to equator of eyeball.

OBLIQUUS SUPERIOR

Origin: Above margin of optic foramen.

Insertion: Into sclera between Rectus superior and Rectus lateralis, posterior to equator of eyeball.

Test Rectus superior and Obliquus inferior: Have the patient look straight upward toward the brow.

Test Rectus inferior and Obliquus superior: Have patient look straight downward toward the mouth. (Not illustrated.)

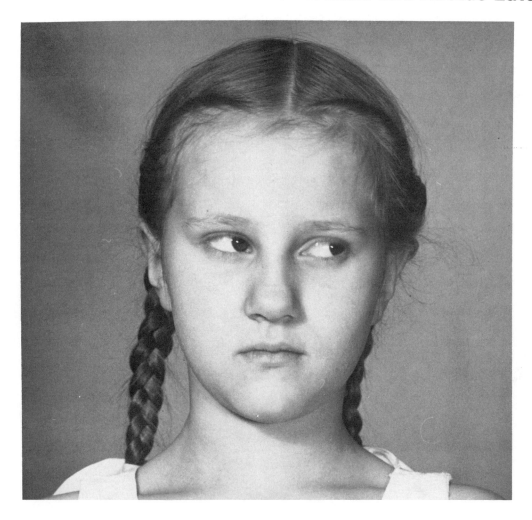

RECTI SUPERIOR, INFERIOR, MEDIALIS, AND LATERALIS

Origins: Fibrous ring which surrounds superior, medial, and inferior margins of optic foramen.

Insertions: Into sclera, anterior to equator of eyeball at the site implied by each name.

Test Rectus medialis: Have the patient look horizontally inward toward the nose. (Right illustrated.)

Test Rectus lateralis: Have the patient look horizontally outward away from the nose. (Left illustrated.)

Orbicularis Oculi

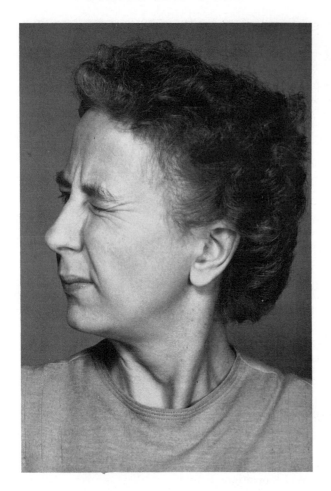

Suprahyoid Muscles

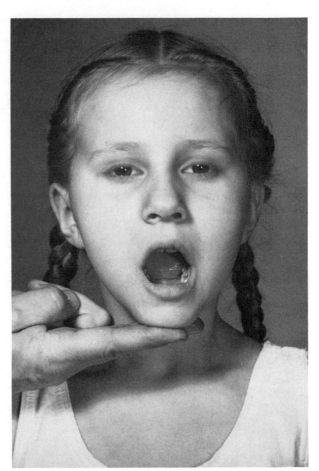

Origin: Nasal part of frontal bone, frontal process of maxilla, anterior surface of medial palpebral ligament.

Insertion: Muscle fibers surround the circumference of the orbit, spread downward on the cheek, and blend with adjacent muscular or ligamentous structures.

Test orbital part: Have the patient close the eyelid firmly, forming wrinkles radiating from the outer angle.

Test palpebral part: Have the patient close the eyelid gently. (Not illustrated.)

Test: Have the patient depress the lower jaw against resistance. The infrahyoid muscles furnish fixation of the hyoid bone during the action of these muscles.

(See p. 266 for origins, insertions, and actions of suprahyoid muscles; and illustration p. 255.)

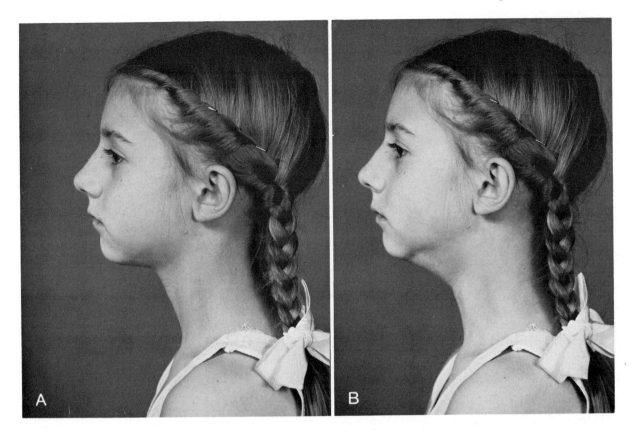

Test: Have the patient depress the hyoid bone (as illustrated). A represents relaxed position; B the test.

(See p. 267 for origins, insertions, and actions of infrahyoid muscles; and illustration p. 255.)

CRANIAL NERVE AND MUSCLE CHART

Name _____   Date _____

Nerve	Region	Grade	SENSORY OR MOTOR TO:	I Olfactory (S)	II Optic (S)	III Oculomotor (M)	IV Trochlear (M)	V Trigeminal (S&M)	VI Abducens (M)	VII Facial (S&M)	VIII Vestibulocochlear (S)	IX Glossopharyngeal (S&M)	X Vagus (S&M)	XI Accessory (M)	XII Hypoglossal (M)
I	NOSE	S	SENSORY—SMELL	●											
II	EYE	S	SENSORY—SIGHT		●										
III	EYELID		LEVATOR PALPEBRAE SUPERIORIS			●									
III	EYE		RECTUS SUPERIOR			●									
			OBLIQUUS INFERIOR			●									
			RECTUS MEDIALIS			●									
			RECTUS INFERIOR			●									
IV	EYE		OBLIQUUS SUPERIOR				●								
V	→	S	SENSORY—FACE & INT. STRUCTURES OF HEAD					●							
	EAR		TENSOR TYMPANI					●							
	PALATE		TENSOR VELI PALATINI					●							
	MASTI-CATION		MASSETER					●							
			TEMPORALIS					●							
			PTERYGOIDEUS MEDIALIS					●							
			PTERYGOIDEUS LATERALIS					●							
	S. HYOID		MYLOHYOIDEUS					●							
			ANTERIOR DIGASTRIC					●							
VI	EYE		RECTUS LATERALIS						●						
VII	TONGUE	S	SENSORY—TASTE, ANTERIOR ⅔ TONGUE							●					
	→	S	SENSORY—EXTERNAL EAR.							●					
	EAR		STAPEDIUS							●					
	S. HYOID		POSTERIOR DIGASTRIC							●					
			STYLOHYOIDEUS							●					
	SCALP		OCCIPITALIS							●					
			INTRINSIC EAR MUSCLES } POST. AURICULAR BRANCH							●					
			AURICULARIS POSTERIOR							●					
	EAR		AURICULARIS ANTERIOR							●					
			AURICULARIS SUPERIOR } TEMPORAL BRANCH							●					
	SCALP		FRONTALIS							●					
	EYEBR.		CORRUGATOR SUPERCILII							●					
	EYELID		ORBICULARIS OCULI } TEMP. & ZYGO. BRANCH							●					
			PROCERUS							●					
	NOSE		DEP. SEPTI & NAS. TRANS.							●					
			NASALIS, ALAR							●					
			ZYGOMATICUS MAJOR							●					
			LEVATOR LABII SUPERIORIS } BUCCAL BRANCH							●					
	MOUTH		BUCCINATOR							●					
			ORBICULARIS ORIS							●					
			LEVATOR ANGULI ORIS							●					
			RISORIUS							●					
			DEPRESSOR ANGULI ORIS							●					
			DEPRESSOR LABII INFERIORIS } MANDIBULAR BRANCH							●					
	CHIN		MENTALIS							●					
	NECK		PLATYSMA CERVICAL BRANCH							●					
VIII	EAR	S	SENSORY—HEARING & EQUILIBRIUM								●				
IX	TONGUE	S	SENSORY—POSTERIOR ⅓ TONGUE									●			
		S	SENSORY—PHARYNX, FAUCES, SOFT PALATE									●			
	PHARYNX		STYLOPHARYNGEUS									●			
		—	STRIATED MUSCLES - PHARYNX									●			
X	→	—	STRIATED MUSCLES—SOFT PALATE, PHARYNX & LARYNX										●		
	→	—	INVOLUNTARY MUSCLES—ALIMENTARY TRACT										●		
	→	—	INVOLUNTARY MUSCLES—AIR PASSAGES										●		
	→	—	INVOLUNTARY CARDIAC MUSCLE										●		
	→	S	SENSORY—AURICULAR										●		
	→	S	SENSORY—ALIMENTARY TRACT										●		
	→	S	SENSORY—AIR PASSAGES										●		
	→	S	SENSORY—ABDOMINAL VISCERA & HEART										●		
XI	NECK		TRAPEZIUS & STERNOCLEIDOMASTOID											●	
	PALATE		LEVATOR VELI PALATINI											●	
	→		STRIATED MUSCLES—SOFT PALATE, PHARYNX, & LARYNX											●	
XII	TONGUE		STYLOGLOSSUS												●
			HYOGLOSSUS												●
			GENIOGLOSSUS												●
			TONGUE INTRINSICS												●

(Leftmost column: MUSCLE STRENGTH GRADE)

SENSORY

(head diagrams labeled C2, C4)

DERMATOMES

(head diagram labeled C3)

CUTANEOUS DISTRIBUTION OF CRANIAL NERVES

OPHTHALMIC — MAXILLARY — MANDIBULAR — CERVICAL NERVES

vent.   dorsal
primary rami

Ophthalmic
1. Supratrochlear N.
2. Supraorbital N.
3. Lacrimal N.
4. Infratrochlear N.
5. Nasal N.

Maxillary
6. Zygomatico-temporal N.
7. Infraorbital N.
8. Zygomatico-facial N.

Mandibular
9. Auriculotemporal N.
10. Buccal N.
11. Mental N.

Cervical Nerves
12. Greater Occipital N.
13. Lesser Occipital N.
14. Great Auricular N.

Redrawn from *Gray's Anatomy of the Human Body*, 28th ed

CRANIAL NERVE AND MUSCLE CHART

The cranial nerve and muscle chart (p. 250) is designed for use primarily as a reference sheet, and secondarily as a chart to record examination findings for the muscles of expression.

Because of the dual purpose of the chart, it contains some material that would not be included if used only for recording muscle examinations. For example, all the cranial nerves are listed, whether they are sensory, motor, or mixed nerves. Some muscles are included which cannot be tested manually either individually or in groups.

On the cranial nerve chart are listed all the cranial nerves with the specific muscles or organs and the general regions which they supply. A column is provided at the left of the muscle names in which to record evaluation of strength of those muscles which can be tested. At the right side of the page are drawings of the head showing areas of dermatomes and distribution of the cutaneous nerves.

On the reverse side of the chart are two drawings of the head. These appear in this chapter as full page illustrations.

The illustration on p. 237 shows the superficial musculature. The one on p. 236 is a sagittal section of the skull at about the center of the left orbit except that the complete eyeball is shown. The muscles depicted in this illustration are mainly those of the tongue, the pharyngeal area, and the eyeball.

The left hemisphere of the brain has been reflected upward to show its inferior surface and the cranial nerve roots. Connecting the nerve roots with the corresponding nerve trunks in the lower section of the drawing are lines bearing the numbers of the respective cranial nerves. Nerve roots I, II, and VIII, which are sensory, are left white. The motor and mixed nerves are shown in yellow except that only the small motor branch of the fifth is yellow.

On the drawings and in the legend, the cranial nerves are denoted by Roman numerals, and the muscles by Arabic numerals.

On the following pages there appear two charts with recordings of muscle examinations of the facial muscles. In the first case, the onset of the Bell's Palsy was one week before the examination date. Three muscles graded zero (0), ten graded trace (T), and two graded poor (P). Three weeks later, the muscles all graded good (G) and approximately three weeks later, the muscles graded normal (N) except three that still graded good. This is an ex-

ample of those facial paralysis cases that make fairly rapid recovery, some within a few days or a week, some, like this case, within a two-month period.

The second case is one in which there was no evidence of any muscle function except slight action in the Corrugator at the time of the first examination which was three weeks after onset. This case showed very little change during the first three and one-half months. By the end of six months, most of the muscles graded fair or better. By the end of eight months, there was further improvement and by the end of nine and one-half months, about one-third of the muscles graded fair and all others either good or normal. This case shows the slow but gradual improvement that occurs in some instances.

This second patient was fitted with a very small plastic hook, contoured to fit in the corner of the mouth and attached by means of a rubber band to the bow on the glasses worn by the patient. She was instructed in giving herself light massage—upward on the affected side and downward and toward the mouth on the unaffected side. At times, transparent scotch tape was used to hold up the side of the mouth and cheek. When she was not using the hook or tape, she was advised to make a habit, when sitting, of resting the right elbow on a table or arm of a chair and placing the right hand with palm under the right chin and fingers along the cheek to hold the right side of the face upward. Also, when speaking, smiling, or laughing, the hand was to be used to push the affected side toward the right and upward to compensate for the weakness, as well as to prevent the unaffected side from distorting the mouth in that direction. She was taught how to exercise the facial muscles, by assisting the weak side and restraining the stronger side.

In some cases of facial paralysis, the Orbicular oculi (which closes and squeezes the eye shut) may be slower to respond than some other muscles. During the period of recovery when there is a problem with closing the eye, exercising the Epicranius is discouraged because it acts in opposition to the Orbicularis oculi. The reason for this may be illustrated by the following: Lift the eyebrow by contracting the Epicranius. With the fingertips placed just above the eyebrows, keep the eyebrow held upward and (1) try to close the eye; (2) try to squeeze the eye tightly shut. The difficulty in doing both is readily demonstrated.

CRANIAL NERVE AND MUSCLE CHART

Name: *Case #1*   Date: 1 week after onset

SENSORY

	Region	Muscle Strength Grade (Feb. 27)	SENSORY OR MOTOR TO: Left	Mar. 20	Apr. 13	I Olfactory (S)	II Optic (S)	III Oculomotor (M)	IV Trochlear (M)	V Trigeminal (S&M)	VI Abducens (M)	VII Facial (S&M)	VIII Vestibulocochlear (S)	IX Glossopharyngeal (S&M)	X Vagus (S&M)	XI Accessory (M)	XII Hypoglossal (M)
I NOSE		S	SENSORY—SMELL			•											
II EYE		S	SENSORY—SIGHT				•									
III EYELID			LEVATOR PALPEBRAE SUPERIORIS					•								
III EYE			RECTUS SUPERIOR					•								
			OBLIQUUS INFERIOR					•								
			RECTUS MEDIALIS					•								
			RECTUS INFERIOR					•								
IV EYE			OBLIQUUS SUPERIOR						•							
V →		S	SENSORY—FACE & INT. STRUCTURES OF HEAD							•						
EAR			TENSOR TYMPANI							•						
PALATE			TENSOR VELI PALATINI							•						
V MASTICATION			MASSETER							•						
			TEMPORALIS							•						
			PTERYGOIDEUS MEDIALIS							•						
			PTERYGOIDEUS LATERALIS							•						
S. HYOID			MYLOHYOIDEUS							•						
			ANTERIOR DIGASTRIC							•						
VI EYE			RECTUS LATERALIS								•					
VII TONGUE		S	SENSORY—TASTE, ANTERIOR ⅔ TONGUE									•				
→		S	SENSORY—EXTERNAL EAR									•				
EAR			STAPEDIUS									•				
S. HYOID			POSTERIOR DIGASTRIC									•				
			STYLOHYOIDEUS									•				
SCALP			OCCIPITALIS									•				
EAR			INTRINSIC EAR MUSCLES — POST. AURICULAR BRANCH									•				
			AURICULARIS POSTERIOR									•				
			AURICULARIS ANTERIOR									•				
			AURICULARIS SUPERIOR — TEMPORAL BRANCH									•				
SCALP		T	FRONTALIS	G	N							•				
EYEBR.		T	CORRUGATOR SUPERCILII — TEMP & ZYGO BRANCH	G	N							•				
EYELID		P	ORBICULARIS OCULI	G	N							•				
VII NOSE		P	PROCERUS	G	N							•				
		−	DEP. SEPTI & NAS, TRANS.	−	−							•				
		T	NASALIS, ALAR	G	N							•				
		T	ZYGOMATICUS MAJOR	G	N							•				
			LEVATOR LABII SUPERIORIS — BUCCAL BRANCH	G	N							•				
MOUTH		T	BUCCINATOR	G	N							•				
		T	ORBICULARIS ORIS	G	G							•				
		T	LEVATOR ANGULI ORIS	G	G							•				
		T	RISORIUS	G	N							•				
		O	DEPRESSOR ANGULI ORIS	G	N							•				
		O	DEPRESSOR LABII INFERIORIS — MANDIBULAR BRANCH	G	N							•				
CHIN		T	MENTALIS	G	G							•				
NECK		O	PLATYSMA — CERVICAL BRANCH	G	N							•				
VIII EAR		S	SENSORY—HEARING & EQUILIBRIUM										•			
IX TONGUE		S	SENSORY—POSTERIOR ⅓ TONGUE											•		
		S	SENSORY—PHARYNX, FAUCES, SOFT PALATE											•		
PHARYNX			STYLOPHARYNGEUS											•		
		−	STRIATED MUSCLES - PHARYNX											•		
X →		−	STRIATED MUSCLES—SOFT PALATE, PHARYNX & LARYNX												•	
→		−	INVOLUNTARY MUSCLES—ALIMENTARY TRACT												•	
→		−	INVOLUNTARY MUSCLES—AIR PASSAGES												•	
→		−	INVOLUNTARY CARDIAC MUSCLE												•	
→		S	SENSORY—AURICULAR												•	
→		S	SENSORY—ALIMENTARY TRACT												•	
→		S	SENSORY—AIR PASSAGES												•	
→		S	SENSORY—ABDOMINAL VISCERA & HEART												•	
XI NECK			TRAPEZIUS & STERNOCLEIDOMASTOID													•	
PALATE			LEVATOR VELI PALATINI													•	
→			STRIATED MUSCLES—SOFT PALATE, PHARYNX, & LARYNX													•	
XII TONGUE			STYLOGLOSSUS														•
			HYOGLOSSUS														•
			GENIOGLOSSUS														•
			TONGUE INTRINSICS														•

DERMATOMES — C2, C4, C3

CUTANEOUS DISTRIBUTION OF CRANIAL NERVES

vent.   dorsal
primary rami

Ophthalmic
1. Supratrochlear N.
2. Supraorbital N.
3. Lacrimal N.
4. Infratrochlear N.
5. Nasal N.

Maxillary
6. Zygomatico-temporal N.
7. Infraorbital N.
8. Zygomatico-facial N.

Mandibular
9. Auriculotemporal N.
10. Buccal N.
11. Mental N.

Cervical Nerves
12. Greater Occipital N.
13. Lesser Occipital N.
14. Great Auricular N.

Redrawn from *Gray's Anatomy of the Human Body*, 28th ed.

CRANIAL NERVE AND MUSCLE CHART

Name: *Case # 2* Date: 3 weeks after onset

Nerve columns (with sensory/motor designation):

#	Nerve	Type
I	Olfactory	S
II	Optic	S
III	Oculomotor	M
IV	Trochlear	M
V	Trigeminal	S & M
VI	Abducens	M
VII	Facial	S & M
VIII	Vestibulocochlear	S
IX	Glossopharyngeal	S & M
X	Vagus	S & M
XI	Accessory	M
XII	Hypoglossal	M

Main chart — "SENSORY OR MOTOR TO: Right"

Group	Region	Grade	Sensory or Motor To	Nerve
I	NOSE	S	SENSORY—SMELL	I Olfactory ●
II	EYE	S	SENSORY—SIGHT	II Optic ●
III	EYELID		LEVATOR PALPEBRAE SUPERIORIS	III Oculomotor ●
III	EYE		RECTUS SUPERIOR	III ●
			OBLIQUUS INFERIOR	III ●
			RECTUS MEDIALIS	III ●
			RECTUS INFERIOR	III ●
IV	EYE		OBLIQUUS SUPERIOR	IV Trochlear ●
V	→	S	SENSORY—FACE & INT. STRUCTURES OF HEAD	V Trigeminal ●
	EAR		TENSOR TYMPANI	V ●
	PALATE		TENSOR VELI PALATINI	V ●
	MASTICATION		MASSETER	V ●
			TEMPORALIS	V ●
			PTERYGOIDEUS MEDIALIS	V ●
			PTERYGOIDEUS LATERALIS	V ●
	S. HYOID		MYLOHYOIDEUS	V ●
			ANTERIOR DIGASTRIC	V ●
VI	EYE		RECTUS LATERALIS	VI Abducens ●
VII	TONGUE	S	SENSORY—TASTE, ANTERIOR ⅔ TONGUE	VII Facial ●
	→	S	SENSORY—EXTERNAL EAR	VII ●
	EAR		STAPEDIUS	VII ●
	S. HYOID		POSTERIOR DIGASTRIC	VII ●
			STYLOHYOIDEUS	VII ●
	SCALP		OCCIPITALIS	VII ●
	EAR		INTRINSIC EAR MUSCLES (POST. AURICULAR BRANCH)	VII ●
			AURICULARIS POSTERIOR	VII ●
			AURICULARIS ANTERIOR	VII ●
			AURICULARIS SUPERIOR (TEMPORAL BRANCH)	VII ●

VII Facial muscle strength grades (serial examinations). Grade column dates: [8-22-61]. Column headings: [11-3-61], [12-11-61], [2-28-62], [4-17-62], [6-6-62]

Region	Grade	Muscle (branch)	11-3-61	12-11-61	2-28-62	4-17-62	6-6-62	VII
SCALP	O	FRONTALIS	T	T	P+	F	F	●
EYEBR.	P	CORRUGATOR SUPERCILII (TEMP. & ZYGO. BRANCH)	P	–	G–	G	G	●
EYELID	O	ORBICULARIS OCULI	P–	P	F+	N	N	●
NOSE	O	PROCERUS	O	P	G–	F	G	●
	O	DEP. SEPTI & NAS. TRANS.	–	–	–	–	–	●
	O	NASALIS, ALAR	O	?	F	F	F	●
MOUTH	O	ZYGOMATICUS MAJOR	P–	P	G–	G	G	●
	O	LEVATOR LABII SUPERIORIS (BUCCAL BRANCH)	?	?	F	F	G	●
	O	BUCCINATOR	–	–	F–	F	F	●
	O	ORBICULARIS ORIS	–	T	F	F–	F	●
	O	LEVATOR ANGULI ORIS	T	?	G–	G	G	●
	O	RISORIUS	P–	P	F+	G	G	●
	O	DEPRESSOR ANGULI ORIS	?	–	F	F–	F	●
	O	DEPRESSOR LABII INFERIORIS (MANDIBULAR BRANCH)	?	–	P+	F–	G	●
CHIN	O	MENTALIS	O	?	F+	G	N	●
NECK	O	PLATYSMA (CERVICAL BRANCH)	T	–	F+	G	G	●

Group	Region	Grade	Sensory or Motor To	Nerve
VIII	EAR	S	SENSORY—HEARING & EQUILIBRIUM	VIII Vestibulocochlear ●
IX	TONGUE	S	SENSORY—POSTERIOR ⅓ TONGUE	IX Glossopharyngeal ●
		S	SENSORY—PHARYNX, FAUCES, SOFT PALATE	IX ●
	PHARYNX		STYLOPHARYNGEUS	IX ●
		—	STRIATED MUSCLES - PHARYNX	IX ●
X	→	—	STRIATED MUSCLES—SOFT PALATE, PHARYNX & LARYNX	X Vagus ●
	→	—	INVOLUNTARY MUSCLES—ALIMENTARY TRACT	X ●
	→	—	INVOLUNTARY MUSCLES—AIR PASSAGES	X ●
	→	—	INVOLUNTARY CARDIAC MUSCLE	X ●
	→	S	SENSORY—AURICULAR	X ●
	→	S	SENSORY—ALIMENTARY TRACT	X ●
	→	S	SENSORY—AIR PASSAGES	X ●
	→	S	SENSORY—ABDOMINAL VISCERA & HEART	X ●
XI	NECK		TRAPEZIUS & STERNOCLEIDOMASTOID	XI Accessory ●
	PALATE		LEVATOR VELI PALATINI	XI ●
	→		STRIATED MUSCLES—SOFT PALATE, PHARYNX, & LARYNX	XI ●
XII	TONGUE		STYLOGLOSSUS	XII Hypoglossal ●
			HYOGLOSSUS	XII ●
			GENIOGLOSSUS	XII ●
			TONGUE INTRINSICS	XII ●

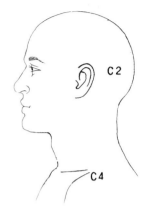

SENSORY

C 2

C 4

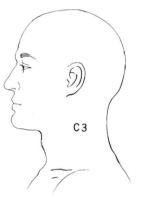

C 3

DERMATOMES

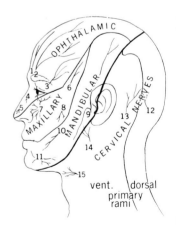

vent. dorsal primary rami

CUTANEOUS DISTRIBUTION OF CRANIAL NERVES

Ophthalamic
1. Supratrochlear N.
2. Supraorbital N.
3. Lacrimal N.
4. Infratrochlear N.
5. Nasal N.

Maxillary
6. Zygomatico-temporal N.
7. Infraorbital N.
8. Zygomatico-facial N.

Mandibular
9. Auriculotemporal N.
10. Buccal N.
11. Mental N.

Cervical Nerves
12. Greater Occipital N.
13. Lesser Occipital N.
14. Great Auricular N.

Redrawn from *Gray's Anatomy of the Human Body*. 28th ed

Anterior and Lateral Neck Muscles

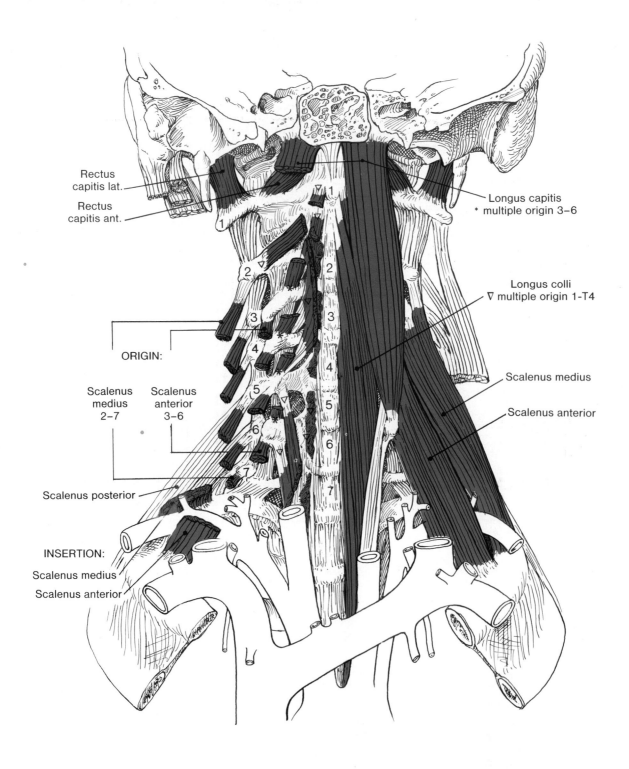

Rectus capitis lat.

Rectus capitis ant.

Longus capitis
* multiple origin 3–6

Longus colli
▽ multiple origin 1–T4

ORIGIN:

Scalenus medius 2–7

Scalenus anterior 3–6

Scalenus medius

Scalenus anterior

Scalenus posterior

INSERTION:

Scalenus medius

Scalenus anterior

Redrawn from Sobotta-Figge.[30]

Suprahyoid and Infrahyoid Muscles

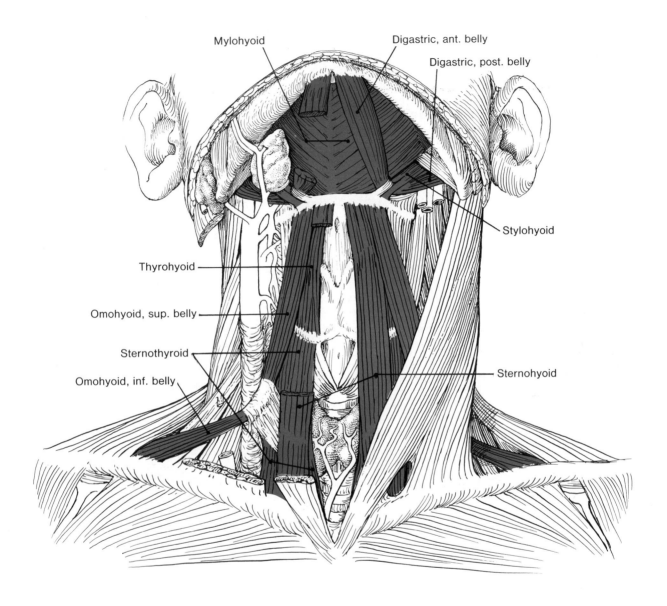

Mylohyoid

Digastric, ant. belly

Digastric, post. belly

Stylohyoid

Thyrohyoid

Omohyoid, sup. belly

Sternothyroid

Sternohyoid

Omohyoid, inf. belly

Redrawn from Sobotta-Figge.[30]

Head and Neck Muscles

	Origin	Insertion
Longus coli*	**Superior oblique portion:** Anterior tubercles of transverse processes of 3rd, 4th, 5th cervical vertebrae.	Tubercle on anterior arch of atlas.
	Inferior oblique portion: Anterior surface of bodies of first 2 or 3 thoracic vertebrae.	Anterior tubercles of transverse processes of 5th and 6th cervical vertebrae.
	Vertical portion: Anterior surface of bodies of first 3 thoracic and last 3 cervical vertebrae.	Anterior surface of bodies of 2nd, 3rd, 4th cervical vertabrae.
Longus capitis*	Anterior tubercles of transverse processes of 3rd through 6th cervical vertebrae.	Inferior surface of basilar part of occipital bone.
Rectus capitis anterior*	Root of transverse process, and anterior surface of atlas.	Inferior surface of basilar part of occipital bone.
Rectus capitis lateralis*	Superior surface of transverse process of atlas.	Inferior surface of jugular process of occipital bone.
Scalenus anterior*	Anterior tubercles of transverse processes of 3rd through 6th cervical vertebrae.	Scalene tubercle and cranial crest of 1st rib.
Scalenus medius*	Posterior tubercles of transverse processes of 2nd through 7th cervical vertebrae.	Cranial surface of 1st rib between tubercle and subclavian groove.
Scalenus posterior*	By 2 or 3 tendons from posterior tubercles of transverse processes of last 2 or 3 cervical vertebrae.	Outer surface of 2nd rib.
Platysma**	Fascia covering superior parts of Pectoralis major and Deltoid.	Inferior margin of mandible, and skin of lower part of face and corner of mouth.
Sternocleidomastoid**	**Medial or sternal head:** Cranial part of manubrium sterni.	Lateral surface of mastoid process, lateral ½ of superior nuchal line of occipital bone.
	Lateral or clavicular head: Medial ⅓ of clavicle.	
Rectus capitis post. major	Spinous process of axis.	Lateral part of inferior nuchal line of occipital bone.
Rectus capitis post. minor	Tubercle on posterior arch of atlas.	Medial part of inferior nuchal line of occipital bone.
Obliquus capitis inferior	Apex of spinous process of axis.	Inferior and posterior part of transverse process of atlas.
Obliquus capitis superior	Superior surface of transverse process of atlas.	Between superior and inferior nuchal lines of occipital bone.

Trapezius, upper	See p. 112.
Splenius capitis Splenius cervicis Iliocostalis cervicis Longissimus cervicis Longissimus capitis Spinalis cervicis Spinalis capitis Semispinalis cervicis Semispinalis capitis Multifidi, cervical Rotatores, cervical Interspinales, cervical Intertransversarii, cervical	See p. 228.

* See illustration, p. 254.
** See illustration, p. 237.

ACTION OF HEAD and NECK MUSCLES

Muscles	Acting bilaterally		Acting unilaterally		
	Extension	Flexion	Lat. flexion	Rotation	
				To same side	To opposite side
Longus colli		×	×	×	
Longus capitis		×		×	
Rectus capitis anterior		×		×	
Rectus capitis lateralis			×		
Scalenus anterior		×	×		×
Scalenus medius			×		×
Scalenus posterior			×		×
Platysma		×			
Sternocleidomastoid	×	×	×		×
Rectus capitis posterior major	×			×	
Rectus capitis posterior minor	×				
Obliquus capitis inferior				×	
Obliquus capitis superior	×		×		
Splenius cervicis	×		×	×	
Splenius capitis	×		×	×	
Trapezius, upper	×		×		×
Iliocostalis cervicis	×		×		
Longissimus cervicis	×				
Longissimus capitis	×		×	×	
Spinalis cervicis	×				
Spinalis capitis	×				
Semispinalis cervicis	×				×
Semispinalis capitis	×				
Multifidi, cervical	×				×
Rotatores, cervical	×				×
Interspinales, cervical	×				
Intertransversarii, cervical			×		

Anterior Neck Flexors

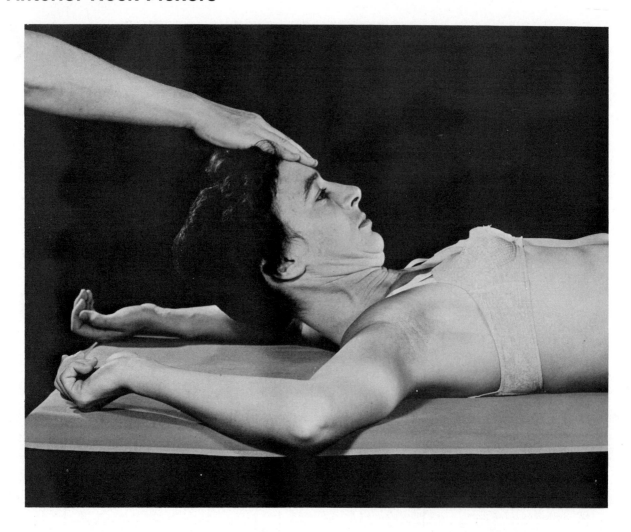

Patient: Supine with elbows bent and hands over head, resting on table.

Fixation: Anterior abdominal muscles must be strong enough to give anterior fixation of thorax to pelvis before the head can be raised by the neck flexors. If abdominal muscles are weak, the examiner can give fixation by firm downward pressure on the thorax. Children approximately five years and under should have fixation of the thorax by the examiner.

Test: Flexion of the cervical spine by lifting the head from the table with the chin depressed and approximated toward the sternum.

Pressure: Against the forehead in a posterior direction.

Modified test: In cases of marked weakness have patient make an effort to flatten the cervical spine on the table, approximating the chin toward the sternum.

Pressure: Against the chin in the direction of neck extension.

Note: The anterior vertebral flexors of the neck are the Longus capitus and colli and the Rectus capitus anterior. In this movement they are aided by the Sternocleidomastoid, Anterior scaleni, suprahyoids, and infrahyoids. The Platysma will attempt to aid when flexors are very weak.

Weakness: Hyperextension of the cervical spine resulting in a forward head position.

Contracture: A neck flexion contracture is rarely seen except unilaterally as in torticollis.

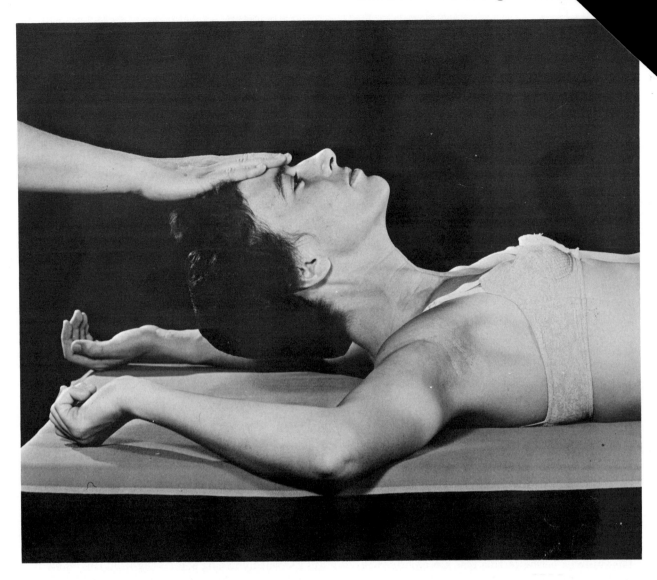

If anterior vertebral neck flexors are weak and the Sternocleidomastoid muscles strong, an individual can raise the head from the table (as illustrated), and can hold against pressure, but this is not an accurate test for neck flexors. Action is accomplished chiefly by the Sternocleidomastoids aided by the anterior Scaleni and the clavicular portions of the upper Trapezius.

Neck Flexors

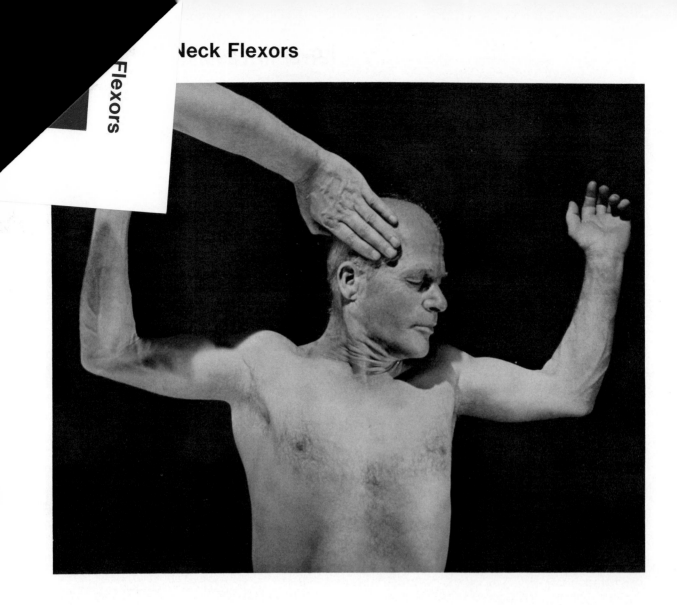

The muscles acting in this test are chiefly Sternocleidomastoid and Scaleni.

Patient: Supine with elbows bent and hands overhead, resting on table.

Fixation: If anterior abdominal muscles are weak, the examiner can give fixation by firm downward pressure on the thorax.

Test: Anterolateral neck flexion.

Pressure: Against the temporal region of the head, in an obliquely posterior direction.

Note: With neck muscles just strong enough to hold, but not strong enough to flex completely, a patient can lift the head from the table by raising the shoulders. He does so especially on the tests for right and left neck flexors because he attempts to

aid himself by taking some weight on the elbow or hand in order to push the shoulder from the table. To avoid this, hold the patient's shoulder flat on the table.

Contracture and weakness: A contracture of the right Sternocleidomastoid produces a right torticollis. The face is turned toward the left and the head is tilted toward the right. Thus, a right torticollis produces a cervical scoliosis convex toward the left. The left Sternocleidomastoid is elongated and weak. Contracture of the left Sternocleidomastoid, with weakness of the right, produces a left torticollis with a cervical scoliosis convex toward the right.

In an habitually faulty posture with forward head, the Sternocleidomastoid muscles remain in a shortened position, and tend to develop shortness.

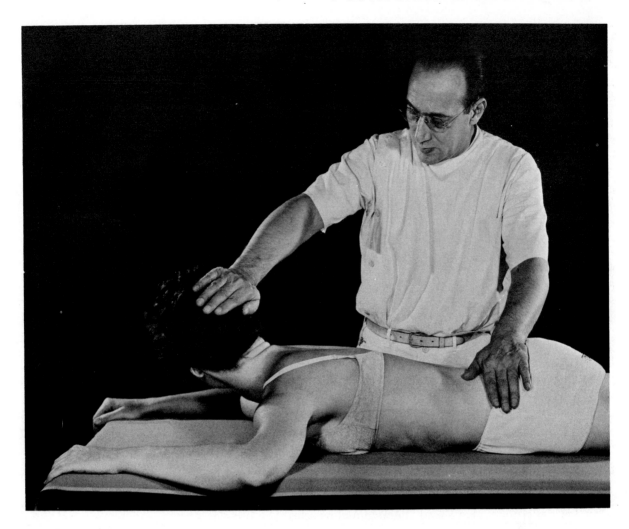

The muscles included in this test are chiefly the Splenius capitis and cervicis, Semispinalis capitis and cervicis, and cervical Erector spinae. (See pp. 228, 229.)

Patient: Prone with elbows bent and hands overhead, resting on table.

Fixation: None necessary.

Test: Posterolateral neck extension with face turned toward the side being tested. (See Note.)

Pressure: Against the posterolateral aspect of the head, in an anterior direction.

Shortness: The right Splenius capitis and left upper Trapezius are usually short along with the Sternocleidomastoid in a left torticollis. The opposite muscles are short in a right torticollis.

Note: The upper Trapezius, which is also a posterolateral neck extensor, is tested with the face turned away from the side being tested. (See p. 117.)

RESPIRATORY MUSCLE CHART

Patient's Name _____ Clinic # _____

Left					Right			
				Examiner				
				Date				
				Inspiratory Muscles Primary				
				Diaphragm				
· · · ·	· · · ·	· · · ·	· · · ·	Levator costarum	· · · ·	· · · ·	· · · ·	· · · ·
				External intercostals				
				Internal intercostals, anterior				
				Accessory				
				Scaleni				
				Sternocleidomastoid				
				Trapezius				
				Serratus ant. & post. superior				
				Pectoralis major & minor				
				Latissimus dorsi				
				Thoracic spine extensors				
· · · ·	· · · ·	· · · ·	· · · ·	Subclavius	· · · ·	· · · ·	· · · ·	· · · ·
				Expiratory Muscles Primary				
				Abdominal muscles				
				Internal oblique				
				External oblique				
				Rectus abdominis				
				Transversus abdominis				
				Internal intercostals, posterior				
· · · ·	· · · ·	· · · ·	· · · ·	Transversus thoracis	· · · ·	· · · ·	· · · ·	· · · ·
				Accessory				
				Latissimus dorsi				
· · · ·	· · · ·	· · · ·	· · · ·	Serratus posterior inferior	· · · ·	· · · ·	· · · ·	· · · ·
				Quadratus lumborum				
				Iliocostalis lumborum				

Notes: _____

_____

_____

_____

_____

_____

_____

Diaphragm

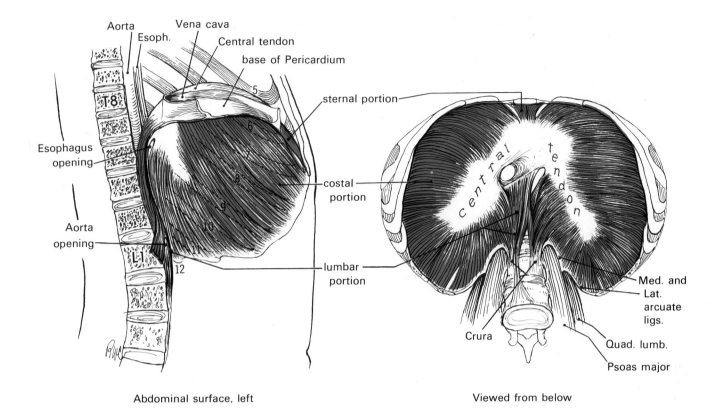

Aorta
Esoph.
Vena cava
Central tendon
base of Pericardium

T8

Esophagus opening

Aorta opening

L·1

12

5
6
7
8
9
10

sternal portion

costal portion

lumbar portion

Abdominal surface, left

central tendon

Med. and Lat. arcuate ligs.

Crura

Quad. lumb.

Psoas major

Viewed from below

Origin of sternal part: Two fleshy slips from dorsum of xiphoid process.

Origin of costal part: Inner surfaces of lower six costal cartilages and lower six ribs on either side, interdigitating with Transversus abdominis.

Origin of lumbar part: By two muscular crura from the bodies of the upper lumbar vertebrae and by two fibrous arches on either side, known as medial and lateral arcuate ligaments, which span from the vertebrae to the transverse processes and from the latter to the twelfth rib.

Insertion: Into central tendon. This tendon is a thin strong aponeurosis with no bony attachment. Since the anterior muscular fibers of the Diaphragm are shorter than the posterior muscular fibers, the central tendon is situated closer to the ventral part of the thorax than to the dorsal.

Action: The dome-shaped Diaphragm separates the thoracic and abdominal cavities and is the principal muscle of respiration. During inspiration, the muscle contracts and the dome descends increasing the volume and decreasing the pressure of the thoracic cavity, while decreasing the volume and increasing the pressure of the abdominal cavity. The descent of the dome or central tendon of the Diaphragm is limited by the abdominal viscera; and when this occurs the central tendon becomes the more fixed portion of the muscle. With continued contraction, the vertical fibers attached to the ribs elevate and evert the costal margin. The dimensions of the thorax are consequently enlarged craniocaudally, anteroposteriorly, and transversely. During expiration, the Diaphragm relaxes and the dome ascends decreasing the volume and increasing the pressure of the thoracic cavity, while increasing the volume and decreasing the pressure of the abdominal cavity.

Note: In cases of pulmonary pathology such as emphysema, the dome of the Diaphragm is so depressed that the costal margin or base of the thorax cannot be expanded.

Nerve: Phrenic, C3, **4**, 5.

Muscles of Respiration

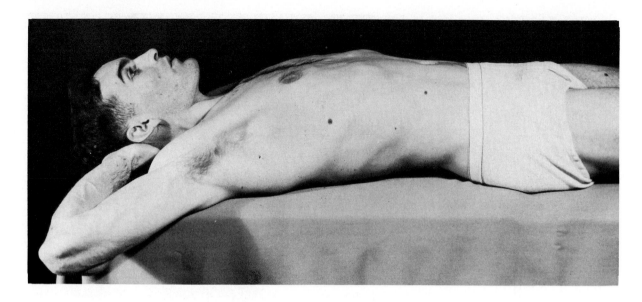

Normal inspiration: Intercostal and diaphragmatic.

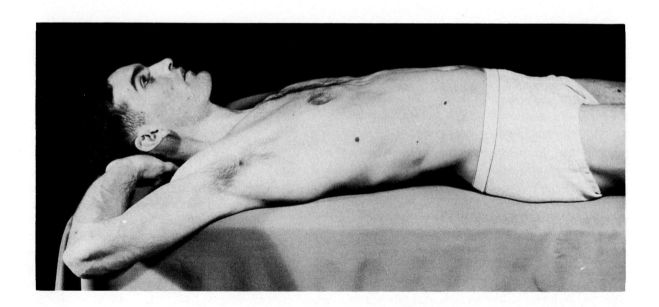

Inspiration: Diaphragmatic only.

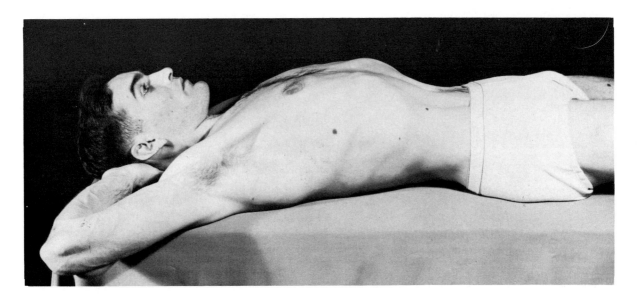

Inspiration: Intercostal.

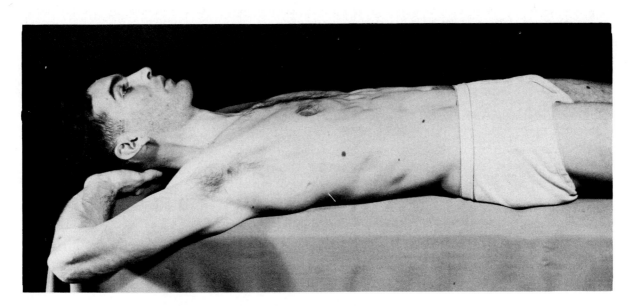

Forced expiration: Intercostal, abdominal, and accessory muscles.

Muscles of Deglutition

MUSCLE	ORIGIN	INSERTION	ACTION	INNERVATION Motor	INNERVATION Sensory	ROLE IN DEGLUTITION
TONGUE						
Sup. longitudinal	Intrinsic	Intrinsic	Shortens tongue; Raises sides and tip of tongue	Hypoglossal XII	General sensation Ant. ⅔ — Trigeminal V; Post. ⅓ — Glossopharyngeal IX; Base — Vagus X	**Bolus Preparation** — During this phase the tongue and the buccinator muscles keep the food between the molar teeth where it is crushed and ground by the action of muscles of mastication. Alternate side to side movements and twisting of the tongue, performed chiefly by the intrinsic muscles and by the styloglossi acting unilaterally, aid in mixing the food with saliva and in sorting larger particles from the sufficiently ground portion which is ready to be rolled into a bolus and swallowed.
Transverse	Intrinsic	Intrinsic	Lengthens and narrows tongue			
Vertical	Intrinsic	Intrinsic	Flattens and broadens tongue			
Inf. longitudinal	Intrinsic	Intrinsic	Shortens tongue; Turns tip of tongue downward			
Genioglossus	Mental spine	Tongue & body of hyoid	Depresses tongue: protrudes & retracts tongue: elevates hyoid	Hypoglossal XII	Special sensation (taste) Ant. ⅔ — Facial VII; Post. ⅓ — Glossopharyngeal IX; Base — Vagus X	
Hyoglossus	Greater horn of hyoid	Tongue	Depresses and pulls tongue posteriorly			
Styloglossus	Styloid process	Tongue	Elevates and pulls tongue posteriorly			
Palatoglossus	Aponeurosis of soft palate	Tongue	Elevates and pulls tongue posteriorly: narrows fauces.	Pharyngeal plexus IX, X, XI		
SOFT PALATE						
Tensor veli palatini	Scaphoid fossa, spine of sphenoid, lateral auditory tube	Aponeurosis of soft palate	Tenses soft palate	Trigeminal V	Trigeminal V Glossopharyngeal IX	**Voluntary Stage** — The tongue depressor muscles contract and form a groove in the posterior portion of the dorsum of the tongue which cradles the bolus. A movement initiated by the intrinsic muscles raises the anterior portion and then the posterior portion of the tongue to the hard palate. This sequential movement dislodges the bolus and squeezes it toward the fauces. In turn the base of the tongue is elevated and pulled posteriorly mainly by the action of the styloglossi muscles forcing the bolus through the fauces into the pharynx. Occurring simultaneously with this elevation of the base of the tongue is a moderate elevation of the hyoid bone and the larynx.
Levator veli palatini	Petrous portion, temporal bone; medial auditory tube	Soft palate	Elevates soft palate	Pharyngeal plexus IX, X, XI		
Uvulae	Posterior nasal spine: aponeurosis of palate	Uvula	Shortens soft palate			
FAUCES						
Palatoglossus	See above					
Palatopharyngeus	Aponeurosis of soft palate	Posterior thyroid cartilage; Posterolateral pharynx	Narrows fauces: ; Elevates larynx & pharynx	Pharyngeal plexus IX, X, XI	Glossopharyngeal IX	
SUPRAHYOID						
Digastric Ant. belly	Inferior border of mandible near symphysis	Intermediate tendon to body and cornu of hyoid	Elevates and pulls hyoid anteriorly; Assist in depressing the mandible	Trigeminal V		**Involuntary (Reflex) Stage** — As the bolus passes through the fauces to the pharynx, branches of cranial nerves V, IX and X are stimulated producing impulses in the afferent limb of the swallow reflex. Upon reaching the brainstem, these impulses are transmitted across synapses to efferent fibers of cranial nerves IX, X and XI completing the reflex arc and effecting the following automatic events:
Post. belly	Mastoid process		Elevates and pulls hyoid posteriorly	Facial VII		
Mylohyoid	Mylohyoid line of mandible	Body of hyoid & median raphe	Elevates hyoid & tongue; depresses mandible	Trigeminal V		
Geniohyoid	Median ridge of mandible	Body of hyoid	Elevates hyoid & tongue; depresses mandible	Ansa cervicalis C1, 2		
Stylohyoid	Styloid process of temporal bone	Body of hyoid	Elevates and pulls hyoid posteriorly	Facial VII		

INFRAHYOID

Muscle	Origin	Insertion	Action	Nerve
Thyrohyoid	Oblique line of thyroid cartilage	Greater horn of hyoid	Elevates the thyroid cartilage; depresses the hyoid	Ansa cervicalis C1, 2
Sternohyoid	Manubrium sterni; medial end of clavicle	Body of hyoid, inf. border	Depresses hyoid	Ansa cervicalis C1, 2, 3
Sternothyroid	Manubrium sterni; costal cartilage of 1st rib	Oblique line of thyroid cartilage	Depresses thyroid cartilage	Ansa cervicalis C1, 2, 3
Omohyoid — Sup. belly	Superior border of scapula near scapular notch	Intermediate tendon, by fascia to clavicle		Ansa cervicalis C1, 2, 3
Omohyoid — Inf. belly	Intermediate tendon by fascia to clavicle	Body of hyoid, inf. border	Depresses the hyoid	Ansa cervicalis C1, 2, 3

The soft palate is elevated and brought into contact with the posterior pharyngeal wall by the contraction of the tensor and levator veli palatini muscles. This action closes off the nasopharynx ensuring passage of the bolus into the lumen of the laryngopharynx. This passage is facilitated when the lumen is expanded by the elevation of the pharyngeal wall and the cranial and anterior movement of the hyoid bone and the larynx. When the last of the bolus leaves the oral cavity, the oropharynx opening is closed by contraction of the palato-pharyngeal muscles and the descent of the soft palate.

LARYNX

Muscle	Origin	Insertion	Action	Nerve
Aryepiglottic	Apex of arytenoid cartilage	Lateral margin of epiglottis	Assists in closing inlet of larynx	Vagus X; Mainly accessory XI, cranial root
Thyroepiglottic	Medial surface of thyroid cartilage	Lateral margin of epiglottis	Assists in closing inlet of larynx	Vagus X; Mainly accessory XI, cranial root
Thyroarytenoid	Medial surface of thyroid cartilage	Muscular process of arytenoid cartilage	Assists in closing glottis; shortens vocal folds	Vagus X; Mainly accessory XI, cranial root
Arytenoid — Oblique	Base of one arytenoid cartilage	Apex of opposite arytenoid cartilage		Vagus X; Mainly accessory XI, cranial root
Arytenoid — Transverse	Posterior surface and lateral border of one arytenoid cartilage	Posterior surface and lateral border of opposite arytenoid cartilage	Assist in closing glottis by adducting arytenoid cartilages	Vagus X; Mainly accessory XI, cranial root
Lat. cricoarytenoid	Upper border of arch of cricoid cartilage	Muscular process of arytenoid cartilage	Adducts and medially rotates arytenoid cartilage assisting in closing glottis	Vagus X; Mainly accessory XI, cranial root
Vocalis	Medial surface of thyroid cartilage	Vocal process of arytenoid cartilage	Regulates tension of vocal folds	Vagus X; Mainly accessory XI, cranial root
Post. cricoarytenoid	Posterior surface of lamina of cricoid cartilage	Muscular process of arytenoid cartilage	Abducts arytenoid cartilage widening glottis	Vagus X; Mainly accessory XI, cranial root
Cricothyroid — Straight / Oblique	Anterior and lateral part of arch of cricoid cartilage	Anterior border, inferior horn of thyroid cartilage; Lower border of lamina of thyroid cartilage	Elevates cricoid arch and elongates vocal folds	Vagus X

The cranial movement of the thyroid cartilage toward the hyoid bone and of these two structures, in turn, toward the base of the tongue results in tilting the epiglottis posteriorly. The weight of the bolus as it contacts the anterior surface of the epiglottis assists in increasing this posterior tilt. The change of position of the epiglottis aids in directing the bolus material around the sides of the larynx through the piriform sinuses and over the tip of the epiglottis into the hypopharynx. It also aids in preventing foodstuffs from entering the larynx. The major mechanism for protecting the larynx, however, is the concurrent sphincter-like closure of the laryngeal inlet to the vestibule and the closure of the vestibular and vocal folds of the glottis.

PHARYNX

Muscle	Origin	Insertion	Action	Nerve
Salpingopharyngeus	Auditory tube	Pharyngeal wall	Elevates pharynx	Pharyngeal plexus IX, X, XI
Palatopharyngeus	See above	Pharyngeal wall	Elevates pharynx	Pharyngeal plexus IX, X, XI
Stylopharyngeus	Styloid process	Posterior border of thyroid cartilage; posterolateral wall of pharynx	Elevates pharynx and larynx	Glossopharyngeal IX
Superior constrictor	Medial pterygoid plate; pterygomandibular raphe; mandible	pharyngeal tubercle; pharyngeal raphe	Constrict, sequentially, nasopharynx,	Pharyngeal plexus IX, X, XI
Middle constrictor	Horns of hyoid	pharyngeal raphe	oropharynx	Pharyngeal plexus IX, X, XI
Inferior constrictor	Thyroid and cricoid cartilages	pharyngeal raphe	laryngopharynx	Pharyngeal plexus IX, X, XI
Cricopharyngeus	Arch of cricoid cartilage	Arch of cricoid cartilage	Acts as sphincter to prevent air entering esophagus; relaxes during swallowing	Pharyngeal plexus IX and X

Occurring simultaneously with the above events is a sequential contraction of the superior, middle and inferior constrictors which strips the pharynx forcing the bolus toward the esophagus. Horizontally oriented fibers found between the inferior constrictor and the esophagus have been named the cricopharyngeus muscle. This muscle acts as a sphincter and functionally is related more to the esophagus then to the pharynx. It relaxes when the bolus reaches the caudal extent of the hypopharynx permitting the foodstuff to enter the esophagus.

chapter 8

Muscle Function in Relation to Posture

Introduction

Effect of muscle imbalance on posture

Exercise programs

The standard posture

Ideal plumb alignment: side view

Faulty plumb alignment: side view

Ideal segmental alignment: side view

Faulty segmental alignment: side view

Faulty posture, side view: analysis and treatment

Ideal alignment: posterior view

Faulty alignment: posterior view

Handedness: Effect on posture

Scoliosis

Faulty posture, back view: analysis and treatment

Shoulders and scapulae

Faulty head and neck positions

Faulty head and shoulder positions: analysis and treatment

Good and faulty posture of feet, knees, and legs

Faulty leg, knee, and foot positions: analysis and treatment

Procedure for postural examination

Charts

Muscle Function in Relation to Posture

Posture is a composite of the positions of all the joints of the body at any given moment. If a position is habitual, there will be a correlation between alignment and muscle test findings. If a reasonable assessment of the joint positions is made, then an assessment also can be made regarding which muscles are in elongated positions and which are in shortened positions. In faulty posture, those muscles in slightly shortened positions tend to be stronger, and those in slightly elongated positions tend to be weaker than the muscles that work in opposition to them. To help portray and analyze the concepts related to joint positions and muscle function, this chapter provides skeletal and muscle drawings along with photographs of good and faulty postures. In addition, useful exercises are recommended, along with supports and shoe alterations, as indicated on the basis of *tests for alignment and tests for muscle length and strength.* Together, these tests constitute postural analysis.

In Chapters 4, 5, and 6, muscle testing and function of extremity and trunk muscles is described in detail. For those muscles in which weakness or shortness has a direct effect on postural alignment, the effect is noted in the text but not illustrated. The purpose of Chapter 8 is to take much of the information contained in other chapters regarding the effect on posture and put it together, with illustrations, in a way that will assist an examiner in evaluating and treating postural problems.

While the alignment of the body in a standing position can indicate specific muscle problems, a definitive assessment should not be made on the basis of the disturbance of alignment alone. Specific muscle tests as described elsewhere in this text are necessary to determine the extent of weakness or shortness.

EFFECT OF MUSCLE IMBALANCE ON POSTURE

Actions of muscles may be demonstrated by muscle testing movements, by electrically stimulating muscles, and by ascertaining the presence of electrical activity generated within muscles during movements in which the muscles are contracting. However, some of the most dramatic evidence of muscle function comes from observing the effects of loss of ability to contract as seen in paralyzed muscles, or the effect of excessive shortening as seen in a muscle contracture. Both paralysis and contractures lead to loss of movement, loss of stability, and deformities. Between these extremes are varying degrees of weakness and muscle shortness which affect the alignment of the body segments and the posture of the body as a whole. Weakness of a muscle allows separation of the parts to which the muscle is attached because the strength to maintain good alignment is lacking. Muscle shortness holds the parts to which the muscle is attached closer, and prevents the parts from returning to a position of good alignment.

Muscle weakness or shortness may cause faulty alignment, and faulty alignment may give rise to *stretch-weakness* or *adaptive shortness* of muscles. The appearance of the fault is the same in either case, making it impossible to distinguish cause and effect when dealing with established postural faults. Stretch-weakness may be defined as the effect on muscles of *remaining* in a lengthened condition, however slight, beyond the neutral (physiological rest) position. The concept is related to the duration of the faulty alignment rather than to the severity of it. It does not refer to *overstretch* which means beyond the range afforded by muscle length. In standing, the ideal alignment may be taken as the neutral position. Persistent postural deviation from this alignment may result in stretch-weakness. Such weakness frequently is found in the middle and lower Trapezius muscles in persons with kyphosis and forward shoulders, and in hip abductor muscles on the side where the hip is high or prominent.

Important to an understanding of adaptive shortening of muscles is this basic physiological concept as stated by Ralston,[31] "After a muscle has been caused to shorten by stimulation, there is no appreciable spontaneous lengthening of the muscle during relaxation. Muscles are caused to lengthen in the intact body by the pull of antagonistic muscles, by the action of gravity and the like. The lengthening of inactive muscle is a passive, not an active process." Consequently, unless the opposing muscle is able to pull the part back to neutral position, or some outside force is exerted to lengthen the short muscle, there will be a tendency for the shortened muscle to remain in a somewhat shortened condition.

In muscle problems associated with habitual faulty posture, *weakness* and *stretch* are so closely related that, interchangeably, they represent cause and effect. On the other hand, muscle *shortness* is invariably associated with muscle *strength*. The relationship of muscle strength with shortness, and muscle weakness with stretch is nowhere more apparent than in the distribution of these findings in handedness patterns. The postural deviation may appear to be slight, the corresponding changes in the postural positions of the joints may appear to be minor, but the muscles will show significant differences related to the habitual postural position. (See pp. 291 and 294.)

When tested for strength, a muscle may be only slightly weak. However, the alignment problem associated with the weakness often provides more substantial evidence of the functional importance than the muscle test. Weakness or shortness of neck muscles, of upper back and shoulder girdle muscles, of lower back and abdominal muscles, and all lower extremity muscles can directly affect the postural alignment of body segments and the body as a whole. For example, weakness of the muscles that pull the scapulae (shoulder blades) toward the spine may result in a position of abducted scapulae (forward shoulders). Short pectoral muscles will hold the shoulders forward. Weakness of muscles that hold the upper back straight may result in a kyphosis (round upper back). Weakness of abdominal muscles may result in forward pelvic tilt and a low back lordotic ("hollow-back") position. Short Iliopsoas (hip flexor) muscles will hold the low back in a lordotic position and the pelvis in a position of forward tilt in the standing posture.

Pain in relation to faulty posture. That faulty posture can cause painful conditions is a generally accepted concept in the health care field. There are, however, those who are dubious. They often ask the very pertinent question why it is that many cases of faulty posture exist without symptoms of pain. They also question why seemingly mild postural defects give rise to symptoms of mechanical and muscular strain. The answer to both of these questions depends on the constancy of the fault. A posture may appear to be very faulty, yet the individual may be flexible and the position of the body may change readily. A posture may appear to be good, but the stiffness or tightness that is present may so limit mobility that position cannot be changed easily. The lack of mobility, which is not apparent as an alignment fault but is observed only in tests for flexibility, may be the more significant fault.

Basic to an understanding of pain in faulty body mechanics is the concept that the cumulative effects of constant or repeated small stresses over a long period of time can give rise to the same difficulties as a sudden severe stress.

Cases of postural pain are extremely variable in the manner of onset, in the severity of symptoms, and in the nature of associated faulty mechanics. There are cases in which only acute symptoms appear, usually as a result of some unusual stress or injury, and in which there are no predisposing faulty body mechanics. Some cases have an acute onset and develop chronically painful symptoms. Some exhibit chronic symptoms which later become acute; others remain chronic.

Symptoms associated with an acute onset are often widespread and the presence of severe pain makes it impossible to do a detailed analysis of the faulty mechanics. Only after acute symptoms have subsided can tests for underlying faults in alignment and muscle balance be done and specific therapeutic measures be instituted.

Effects of faulty alignment and lack of mobility on skeletal structures. From a mechanical standpoint, faults in alignment and mobility create two types of problems: Undue compression on articulating surfaces of bone; and undue tension on bones, ligaments, or muscles. Eventually two types of bony changes may occur. Undue compression may result in a "wearing away" of the articulating surface, while undue traction may result in an increase in bony growth at the point of attachment.

It is a fault in alignment if a deviation is persistent or if it is severe.

It is a fault in mobility if motion is restricted, or if it is excessive.

Persistence of faulty alignment results in undue compression on the parts of the articulating surfaces which bear constant or repeated stress. The ability to tolerate ordinary stresses decreases as the degree or duration of the fault increases. As a fault in alignment progresses toward the extreme, there is added to the factor of undue compression on bony structure the factor of undue tension on ligaments, or by ligaments on bone.

If a postural deviation exceeds the limit of motion permitted by the articulating surfaces, it is at once a fault, whether it be momentary or persistent. When such a fault occurs, it is usually responsible for a sudden onset of acute pain. An abnormal deviation usually is labeled as a slipping, subluxation, luxation, or dislocation. These terms, although correctly applied to some faults in alignment, are often inappropriately used.

Lack of mobility is closely associated with persistent faulty alignment as a factor in causing undue compression. When mobility is lost, there is stiffness and a certain alignment remains constant. The lack of mobility may be due to restriction of motion by *short muscles* or due to the inability of *weak muscles* to move the part through the arc of motion. Muscle shortness is a constant factor tending to maintain the part in faulty alignment regardless of the position of the body. Muscle weakness is a less constant factor because changing the body position can bring about a change in alignment of the part. When there is normal movement in joints, wear and tear on joint surfaces tend to be distributed. If there is limitation of range, the wear will take place only on the joint surfaces that represent the "arc of use."

If limitation of range or stiffness occurs with the part in *good alignment* for weight-bearing, there may be a minimum of pain or disability in standing or sitting, but muscle strain may occur on motion. If the part that is restricted by muscle shortness is protected against any movement that might cause strain, other parts that must compensate for such restriction may suffer the strain instead.

When limitation of motion holds the body in *faulty alignment*, the strain and painful symptoms will tend to be constant whether the individual is quiet or in motion. There is also the factor of undue compression on the margins of the articulating surfaces in cases in which the excessive range is maintained for a prolonged period of time.

Excessive joint mobility results in stretch of the ligaments which normally limit the range of motion. Furthermore, according to Steindler,[26] "Repeated forceful stretching of ligaments impairs their elasticity and permanent lengthening and relaxation results."

EXERCISE PROGRAMS

To maintain a position of good alignment, there must be adequate muscle balance between opposing sets of muscles. Muscle imbalance occurs when muscles are exercised persistently to develop strength, and opposing muscles are not also strengthened; or if muscles remain habitually in a shortened position while the opposing muscles remain lengthened.

Exercises which strengthen weak muscles, and exercises that stretch short muscles, are the means by which muscle balance is restored. Programs of exercise should not include those that increase the strength of short, strong muscles, nor those which put undue load or stretch on already stretched, weak muscles.

Most exercise programs aim to improve flexibility and muscle strength, and to increase endurance. Programs must have, also, the restoration of muscle balance as a major objective. A hand surgeon, operating to transplant muscles for the restoration of function, knows that it is more important to obtain muscle balance between opposing involved muscles than to try to secure maximum strength in a specific muscle or muscle group. On a much larger scale, but no less important, is the need to apply this principle in planning exercise programs for the prevention or correction of faulty posture.

It is not enough that a weak muscle be exercised in a manner that will specifically bring it into action; it must be combined with the actions of other muscles in a way that will be therapeutic for the individual. Take as an example, an individual with a kyphosis-lordosis posture (see p. 281) who needs strengthening exercises for the upper back extensors and for the lower abdominal muscles and who should avoid exercise of the low back. The upper back extensors could be exercised by lying prone (face-lying) and raising the head and shoulders up backwards, but the low back muscles would be exercised also because they stabilize the low back in extension during this movement. A curled-trunk sit-up, properly done, would exercise certain abdominal muscles. However, this is a strong hip flexor exercise (whether knees and hips are flexed or extended, see pp. 203 and 204) and the effect of the exercise would be to increase the lordosis. In addition, curling the trunk can increase a tendency toward kyphosis, and, therefore, should be avoided in this type posture. It is just as important to omit undesirable exercises as to include the needed ones. Furthermore, the abdominal muscles involved in trunk-raising are often not weak in a kyphosis-lordosis posture and do not need strengthening exercises, while those involved in posterior pelvic tilt do need to be strengthened.

A simple but unglamorous type of exercise that is therapeutic for the individual with the kyphosis-lordosis posture is the "wall-standing" exercise. (See p. 289.) The upper back and lower abdominal muscles are specifically strengthened, and the low back and hip flexor muscles are elongated by the flattening of the low back and the posterior tilting of the pelvis. By doing the exercise in this manner, the specific exercises needed are combined in a way that makes the overall exercise therapeutic for the individual. Muscle balance involves restoration of both normal strength and normal length. Once restored, an individual may participate in a "well-rounded" program of exercises provided the program is also "well-balanced." Some exercises are neither specific nor therapeutic for the purpose intended. Double leg-raising, as an exercise intended to strengthen abdominal muscles that are very weak, falls into this category. (See p. 213.)

Occupational and recreational activities may be regarded as forms of exercise repeated frequently and persisted in for some time. In dealing with faulty body mechanics, it is important to be cognizant of the role of these activities and to analyze them in prescribing treatment. When a postural defect appears to be related to occupational or recreational habits, the duration or the intensity of the pattern must be altered in some way in order to treat the condition effectively.

Whether adjustment of the occupational factors alone will alleviate the painful symptoms depends primarily on whether the pattern of activity or position has become a fixed pattern in the individual in the form of muscle tightness or weakness. Usually a condition which has reached the painful stage is one in which some muscle imbalance is present, and specific treatment should be given to the individual to strengthen the weak muscles and stretch the tight muscles.

Stretching movements must be done gradually to avoid damage to tissue structures. Tightness that has occurred over a period of time must be given a reasonable time for correction. A period of several weeks is usually necessary for restoration of mobility in muscles exhibiting moderate tightness.

Treatment of muscle weakness requires consideration of the factors of muscle-stretch and disuse that are the underlying faults. In cases of faulty body mechanics, there are numerous instances of muscle stretch-weakness while the element of disuse atrophy is much less common.

To some degree, the lack of opportunity to use the muscles because of occupational restrictions, the lack of necessity to use them to capacity, or the state of semi-immobilization imposed by tightness of an opposing muscle results in weakness from disuse. This type of weakness is usually superimposed on the same muscles affected by stretch-weakness. The muscles that are most often affected by stretch and disuse weakness are the abdominals, upper back erector spinae, middle and lower trapezius, and anterior vertebral neck flexors.

Stretch-weakness, being the result of persistent tension on the muscle, must be treated by relief of tension. Realignment of the part, bringing it into a neutral position, is the essence of this treatment. Use of supportive measures to help restore and maintain such alignment until weak muscles recover strength is a most important factor in treatment. Any opposing tightness which tends to hold the part out of alignment must be corrected in order to relieve tension on weak muscles. Faulty occupational positions which impose continuous tension on certain muscles must also be adjusted or corrected.

When the tension on the weak muscle has been relieved, exercises to help restore strength are indicated. Care must be taken not to overwork a muscle which has been subjected to a prolonged tension stress. In practical application, exercises usually begin with the use of three or four repetitions, increasing to six or eight by the end of ten days or two weeks. It is seldom advocated that specific exercises be done more than once a day, but as soon as the strain is relieved, and the part capable of good realignment, the patient is expected to use the muscles in their normal function of maintaining good alignment. He should know that it is important to try to maintain good alignment *most of the time.* When a faulty position is assumed more often than a good one, the posture tends to become faulty.

THE STANDARD POSTURE

An evaluation of postural faults necessitates a standard by which individual postures can be judged. The alignment used as standard must be consistent with sound scientific principles. It should be the kind of posture which involves a minimal amount of stress and strain and which is conducive to maximal efficiency in the use of the body. The standard must meet these requirements if the whole system of posture training that is built around it is to be sound.

The *Standard Posture* as used and described in this text refers to an "ideal" posture rather than an average posture. It is important that the reader visualize this standard as a basis for comparison, since good alignment or postural faults will be described in terms of this standard.

The standard is one of *skeletal alignment* because posture is basically a matter of alignment. The body contour in the illustrations of the standard posture is included because it shows the relationship of skeletal structures to surface outline when alignment is excellent. It is recognized that there are variations in body type and size among individuals, and that shape and proportions of the body are factors in weight distribution. Nevertheless, the authors believe that a standard of skeletal alignment can be considered a valid standard for evaluating the posture of any individual regardless of body type or size. The importance of body contour lies chiefly in the fact that variations in contour are correlated to some degree with variations in skeletal alignment. Fortunately, this is essentially true of everyone regardless of body build. Thus, an experienced observer is able, by observing the contours of the body, to estimate the position of the skeletal structures.

The standing position may be regarded as the composite alignment of a subject from four views, front, back, right side, and left side. It involves the position and alignment of so many joints and parts of the body that it is not probable that any individual can meet this standard in every respect. As a matter of fact, the authors have not seen an individual who matches the standard in all respects.

The standard posture is illustrated from front, back, and side by line drawings. In the drawings showing the front and back views of the body, the vertical *line of reference* represents a plane which coincides with the midline of the body. It is illustrated as beginning midway between the heels and extending upward midway between the lower extremities, through the midline of the pelvis, spine, sternum, and skull. The right and left halves of the skeletal structures are essentially symmetrical and by hypothesis the two halves of the body exactly counterbalance. (See p. 290.)

In the side view drawing, the vertical *line of reference* represents a plane which hypothetically divides the body into front and back sections of equal weight. These sections are not symmetrical and there is no obvious line of division on the basis of anatomical structures. It is necessary, therefore, to describe the relationship of the body to this plane on the basis of the mechanical and physiological factors involved.

The intersection of these two midplanes of the body forms a line which is analogous to the gravity line. Around this line, the body is hypothetically in a position of equilibrium. This implies a balanced distribution of weight, and a stable position of each joint. Along the course of each line of reference are certain *points of reference*. In the drawings of the standard posture these are shown as deep structures. (See p. 278.)

The lines and points of reference discussed above are put to practical use in *plumb line tests* for postural alignment. For the purpose of testing, the subject steps up to a suspended plumb line. In back or front view he stands so that the feet are equidistant from the line; in side view, so that the point just in front of the lateral malleolus (in adults about an inch in front of the center of the malleolus) is in line with the plumb line. Thus, the *base* points of reference indicated on the drawings of the standard posture are the points in line with which the plumb

line is suspended. It is necessary that the base point be the fixed reference point because the base is the only stationary or fixed part of the standing posture.

A plumb line test is used to determine whether the *points of reference* of the individual being tested are in the same alignment as are the corresponding points in the standard posture. The amount of deviation of the various points of reference from the plumb line reveal the extent to which the subject's alignment is faulty. Whenever deviations from plumb alignment are evaluated they are described as slight, moderate, or marked, rather than in terms of inches or degrees. Because of the fact previously noted that, in a standing position, only the base of the body can be considered stationary, it would be extremely difficult and of no practical value to try to determine more exactly how much each point of reference had deviated from plumb alignment.

In the following pages are drawings of the standard posture showing the deep skeletal structures, and the legends indicate which ones coincide with the points of reference. For comparison, beside each drawing is a photograph showing a subject whose alignment approaches closely that of the standard posture.

The detailed explanations which follow are as simple as possible. A section is written with respect to each part of the body for the purpose of presenting what seems to be a reasonable and logical explanation of ideal alignment.

Lower extremity. As illustrated in the following pages, the line of reference in side view passes slightly in front of the ankle joint, slightly in front of the center of the knee joint, and slightly behind the center of the hip joint. The forward curve in the femur permits this relationship of the knee and hip joints to the line of reference.

Since there is no obvious division of the body into anterior and posterior portions, the conclusions regarding the relationship of the line of reference to the body are based on an analysis of what constitutes a stable position of the joints of the lower extremity in standing.

The bony structures of the human body are designed to support and transmit superimposed weight. From a mechanical standpoint, it might be logical to assume that a line of gravity should pass through the centers of the weight-bearing joints of the extremities. It might also be logical to assume that it should pass through the apex of the arch at the base of support. However, from the standpoint of body mechanics, an "on-center" position is not a stable one. It can be held only momentarily in the presence of normal external stresses. The gravity line in a "normal" posture has been described by some[26] as coinciding with the centers of these joints, but they also acknowledge that the position can be held only momentarily.

The normal limitation of joint motion in certain directions has a postural significance in relation to the stability of the body in the standing position. Dorsiflexion at the ankle with the knee straight is normally about 10° to 15°. This means that standing barefoot, with feet nearly parallel, the lower leg does not sway forward on the foot more than about 10°. The knee joint has a few degrees of extension. In standing, then, the femur and lower leg relationship will permit a few degrees of postural deviation backward. The hip joint has about 10° hyperextension and, in standing, the joint motion of the pelvis on the femur is restricted to about 10° of postural deviation forward.

When the center of the knee joint coincides with the plane through the gravity line there is equal tendency for the joint to flex or to hyperextend. The slightest force exerted in either direction will cause it to move off-center. If the body must call on muscular effort at all times to resist knee flexion, for example, there is an unnecessary expenditure of muscular effort. To offset this necessity, the center of the knee-joint moves slightly behind the line of gravity for stability in standing. Similarly, if the knee joint moved freely backward, the position of slight extension could not be maintained, without constant muscular effort. However, ligamentous structures and strong muscles with tendons that help reinforce the ligaments are the restraining force preventing hyperextension. The thrust of body weight above helps to extend the knee joint, and ligamentous structures accept the function of counteracting this thrust, all with a minimum of muscular effort.

At the hip joint, the same principles apply but it is forward deviation at this joint that is limited as it is a backward deviation at the knee. The hip joint is most stable when the center of this joint is slightly in front of the gravity line. The strong ligaments that cross the hip joint anteriorly restrict additional forward sway of the pelvis.

There should be careful scrutiny of exercises or manipulations which allow hyperextension of the knee or hip joint, or which excessively stretch such muscles as Hamstrings. The normal restraining influence of the ligaments and muscles helps to maintain good postural alignment with a minimum of muscular effort. When muscles and ligaments fail to offer adequate support, the joints exceed their normal range and posture becomes faulty with respect to positions of knee and hip hyperextension.

At the ankle, the line of gravity passes slightly in front of the joint, and the slightly forward deviation of the body is checked by the restraining tension of strong posterior muscles and ligaments. In this relationship to the ankle joint, the line of gravity passes approximately through the apex of the arch, designated laterally by the calcaneo-cuboid joint.

Pelvis. Without minimizing the importance of proper foot positions which establish the base of support, it may be said that the position of the pelvis is the key to good or faulty postural alignment. The muscles that maintain good alignment of the pelvis, both anteroposteriorly and laterally, are of utmost importance in maintaining good overall alignment. Imbalance between muscles that oppose each other in standing changes the alignment of the pelvis and adversely affects the posture of the parts of the body above and below.

In the side view drawing of the standard posture, the pelvis is a composite of male and female pelves, and shows an average in regard to shape, length of sacrum, coccyx, and other measurements.

The relation of the pelvis to the line of reference is determined to a great extent by the relationship of this structure to the hip joint. Since the side-view line of reference represents the plane passing slightly behind the hip joint, the pelvis will be intersected through the acetabulum. It is necessary to state what is considered a neutral position of the pelvis in describing the standard posture, and this one point of reference through the pelvis is not sufficient because the pelvis tilts forward or backward about the axis that passes through the hip joints.

Because of structural variations of the pelvis, it appears almost impossible to say that any designated point anteriorly is on the same horizontal plane as a posterior landmark when the pelvis is in a neutral position.

The neutral position used as standard by some anatomists[22, 27] is that in which the anterior superior spines and the symphysis pubis are in the same vertical plane. To the authors of this text, such a position of the pelvis is most acceptable as a neutral position in regard to the standard alignment in the upright posture. It is logical from the standpoint of the action of muscles attached to these two points. The Rectus abdominis with its attachment on the pubis extends upward to the sternum, and the Rectus femoris, Sartorius, and the Tensor fasciae latae with their attachments on the anterior superior spines extend downward to the thigh. In the neutral position, these opposing groups of muscles have essentially an equal mechanical advantage in a straight line of pull.

Feet—a consideration of whether or not they should be parallel. In the standard posture, the position of the feet is given as one in which the heels are separated about three inches, and the forepart of each foot is abducted about 8° to 10° from the midline.

This position of the feet refers only to the static and barefoot position. Elevation of the heels and motion affect the foot position. Since the usual standing position for the average individual is one in which the heels are elevated to some degree, it is also important to understand the foot position in relation to heel height. A rather detailed analysis follows. The conclusion drawn from the material presented is that a parallel position of the feet is indicated for the average individual who wears an average heel.

Much has been written about the angle of the feet in relation to the standing position and to gait. There is a wealth of information available regarding the mechanics of the foot and the relationship to other joints of the lower extremity. However, the interpretation of the mechanics in terms of postural significance seems inadequate.

A degree of out-toeing used to be considered correct for both standing and walking. Various authorities, in the field of orthopedic surgery especially, have observed that such a position of the feet is conducive to a weakening of the structures of the foot, and have advised that the feet should be parallel in walking. Some advise, also, that the feet should be parallel in the standing position, but there is more variance in this opinion.

Morton,[28] who is considered by many as an outstanding authority on the foot, advocates, however, a degree of out-toeing. He states that "Physiological analysis indicates that the most natural position of the feet in standing is that in which the heels are together while the forepart of the feet are separated sufficiently to give proper security to the lateral balance of the body. Thus the angle of out-toeing should be about 30° to 40° in order that the width of the area of foot contact be equal to its length. This angle is diminished during locomotion."

On the assumption that the most natural position of the feet in standing is with heels together, the argument that there should be out-toeing for lateral stability is convincing. The authors question the basic assumption and regard the natural standing position as one in which the feet are separated at the heel as well as at the forepart of the foot. Stability is increased when the base is square rather than tripod. Though separation at the front of the feet with heels together may result in the addition of lateral stability, it also results in the loss (proportionately to the angle of separation) of stability in the forward and backward direction. Laterally there are two bases of support so that in any tendency to sway off balance either may assume full support. In the anteroposterior direction there is, functionally, one base of support and it is not logical to decrease its efficiency in regard to balance as occurs with the tripod base position.

To further establish the conclusions drawn regarding the position of the feet it is necessary to consider the relationship of the foot to the rest of the lower extremity with particular reference to where, if at all, rotation should occur. It should not occur at the knee joint. On the basis of anatomical construction of the knee joint, there is no rotation in extension. It may be assumed that the axis of the extended knee joint is in a frontal plane, and that it should be so from the standpoint of forward progression as in walking.

Assuming this relationship of the knee joint to the frontal plane, the possibility that outward deviation of the foot should take place from hip joint level is eliminated since rotation of the hip would affect the knee.

This makes the question of whether there should be rotation of the foot into an out-toe position dependent on an analysis of the relationship of the foot and ankle to the lower leg. As a movement, rotation is not present in the ankle joint. In regard to position, the question is whether the ankle joint, like the knee joint, is in a frontal plane. The answer, according to anatomists, is that it is not but that it is slightly oblique. The line of obliquity is such that it extends from slightly anterior at the medial malleolus to slightly posterior at the lateral malleolus. The angle at which the axis of the ankle joint deviates from the frontal plane suggests that the foot is normally in a position of slight abduction (or out-toeing) in relation to the lower leg. But the position of the axis of the joint also suggests that movement about this axis would result in the foot moving inward as it moved upward, and outward as it moved downward. Functionally, this does not occur.

Passive or active movements of the foot reveal that the foot tends to abduct and evert in full dorsiflexion, and that the range of motion in plantar flexion is more free in adduction than in abduction. In other words, the foot tends to move *out* as it moves upward, and *in* as it moves downward, the reverse of the direction suggested by the axis of ankle joint motion.

The answer to much of the discussion regarding proper foot position in standing and walking revolves, then, about the following:

1. What is the anatomical relationship of the foot to the leg in standing barefoot as compared with standing in shoes with heels?
2. How does speed of progression affect foot position?

The relationship of the foot to the lower leg must be analyzed with regard to the arc of motion and the position within that arc of motion which is the equivalent of the upright position, with and without heels.

According to Whitman,[29] "Extreme abduction is attained in the attitude of dorsal flexion, its extent being about one-half that of adduction; the entire range of motion between the two extremes being about 45°." If this figure is acceptable, the degree of abduction of each foot in the fully dorsiflexed position is about 15° degrees and that of adduction about 30° degrees in the plantar flexed position. If an angle of 30° to 40° between the feet were permitted in the standing position, the feet would have assumed the position of maximum abduction. The movement of abduction is so closely related to pronation (or eversion) that many authorities say these terms may be considered synonymous. This being true, the question of what constitutes a good standing position in relation to the degree of abduction must necessarily involve a consideration of the degree of pronation that is advisable.

The foot in full abduction actually represents, in the weight-bearing position, a position of potentially full pronation. Since the standing position is not one in which the foot is fully dorsiflexed on the lower leg, it does not follow that a position of full abduction is anatomically correct. From the standpoint of stress and strain on the foot itself, a position which permits full pronation is most inadvisable.

As abduction and pronation are related, so also are adduction and supination (or inversion). As abduction is normally related to dorsiflexion, so adduction is closely related to plantar flexion.

It is difficult, if not impossible, to determine exactly what degree of abduction or adduction of the foot corresponds with each degree of dorsal or plantar flexion. The two are not necessarily so correlated that an exact relationship exists, but it may be assumed that the movement from abduction in the dorsiflexed position to adduction in the plantar flexed position is relatively uniform.

When wearing shoes with heels, the standing position represents varying degrees of plantar flexion of the foot based on the heel height as related to foot length, or more precisely, to the length of the foot from the heel to the ball of the foot. It would be futile to make an effort to compute the degree of out-toeing that corresponds to each heel height according to foot length.

It should be sufficient to know that as the heel height is increased the tendency toward parallel position or in-toeing increases.

The relationship of heel height to out-toeing or in-toeing of the foot is interestingly analogous to the relationship of the foot to standing, walking, and running. Standing barefoot, a slight degree of out-toeing is natural. Standing with heels raised or walking fast, the feet tend to become parallel. In running, the heels do not contact the ground, and the weight is borne on the anterior part of the foot entirely. There is then a tendency for the print of the forefoot to show in-toeing.

Ideal Plumb Alignment: Side View

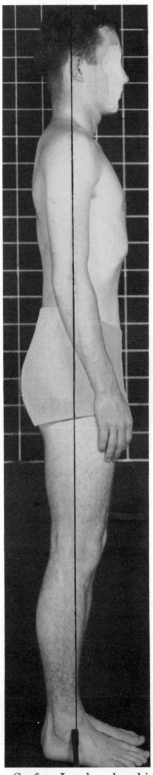

Slightly posterior to the apex of the coronal suture.

Through the lobe of the ear. (Head is slightly forward.)

Through the external auditory meatus.
Through the odontoid process of the axis.
Through the bodies of the cervical vertebrae.

Through the shoulder joint (providing the arms hang in normal alignment in relation to the thorax).

Approximately midway through the trunk.

Through the bodies of the lumbar vertebrae.

Through the sacral promontory.

Slightly posterior to the center of the hip joint.
Approximately through the greater trochanter of the femur.

Slightly anterior to the center of the knee joint.
Slightly anterior to a midline through the knee.

Through the calcaneo-cuboid joint.
Slightly anterior to the lateral malleolus.

Surface Landmarks which Coincide with the Plumb Line. (This subject shows excellent alignment except that the head is slightly forward.)

Anatomical Structures which Coincide with the Line of Reference

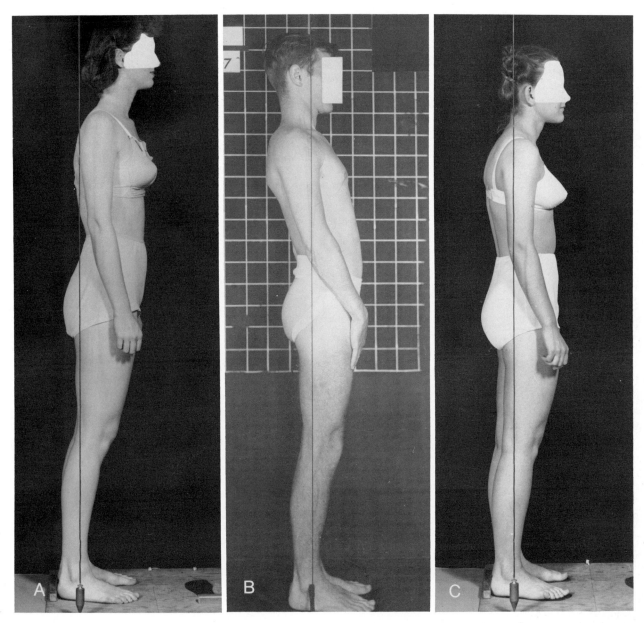

Fig. A shows a marked anterior deviation of the body in relation to the plumb line with body weight carried forward over the balls of the feet. Subjects who stand this way habitually may exhibit strain on the anterior part of the foot with calluses under the ball of the foot and even under the great toe. Metatarsal arch supports may be indicated along with correction of the overall alignment. The ankle joint is in slight dorsiflexion because of the forward inclination of the leg.

Posterior muscles of the trunk and lower extremities tend to remain in a state of constant contraction and the alignment must be corrected to achieve relaxation of these muscles effectively.

This type of faulty alignment is seen most frequently among tall slender individuals.

Fig. B shows a marked posterior deviation of the upper trunk and head. The knees and pelvis are displaced anteriorly to counterbalance the posterior thrust of the upper part of the body.

Fig. C shows a counterclockwise rotation of the body from the ankles to the cervical region.

The deviation of the body from the plumb line appears different from the right and left sides in subjects which have such rotation. The body is anterior from the plumb line as seen from the right, but would show fairly good alignment from the left. From both sides, the head would appear forward.

Ideal Segmental Alignment: Side View

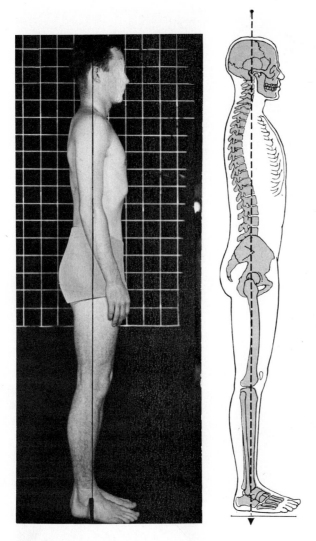

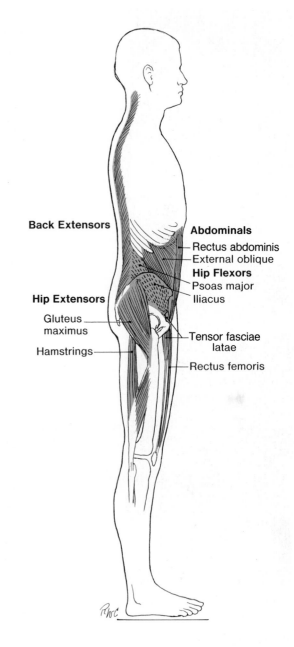

Head: Neutral position, not tilted forward or back. (Slightly forward in photograph.)

Cervical spine: Normal curve, slightly convex anteriorly.

Scapulae: As seen in the photograph, appear to be in good alignment, flat against upper back.

Thoracic spine: Normal curve, slightly convex posteriorly.

Lumbar spine: Normal curve, slightly convex anteriorly.

Pelvis: Neutral position, anterior superior spines in same vertical plane as symphysis pubis.

Hip joints: Neutral position, neither flexed nor extended.

Knee joints: Neutral position, neither flexed nor hyperextended.

Ankle joints: Neutral position, leg vertical and at right angle to sole of foot.

In lateral view, the anterior and posterior muscles attached to the pelvis maintain it in ideal alignment. Anteriorly, the abdominal muscles pull upward and the hip flexors pull downward; posteriorly, the back muscles pull upward and the hip extensors pull downward. Thus, the anterior abdominal and hip extensor muscles work together to tilt the pelvis posteriorly; the low back and hip flexor muscles work together to tilt the pelvis anteriorly.

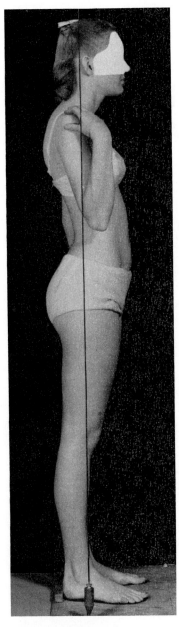

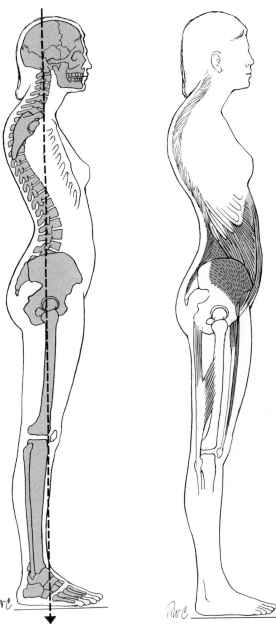

Head: Forward.

Cervical spine: Hyperextended.

Scapulae: Abducted.

Thoracic spine: Increased flexion (kyphosis).

Lumbar spine: Hyperextended (lordosis).

Pelvis: Anterior tilt.

Hip joints: Flexed.

Knee joints: Slightly hyperextended.

Ankle joints: Slight plantar flexion because of the backward inclination of the leg.

In a kyphosis-lordosis posture with forward head, weakness usually exists in the anterior neck and upper back muscles and the muscles of the lower abdomen. Muscle shortness frequently is present in the hip flexors and may or may not be present in the low back.

Both the lordotic posture in standing and the sitting posture place the one-joint hip flexors in a shortened position. However, sitting allows the low back muscles to elongate as the back flattens. Undoubtedly this combination of circumstances has a bearing on the fact that low back muscle shortness is less prevalent than hip flexor shortness in this type of posture.

"Military-Type" Posture

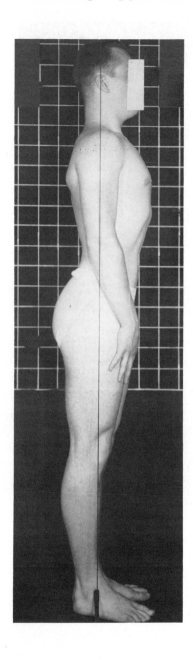 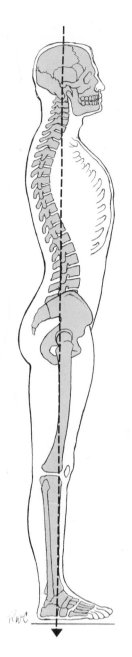 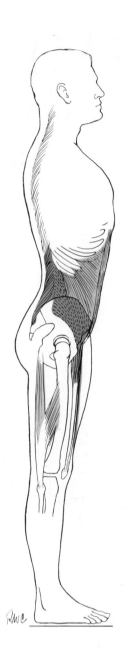

Head: Slight posterior tilt.

Cervical spine: Normal curve, slightly anterior.

Thoracic spine: Normal curve, slightly posterior.

Lumbar spine: Hyperextended (lordosis).

Pelvis: Anterior tilt.

Knee joints: Slightly hyperextended.

Ankle joints: Slightly plantar flexed.

In this "military-type posture", the chest is elevated and the pelvis is tilted forward putting the Rectus abdominis in a somewhat elongated position. The low back and hip flexor muscles are in a shortened position. Due to the anterior pelvic tilt, the Hamstrings are in a slightly longer position than in ideal alignment.

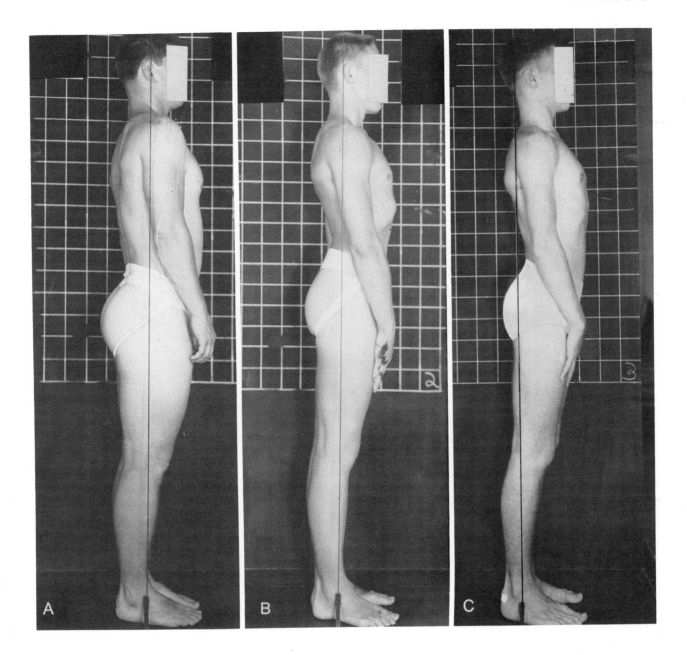

Fig. A shows a marked anterior pelvic tilt and a curve that is sharply convex forward in the lumbosacral area. This degree of tilt and lordosis is often associated with marked shortness of the Iliopsoas (hip flexor) muscles. There is slight counterclockwise rotation of the pelvis and trunk.

Fig. B shows a high and rather marked lordosis. The lumbar spine is inclined forward to the level of about the second lumbar vertebra. Above this level, there is a sharp deviation backward. This type of posture suggests weakness of the anterior abdominal muscles and shortness of the hip flexors.

Fig. C shows an anterior deviation from the plumb line in addition to a marked anterior pelvic tilt and lordosis. This forward deviation from the plumb line compounds the problem of muscle imbalance associated with the segmental alignment faults, and puts strain on the forefoot. (Note the difference in appearance of the feet in fig. C compared to A and B. See also fig. B, p. 293.)

Sway-Back Posture

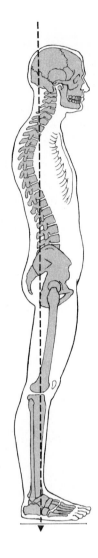

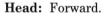

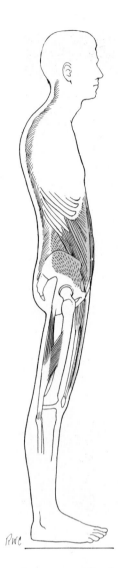

Head: Forward.

Cervical spine: Slightly extended.

Thoracic spine: Increased flexion (long kyphosis) with posterior displacement of upper trunk.

Lumbar spine: Increased flexion with flattening of low lumbar area.

Pelvis: Posterior tilt.

Hip joints: Hyperextended.

Knee joints: Hyperextended.

Ankle joints: Neutral. Knee joint hyperextension usually results in plantar flexion of the ankle joint but does not occur here because of anterior deviation of the pelvis and thighs.

The pelvis and upper end of the thighs sway forward in relation to the feet, which remain stationary. This results in hyperextension of the hip joint in the same manner as does extending the leg backward when the pelvis is fixed. The pelvis tilts posteriorly. There is no increased anterior curve in the lumbar spine, and hence there is no lordosis. The long curve in the thoracolumbar region which is due to the backward deviation of the upper trunk is sometimes mistakenly referred to as a lordosis in this type of posture. The term "sway back" is more appropriately applied to this type of posture.

This position places stretch on the anterior hip joint ligaments, the one-joint hip flexor muscles, and the External oblique abdominal muscles. These muscle groups usually show stretch-weakness. The lower back and Hamstring muscles are likely to be strong and somewhat short. In the upper trunk, the opposite conditions prevail. The upper back muscles show stretch weakness and the "upper" abdominals are likely to be strong.

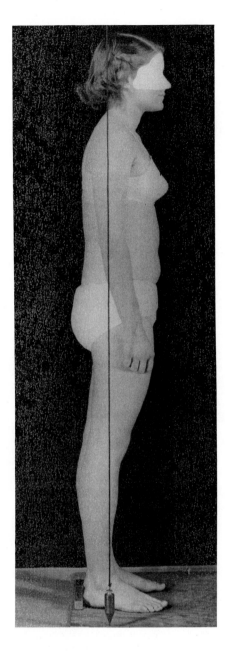 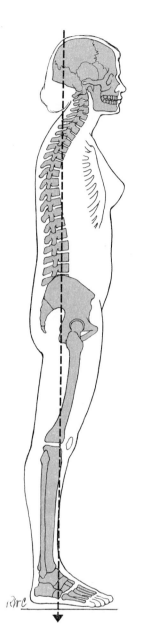 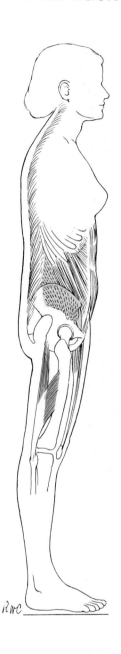

Head: Forward.

Cervical spine: Slightly extended.

Thoracic spine: Upper part increased flexion; lower part, straight.

Lumbar spine: Flexed (straight).

Pelvis: Posterior tilt.

Hip joints: Extended.

Knee joints: Extended.

Ankle joints: Slight plantar flexion.

Muscle findings in this flat-back type of posture are less constant than in other types. The most constant finding is a tightness of Hamstring muscles which pulls the pelvis into posterior tilt, and weakness of hip flexors. Back muscles may be very strong and inflexible even though they are in a slightly elongated position. Abdominal muscles may or may not be strong. Slight knee flexion may be found with this posture due to tightness of the Hamstrings.

Posture of Children

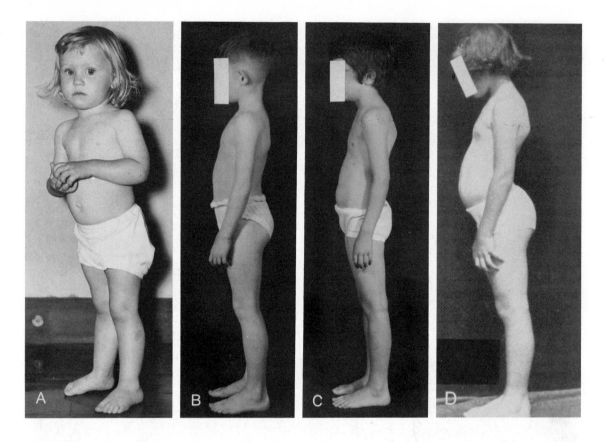

Fig. A shows the posture of a small child. The flexed hips and wide stance of this eighteen-month-old child suggest the uncertain balance associated with this age. Although it is not very evident in the picture, the subject had at this time a mild degree of knock-knees. (This deviation gradually decreased without any corrective measures so that at the age of six, this child's legs were in good alignment.)

The development of the longitudinal arch in this subject is very good for a child of her age.

Fig. B shows a seven-year-old child who has very good posture for his age.

Fig. C shows poor posture in a six and one-half-year-old child. There is forward head, kyphosis, depressed chest and lordosis. Prominence of the scapulae is evident in side view (See p. 299, right column, lower figure for back view of this subject.)

Fig. D shows a marked lordosis in an eight-year-old child. A corset to hold the back in good alignment and to support the abdomen is needed along with therapeutic exercises when alignment is this faulty.

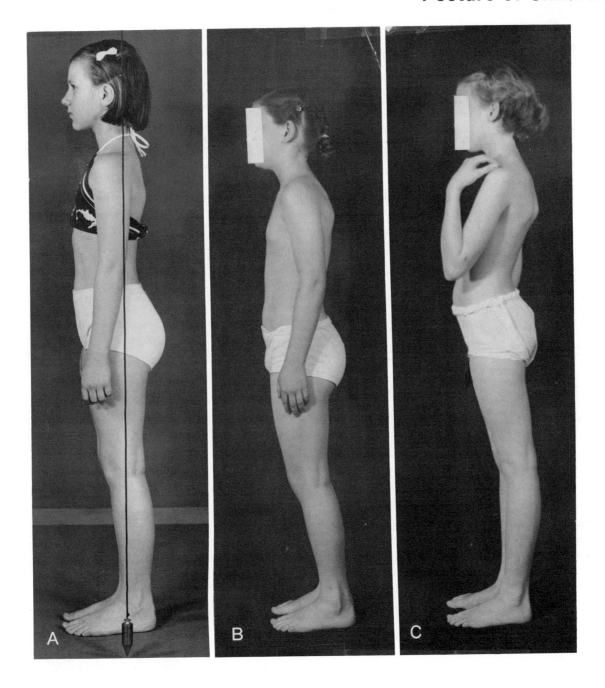

Fig. A shows a ten and one-half-year-old child who has very good posture for this age. The posture resembles the normal adult posture more than does that of the smaller child. The curves of the spine are nearly normal and the scapulae are less prominent. It is characteristic of small children to have a protruding abdomen, but there is a noticeable change about the age of ten or twelve when the waistline becomes relatively smaller and the abdomen no longer protrudes.

Fig. B shows a nine-year-old child whose posture is about average for this age.

Fig. C is an eleven-year-old child whose posture is very faulty with forward head, kyphosis, lordosis, anterior pelvic tilt, and hyperextended knees.

Faulty Posture, Side View: Analysis and Treatment

Postural fault	Anatomical position of joints	Muscles in shortened position	Muscles in lengthened position	Treatment procedures, if indicated on the basis of tests for alignment and muscle length & strength tests
Lordosis posture	Lumbar spine hyperextension Pelvis; anterior tilt	Low back Erector spinae	Abdominals, esp. External oblique	Stretch low back muscles, if tight. Strengthen abdominals by posterior pelvic tilt exercises, and if indicated, by trunk curl. Avoid situps because they shorten hip flexors. Stretch hip flexors, when short. Strengthen hip extensors, if weak.
	Hip joint flexion	Hip flexors	Hip extensors	Instruct regarding proper body alignment. Depending upon degree of lordosis and extent of muscle weakness and pain, use support (corset) to relieve strain on abdominals, and to help correct the lordosis.
Flat-back posture	Lumbar spine flexion	Anterior abdominals	Low back Erector spinae	Low back muscles seldom are weak, but if they are weak, do exercise to strengthen them and to restore normal anterior curve. Tilt pelvis forward, bringing low back into anterior curve. *Avoid* prone hyperextension because it increases posterior pelvic tilt and stretches hip flexors.
	Pelvis, posterior tilt			Instruct in proper body alignment. If back is painful and in need of support, apply corset that holds the back in a position of normal anterior lumbar curve.
	Hip joint extension	Hip extensors	Hip flexors	Strengthen hip flexors to help produce normal anterior lumbar curve. Stretch hamstrings, if tight.
Sway-back posture (Pelvis displaced forward, upper trunk back)	Lumbar spine position depends on level of posterior displacement of upper trunk	Upper anterior abdominals esp. upper Rectus and Internal oblique	Lower anterior abdominals esp. External oblique	Strengthen lower abdominals (stress External oblique). Stretch arms overhead and do deep breathing to stretch tight intercostals and upper abdominals. Instruct in proper body alignment. Wall-standing exercise particularly useful.
	Pelvis, posterior tilt	Hip extensors	Hip flexors	Stretch Hamstrings, if tight. Strengthen hip flexors, if weak, using alternate hip flexion in sitting, or alternate leg raising from supine position. *Avoid* double leg-raising exercises because of strain on abdominals.
	Hip joint extension			

Common painful conditions associated with imbalance of anteroposterior trunk and hip joint muscles:
Low back pain and upper back pain.

Low Back Stretching.

In face-lying position, place a firm pillow under the abdomen and a rolled blanket under the ankles.

Have heat or cold and massage applied to low back while lying in this position. (About . . . . . min. and . . . . . min. massage.)

Low Back Stretching.

Sit with legs extended forward. Place a rolled blanket under the knees to allow slight knee-bend. Pull in with the abdominal muscles, keep the pelvis tilted back, and reach forward toward the toes. The stretch should be felt in the low back, not under the knees or in the upper back or neck.

Hamstring Stretching.

In the face-lying position have heat or cold and massage applied to the back of the thighs. (About . . . . . min., ..., and . . . . . min. massage.)

To stretch right Hamstrings, lie supine with legs extended, hold left leg down, and gradually raise the right with the knee straight. Exercise may be done actively or with assistance.

To stretch left Hamstrings apply the same procedures to the left leg.   Or

Raise the leg and rest the heel against the back of a chair or other support with Hamstrings in stretched position.   Or

Sit on a stool with back against a wall. Keep back straight and buttocks against the wall. With one knee bent, straighten the other. A stretch should be felt under the knee and along the Hamstring muscles. (See p. 153.)

Hip Flexor Stretching.

Have heat or cold and massage applied to front and outer side of thighs. (About . . . . . min., ... and . . . . . min. massage.)

To stretch right hip flexors, lie supine with right lower leg hanging over end of table. Pull left knee firmly toward chest and to help press low back flat down on table. (With hip flexor shortness, this position will bring the right thigh up from the table.) Keeping low back flat, stretch right hip flexors by pulling the right thigh downward towards the table by contracting the buttock muscle.

To stretch the left hip flexors, lie with the right knee bent toward the chest and stretch the left thigh as described above for the right. (See pp. 160–163.)

"Lower" Abdominal Exercise.

In back-lying position, place a rolled blanket or small pillow under the knees. With hands up beside the head, tilt the pelvis to flatten the low back on the table by pulling up and in with lower abdominal muscles. Hold low back flat and breathe in and out easily, relaxing upper abdominal muscles. (There should be good chest expansion during inspiration, but the back should not arch.) (See photographs p. 221.)

"Lower" Abdominal Exercise.

In back-lying position, bend the knees and place the feet flat on the table. With hands up beside the head, tilt the pelvis to flatten the low back on the table. Hold low back flat and slide the heels down along the table. Straighten the legs as much as possible with low back held flat. Keep low back flat and return knees to bent position, *sliding one leg back at a time.* (See photographs p. 221.)

"Upper" Abdominal Exercise.

In back-lying position, tilt the pelvis to flatten the low back on the table. With arms extended forward, raise head and shoulders about eight inches up from the table. (Do not attempt to come to a sitting position, but raise the upper trunk, curling it as much as the *back* will bend.) (See photographs p. 209 and p. 211.)

Standing Postural Exercise.

Stand with the back against a wall, heels about three inches from the wall. Place hands up beside head with elbows touching the wall. Tilt pelvis to flatten low back against the wall by pulling up and in the lower abdominal muscles.

Repeat the above exercise with arms overhead.

Ideal Alignment: Posterior View

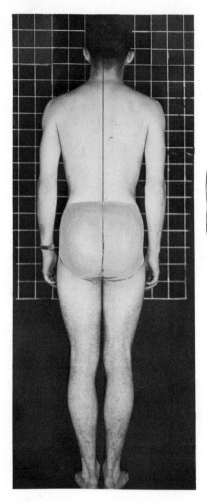

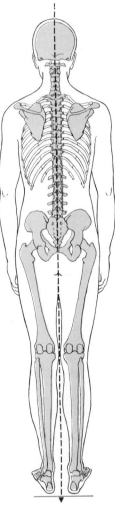

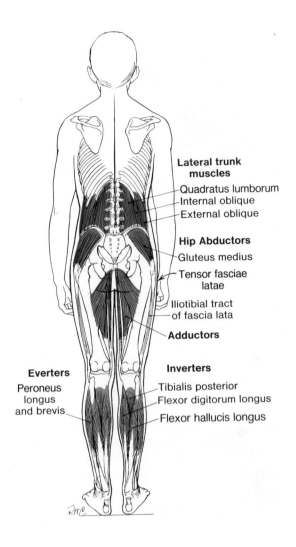

Lateral trunk muscles
- Quadratus lumborum
- Internal oblique
- External oblique

Hip Abductors
- Gluteus medius
- Tensor fasciae latae
- Iliotibial tract of fascia lata

Adductors

Everters
- Peroneus longus and brevis

Inverters
- Tibialis posterior
- Flexor digitorum longus
- Flexor hallucis longus

Head: Neutral position, neither tilted nor rotated. (Slightly tilted toward right in photograph.)

Cervical spine: Straight in drawing. (Slight lateral flexion toward right in photograph.)

Shoulders: Level, not elevated or depressed.

Scapulae: Neutral position, medial borders essentially parallel and about three to four inches apart.

Thoracic and lumbar spines: Straight.

Pelvis: Level, both posterior superior iliac spines in same transverse plane.

Hip joints: Neutral position, neither adducted nor abducted.

Lower extremities: Straight, neither bowed nor knock-kneed.

Feet: Parallel or toeing out slightly. Outer malleolus and outer margin of sole of foot in same vertical plane so that foot is not pronated or supinated. (See p. 302.) Tendo calcaneus should be vertical when seen in posterior view; in photograph, alignment suggests slight pronation.

Laterally, the following muscles work together in stabilizing the trunk, pelvis, and lower extremities:

Right lateral trunk flexors
Right hip adductors
Left hip abductors
Right Tibialis posterior
Right Flexor hallucis longus
Right Flexor digitorum longus
Left Peroneus longus and brevis

Left lateral trunk flexors
Left hip adductors
Right hip abductors
Left Tibialis posterior
Left Flexor hallucis longus
Left Flexor digitorum longus
Right Peroneus longus and brevis

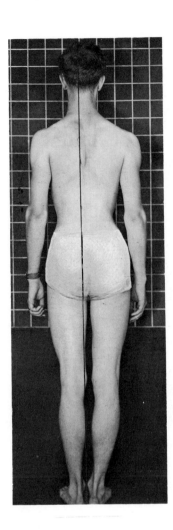 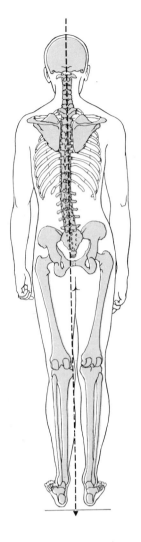 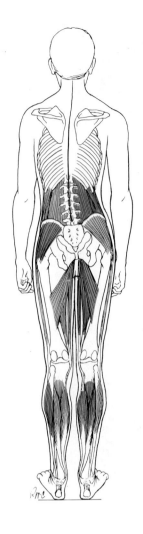

Head: Erect, neither tilted nor rotated.

Cervical spine: Straight.

Shoulder: Right low.

Scapulae: Adducted and right slightly depressed.

Thoracic and lumbar spines: Thoracolumbar curve convex toward *left*.

Pelvis: Lateral tilt, high on right.

Hip joints: Right adducted and slightly medially rotated, left abducted.

Lower extremities: Straight, neither bowed nor knock-kneed.

Feet: In the photograph, the right is slightly pronated as seen in the alignment of the Tendo calcaneus. The left is in a position of slight postural pronation by virtue of the deviation of the body toward the right.

With a deviation of the pelvis toward the right and a lateral tilt down on the left, the right hip is in adduction and the position is referred to as "postural adduction" of the right leg. In this type posture the right Gluteus medius is weaker than the left. The left leg is in a position of "postural abduction" and the left hip abductor group, especially the fascia lata, is often somewhat tight.

This position of the right hip resembles the Trendelenburg sign or gait.

The right lateral trunk muscles, left hip abductors, right hip adductors, left Peroneous longus and brevis, and right invertors are in a shortened position and usually are strong. The left lateral trunk muscles, right hip abductors, left hip adductors, right Peroneus longus and brevis, left Tibialis posterior, left Flexor hallucis longus, and left Flexor digitorum longus are in a somewhat lengthened position and tend to show slight weakness.

Faulty Alignment: Posterior View

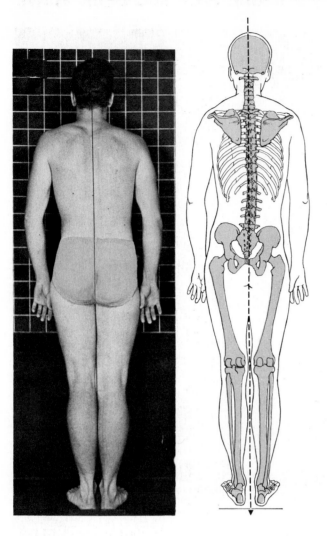

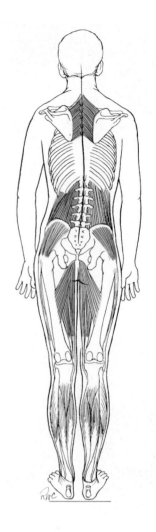

Head: Erect, neither tilted nor rotated.

Cervical spine: Straight.

Shoulders: Elevated and adducted.

Shoulder joints: Medially rotated as indicated by position of hands facing posteriorly.

Scapulae: Adducted and elevated.

Thoracic and lumbar spines: Slight thoraco-lumbar curve convex to *right*.

Pelvis: Lateral tilt, higher on left.

Hip joints: Left adducted and slightly medially rotated, right abducted.

Lower extremities: Straight, neither bowed nor knock-kneed.

Feet: Slightly pronated.

With the scapulae adducted and elevated, the Rhomboid muscles are in a shortened position. The left trunk flexors, right hip abductors, and left hip adductors are also in a shortened position and tend to be strong. Slight weakness tends to be present in right lateral trunk muscles, left hip abductors, right hip adductors, right Tibialis posterior, right Flexor hallucis longus, right Flexor longus digitorum, and left Peroneus longus and brevis.

Faulty Posture: Side and Back Views

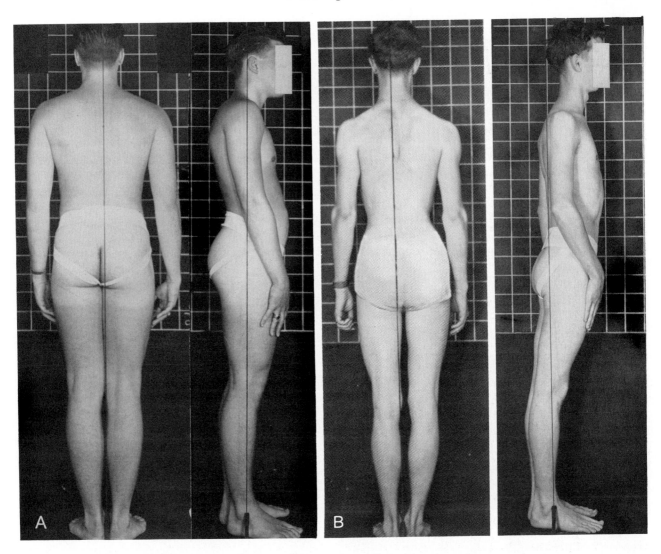

Fig. A is an example of posture which appears good in back view but is very faulty in side view.

The side view posture shows marked segmental faults but the anterior and posterior deviations compensate for each other so that the plumb alignment is quite good. The contour of the abdominal wall almost duplicates the curve of the low back.

Fig. B shows a posture which is faulty in both side and back views. The back view shows a marked deviation of the body to the right of the plumb line, a high right hip, and a low right shoulder.

In side view, the plumb alignment is worse than the segmental alignment. The knees are posterior, the pelvis, trunk, and head are markedly anterior.

Segmentally, the anteroposterior curves of the spine are only slightly exaggerated. The knees, however, are quite hyperextended.

This type of posture might well result from the effort to follow such misguided but common admonitions as "Throw your shoulders back", "Stand with your weight over the balls of your feet."

The result in this subject is so much forward deviation of the trunk and head that the posture is most unstable and requires a good deal of muscular effort to maintain balance.

An individual with this type of fault might appear as one with good posture when fully clothed.

In this, as in the right-hand figure on p. 283, the anterior part of the foot shows evidence of strain.

Handedness: Effect on Posture

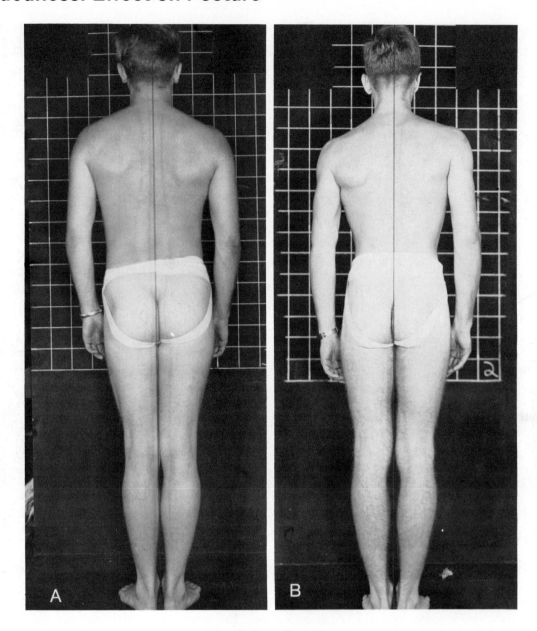

Handedness Patterns

Each of the above figures illustrates a typical pattern of posture as related to handedness. In fig. A, the right shoulder is lower than the left, the pelvis is deviated slightly toward the right, and the right hip appears slightly higher than the left. This pattern is typical of right-handed people. Usually there is a slight deviation of the spine toward the left and the left foot is more pronated than the right. The right Gluteus medius is usually weaker than the left.

Handedness patterns related to posture begin at an early age. The slight deviation of the spine toward the side opposite the higher hip may appear as early as age seven or eight. There tends to be a compensatory low shoulder on the side of the higher hip. In most cases, the low shoulder is less significant than the high hip. Usually shoulder correction tends to follow correction of lateral pelvic tilt, but the reverse does not necessarily occur.

Fig. B shows the opposite pattern which is typical of left-handed individuals. Usually, however, the low shoulder is not quite as marked as in this subject.

294

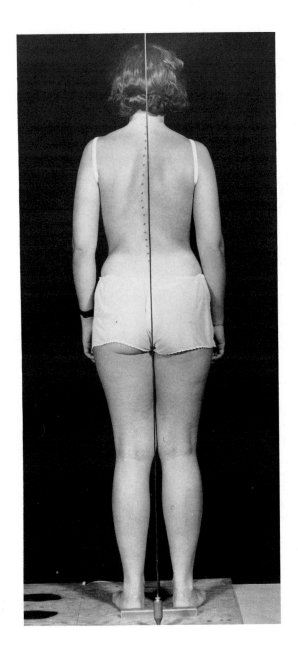

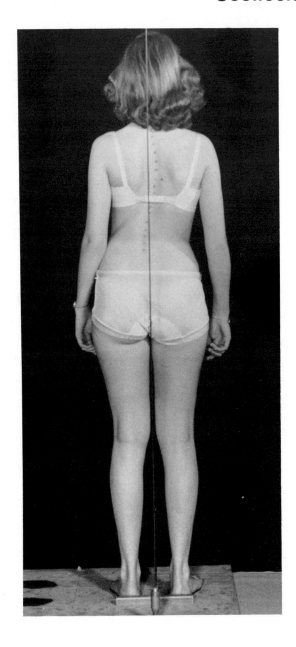

C-type Scoliosis (C-curve)
This subject has a left thoracolumbar curve with low right shoulder but without pelvic deviation or tilt.

S-type Scoliosis (S-curve)
This subject has a right thoracic, left lumbar curve. The shoulder level is good; there is no deviation of the pelvis in relation to the plumb line.

Postural fault	Anatomical position of joints	Muscles in shortened position	Muscles in lengthened position	Treatment procedures, if indicated on basis of tests for alignment and muscle length and strength tests
Slight left C-curve: Thoracolumbar scoliosis	Thoracolumbar spine: Lateral flexion, convex toward left	Right lateral trunk muscles	Left lateral trunk muscles	*If present without lateral pelvic tilt,* stretch right lateral trunk muscles, if short, and strengthen left lateral trunk muscles, if weak. *If present with lateral pelvic tilt,* see below for additional treatment procedures. Correct faulty habits that tend to increase the lateral curve: *Avoid* sitting on left foot in manner that thrusts spine toward left; *Avoid* lying on left side, propped up on elbow, to read or write.
		Left Psoas major	Right Psoas major	Exercise right Iliopsoas, in sitting position. See facing page.
Prominent or high right hip	Pelvis: Lateral tilt, high on right Right hip joint: Adducted Left hip joint: Abducted	Right lateral trunk muscles Left hip abductors and fascia lata Right hip adductors	Left lateral trunk muscles Right hip abductors, esp. Guteus medius Left hip adductors	Stretch right lateral trunk muscles, if short. Strengthen left lateral trunk muscles, if weak. Stretch left lateral thigh muscles and fascia, if short. Specific exercises to strengthen right Gluteus medius *are not required* to correct slight postural weakness; functional activity will suffice if the alignment is corrected and maintained. The subject should be advised to: Wear straight raise on heel of left shoe (usually about 3/16 inch), or pad inside heel of shoe and in bedroom slippers; *Avoid* standing with weight on right leg, causing right hip to be in postural adduction. Stand with weight evenly distributed over both feet, with pelvis level.

[Opposite of above for posture with right C-curve and high left hip.]

Common painful conditions associated with imbalance of lateral trunk and hip muscles:

Low back pain, with pain and tenderness more often unilateral and most pronounced in the area of L5 on the high side of the pelvis;

Pain in the posterolateral buttocks region on the side of the high hip, in the area of the stretched Gluteus medius;

Pain along the lateral thigh on the side of the low hip, in the area of the short hip abductors and the tight iliotibial tract.

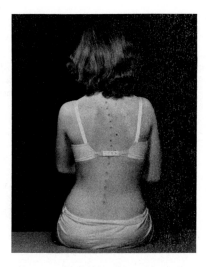

The subject in the illustration above had a moderate right thoracic curve and a slight left lumbar curve. In addition to the imbalance in the anterior, lateral, and oblique abdominal muscles, there was weakness in the right Iliopsoas. This weakness contributed to the deviation of the lumbar spine toward the left.

To be therapeutic, exercises must be combined in a manner that results in overall correction of the deformity.

The exercises being performed by this subject consist of the following:

With the pelvis held level, and the right hand on the right thigh, the subject resists the effort to flex the right hip joint, while, at the same time, reaching diagonally upward and slightly forward with the left arm.

The thigh is not lifted from the table, but the force exerted by the right Psoas major results in a pull on the lumbar spine and a correction of the lumbar curve.

Stretching Fascia Lata and Hip Abductors. To stretch the left fascia lata, lie on the right side with the right hip and knee bent. Relax the left leg on pillows placed between  the thighs and lower legs. Have heat and massage applied to the outer side of the left thigh from the hip to the knee. (About ...... min. heat and ...... min. massage.) Remove the pillows. Have an assistant hold the pelvis firmly with one hand, draw the thigh slightly back, and press it downward toward the table stretching the muscles and fascia between the hip and knee. (The knee should not be allowed to rotate inward, and care should be taken to avoid strain at the knee joint.)

To stretch the right fascia lata, lie on the left side and have heat, massage, and stretching applied to the right thigh in the same manner as described for the left above.

Tightness of the fascia lata is a fairly common finding in cases of pain along the lateral thigh, or in cases of postural deviation with lateral pelvic tilt. Many scoliosis patients have some degree of lateral tilt, and tests for tightness of the fascia lata should be included in the postural examination.

In addition to the treatment described for use in stretching the tight fascia lata, one of the most effective ways of stretching is by the use of a straight raise (usually about 3/16") on the heel of the shoe on the side of tightness.

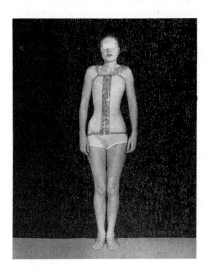

Adequate supports are important in the treatment of scoliosis patients. Supports may be in the form of corsets, jackets (as illustrated), or braces.

Shoulders and Scapulae

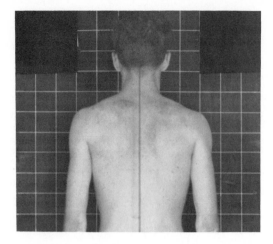

Shoulders and scapulae, good position. This picture illustrates a good position of the shoulders and scapulae.

The scapulae lie flat against the thorax and no angle or border is unduly prominent. Their position is not distorted by unusual muscular development or misdirected efforts at postural correction.

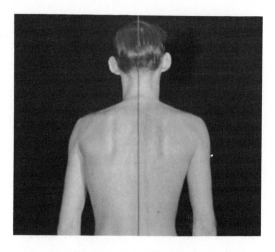

Scapulae, abducted and slightly elevated. In this subject, both scapulae are abducted, the left one more than the right. They are slightly elevated. This kind of elevation goes with round shoulders and round upper back. (For side view of this subject see p. 300, lower left figure.)

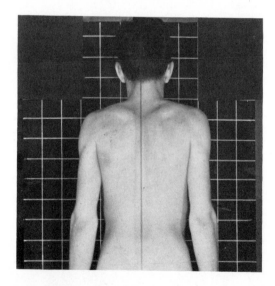

Shoulders elevated, scapulae adducted. In this figure, both shoulders are elevated with the right slightly higher than the left.

The scapulae are adducted.

The upper Trapezius and other shoulder elevators are in a state of contraction.

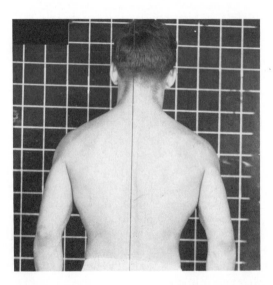

Shoulders depressed, scapulae abducted. In this subject, the shoulders slope downward sharply, accentuating their natural broadness. The marked abduction of the scapulae also contributes to this effect of broadness.

Exercises to strengthen the Trapezius muscles, including upper as well as middle and lower parts, are needed to correct this faulty posture of the shoulders.

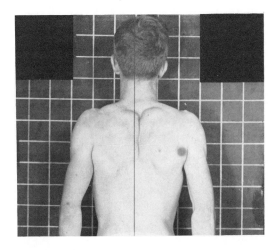

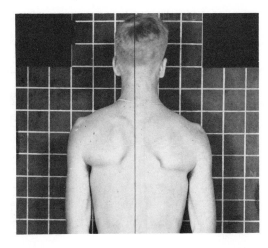

Scapulae, adducted and elevated. In this subject, the scapulae are completely adducted and considerably elevated.

The position illustrated is obviously held by voluntary effort, but when this habit has persisted for some time the scapulae do not return to normal position when the subject tries to relax.

This position is the inevitable end-result of persisting in the military practice of "bracing" the shoulders back.

Scapulae, abnormal appearance. This subject shows abnormal development of some of the scapular muscles with a faulty position of the scapulae.

The Teres major and Rhomboids, which show clearly, form a V at the inferior angle. The scapula is tilted so that the axillary border is more nearly horizontal than normal. The appearance suggests weakness of either the Serratus anterior or the Trapezius or both.

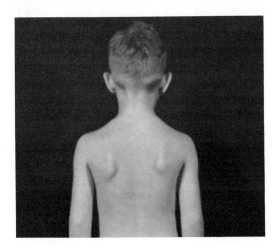

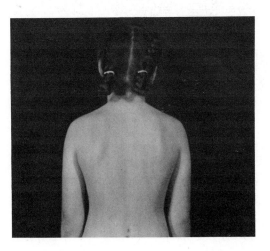

Abducted and slightly "winged" scapulae. This subject shows a degree of scapular prominence that is rather typical of children of this age (eight years). Slight prominence and slight abduction need not be a matter of concern at this age but in this subject there is a difference in level of the scapulae which may indicate some muscle imbalance that should be checked.

Abducted scapulae and forward shoulders. This subject is a nine-year-old girl who is rather mature for her age. The forward position of the shoulders is typical of that assumed by many young girls at the time of beginning development of the breasts. When such a postural habit persists, it may result in a fixed postural fault.

Faulty Head and Neck Positions

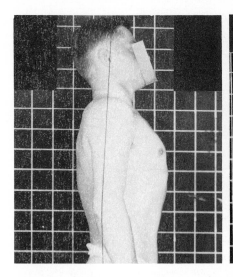
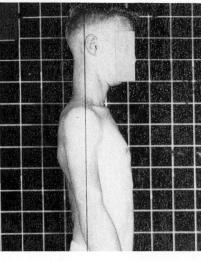
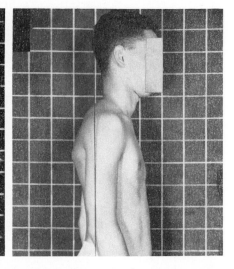

Head, Posterior Tilt. In the left figure, the head tilts backward and there is hyperextension of the cervical spine. The chest and shoulders are held high.

Head, Anterior Tilt. In the center figure, the head is tilted forward and the cervical spine is in flexion.

A posture in which the normal anterior curve of the cervical spine tends to reverse, as it does in this subject, is unusual.

Forward Head with Attempted Correction. The subject in the right figure apparently is trying to correct what is basically a forward head position.

The curve of the neck begins in a typical way in the low cervical region, but a sharp angulation occurs at about the 6th cervical vertebra. Above this level, the curve seems very much decreased. The chin is pressed against the front of the throat. This distorted, rather than corrected, position of the neck results from a failure to correct the related faulty position of the upper trunk.

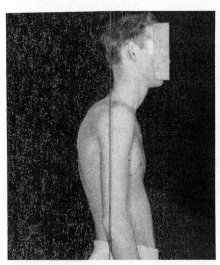
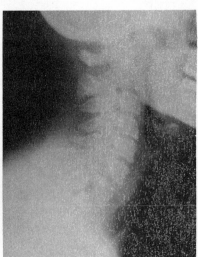
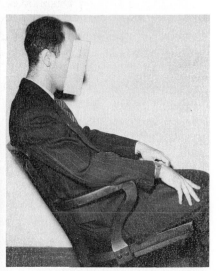

Forward Head, Marked. This subject shows an extremely faulty alignment of the neck and thoracic spine. The degree of deformity in the thoracic spine is suggestive of an epiphysitis. This patient was treated for pain in the posterior neck and occipital region.

In a forward head position, the cervical spine is in hyperextension. Narrowing of the interspaces is most marked between C4 and C5, and between C5 and C6 where C5 and C6 nerve roots emerge.

Sitting in a tilt-back swivel chair or in an automobile seat inclined back is conducive to a forward head position.

Faulty Head and Shoulder Positions: Analysis and Treatment

Postural fault	Anatomical position of joints	Muscles in shortened position	Muscles in lengthened position	Treatment procedures, if indicated on the basis of tests for alignment and muscle length and strength tests
Forward head	Cervical spine hyperextension	Cervical spine extensors Upper Trapezius	Cervical spine flexors	Stretch cervical spine extensors, if short, by trying to flatten the cervical spine. Strengthen cervical spine flexors, if weak. (See A and B below.) A forward head position is usually the result of faulty upper back posture. If neck muscles are not tight posteriorly, the head position will usually correct as the upper back is corrected. Strengthen the thoracic spine extensors. Do deep breathing exercises to help stretch the intercostals and the upper parts of abdominal muscles. Stretch Pectoralis minor. (See C below.) Stretch shoulder adductors and internal rotators, if short. (See below, and pp. 102–104.) Strengthen middle and lower Trapezius. (See exercises below.) Use shoulder support (See D below) when indicated, to help stretch Pectoralis minor and relieve strain on middle and lower Trapezius.
Kyphosis and Depressed chest	Thoracic spine flexion Intercostal spaces diminished	Upper and lateral fibers of Internal oblique Shoulder adductors Pectoralis minor Intercostals	Thoracic spine extensors Middle Trapezius Lower Trapezius	
Forward shoulders	Scapulae abducted & (usually) elevated	Serratus anterior Pectoralis minor Upper Trapezius	Middle Trapezius Lower Trapezius	

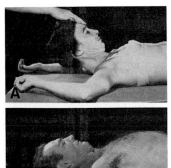

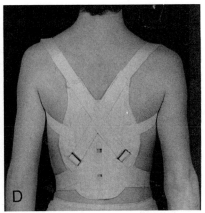

Shoulder Adductor Stretching. Have heat (or other modality) and massage applied to the pectorals and/or teres while in the back-lying position. (About . . . . . . min. heat, and . . . . . . min. massage.)

With knees bent and feet flat on the table, tilt the pelvis to flatten the low back on the table by pulling up and in with muscles in the lower abdomen. Hold the back flat and extend both arms overhead. Try to touch the entire arm to the table with elbows straight. Bring upper arms close to the sides of the head to stretch teres, or diagonally overhead to stretch pectorals.

Wall-sitting. Sit on a stool with back to wall. Place hands up beside the head with elbows touching the wall (for middle Trapezius) or overhead, keeping arms in contact with the wall (for lower Trapezius). Avoid sudden or vigorous movement when placing the arms up beside the head or overhead. Straighten the upper back and hold the head back with chin down and in. Tilt the pelvis to flatten the low back against the wall by pulling up and in with the muscles in the lower abdomen. Hold several seconds, then relax. Repeat several times.

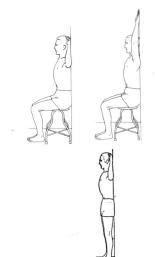

Wall-standing. Stand with back against a wall, heels about three inches from the wall. Place hands up beside the head with elbows touching the wall, or overhead, keeping arms in contact with the wall. Avoid sudden or vigorous movement when placing the arms up beside the head or overhead. Tilt the pelvis to flatten the low back against the wall by pulling up and in with the muscles in the lower abdomen. Hold the position of the body while slowly lowering the arms. Relax. Repeat several times.

Good and Faulty Posture of Feet and Knees

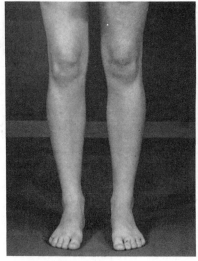

Good Alignment of Feet and Knees. The patellae face directly forward and the feet are neither pronated nor supinated.

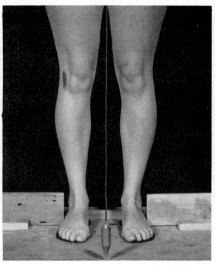

Pronation of Feet and Medial Rotation of Femurs. The distance between the lateral malleolus and the foot board indicates a moderate pronation of the feet, and the position of the knee-caps indicate a moderate degree of medial rotation of the femurs.

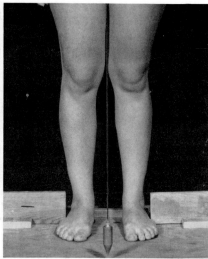

Pronation of Feet and Knock-knees. The feet are moderately pronated; there is slight knock-knee position, but no medial or lateral rotation.

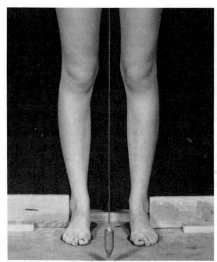

Feet Good, Knees Faulty. The alignment of the feet is very good but there is medial rotation of the femurs as indicated by the position of the patellae. This fault is harder to correct by use of shoe corrections than one in which pronation accompanies the medial rotation.

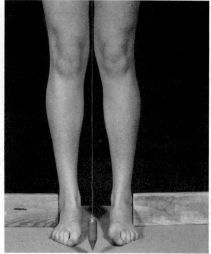

Supinated Feet. In the figure above, the weight is borne on the outer borders of the feet, and the long arches are higher than normal.

The perpendicular foot-board touches the lateral malleolus, but is not in contact with the outer border of the sole of the foot.

The position shown is the natural posture of this subject's feet, but the anterior tibial muscles show so clearly that it appears as if an effort were being made to invert the feet.

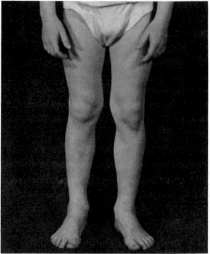

Lateral Rotation of the Legs. The lateral rotation of the legs as seen in this subject is the result of lateral rotation at the hip joint.

This position is more typical of boys than of girls. It may or may not have serious effects, although persistence of such a pattern in walking as well as in standing may put undue strain on the longitudinal arches.

Good and Faulty Posture of Knees and Legs

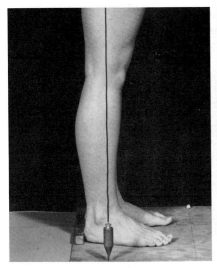

Knees, Good Alignment. In good alignment of the knees in side view, the plumb line passes slightly anterior to a midline through the knee.

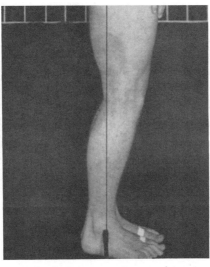

Knee Flexion, Moderate. Flexion of the knees is seen less frequently than hyperextension in cases of faulty posture. The flexed position requires constant muscular effort by the Quadriceps. Knee flexion in standing may result from hip flexor tightness. When hip flexors are tight there must be compensatory alignment faults of the knees or the low back or both. Attempting to reduce a lordosis by flexing the knees in standing is not an appropriate solution when hip flexor stretching is needed.

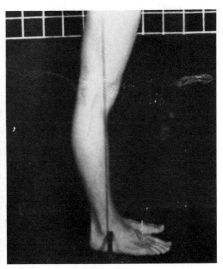

Knee Hyperextension. With marked hyperextension of the knee, the ankle joint is in plantar flexion.

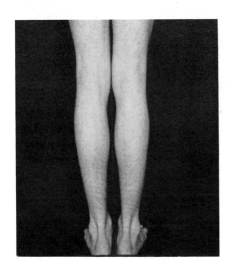

Good Alignment of Legs.

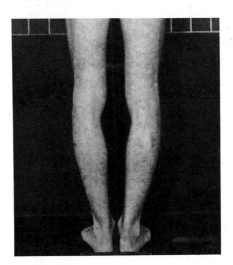

Bowlegs. This figure shows a mild degree of structural bowlegs (genu varum).

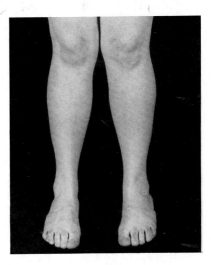

Knock-knees. This figure shows a moderate degree of structural knock-knees (genu valgum).

Postural Bowing Compensating for Knock-knees

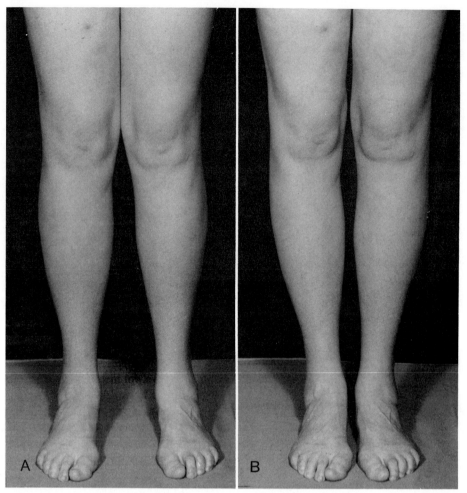

A

B

Mechanism of postural bowing compensatory for knock-knees. Fig. A shows the position of knock-knees which the subject exhibits when the knees are in good anteroposterior alignment.

Fig. B shows that by hyperextending her knees the subject is able to produce enough postural bowing to accommodate for the four and one-half inch separation of her feet in fig. A.

See fig. A on opposite page for the extent of postural bowing which can be produced by hyperextension in an individual who has no knock-knee condition.

Children are often embarrassed by the appearance of knock-knees, and it is not uncommon for them to compensate if the knock-knee condition persists. Sometimes they "hide" the knock-knee position by flexing one knee and hyperextending the other so the knees can be close together. Rotation faults may result if the same knee is habitually flexed while the other is hyperextended.

Postural bowlegs and knock-knees. In contrast to actual changes in the alignment of the bones as seen in structural bowlegs and knock-knees, there may be changes in joint position which give rise to postural bowing or knock-knees.

The knee joint, essentially a hinge joint, allows the knee to flex and extend. There is free flexion through a good arc of motion. When the knee straightens in standing, the motion should stop a few degrees beyond the straight line. However, like some "sprung" hinges, some knees go beyond and curve backward into a position of hyperextension which may be of a mild, moderate, or marked degree.

Patellae should face straight ahead but at times they face medially or laterally. This position of the patella results from rotation of the femur at the hip joint because the knee in extension does not rotate. *Combinations of medial or lateral rotation of the hip joint with hyperextension of the knee joint account for the postural bowlegs or postural knock-knees as seen in figures (A) and (C) on the facing page.*

The axis of the knee joint about which flexion and extension take place is in a coronal plane of the body when patellae face straight ahead, that is, when there is no medial or lateral rotation of the hip joints. From this position, flexion, extension, or even hyperextension will occur directly anteriorly or posteriorly.

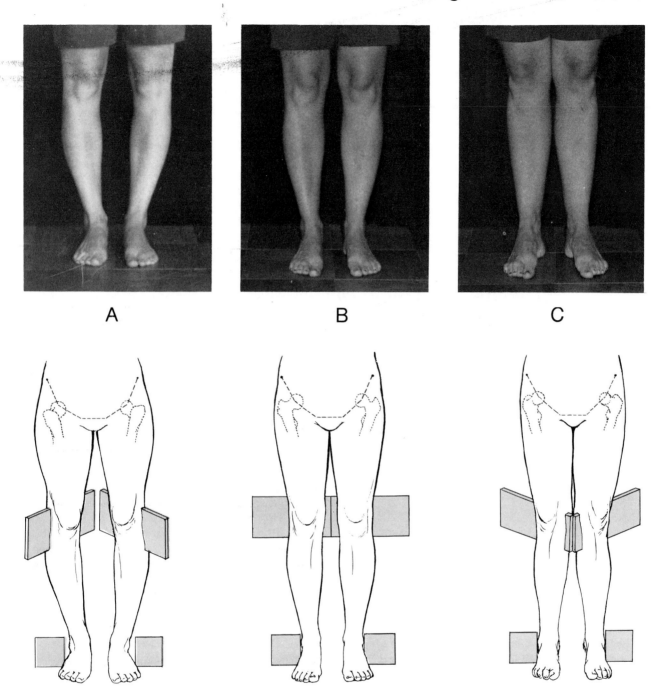

With medial rotation, the axis of the knee joint is oblique to a coronal plane of the body and flexion will occur in an anteromedial direction, and extension or hyperextension in a posterolateral direction. As a result, there will be an apparent bowing of the legs in hyperextension.

With lateral rotation, the axis of the knee joint is also oblique to a coronal plane of the body, but in this case flexion will occur in an anterolateral direction and extension or hyperextension in a posteromedial direction. As a result there will be an apparent knock-kneed position in hyperextension.

The appearance of postural bowlegs and postural knock-knees also may result from the combination of knee flexion with rotation, but is not illustrated. With lateral rotation and slight flexion, legs will appear slightly bowed, and with medial rotation and slight flexion, there will appear to be a position of knock-knees. These variations associated with flexion are of less concern than those associated with hyperextension because flexion is a normal movement while hyperextension is an abnormal one.

305

X-rays of Legs in Good and Faulty Alignment

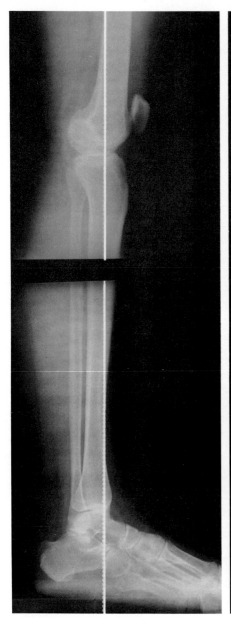 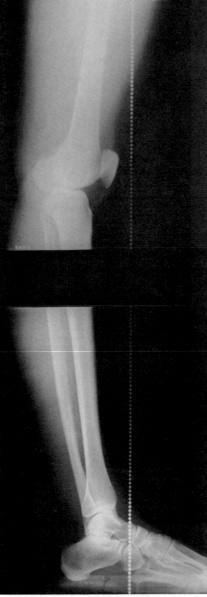 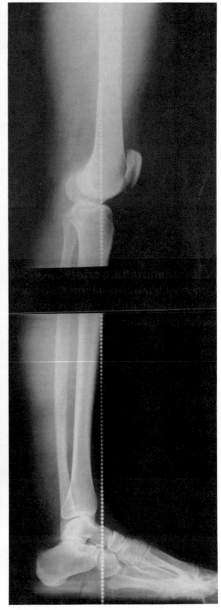

The above illustration shows the relationship of the plumb line to the bones of the foot and lower leg, with the subject standing in a position of good alignment. The beaded metal plumb line was suspended beside the subject when the x-ray was taken. Two x-ray films were in position for the single exposure.

This x-ray is of a subject who had a habit of standing in hyperextension as a child. The plumb line was suspended in line with the standard base point while the x-ray was taken. The cassettes were placed between the lower legs and thighs during the single x-ray exposure.

The above x-ray is of the same subject shown in center figure. As an adult she has attempted to correct her hyperextension fault. The alignment through the knee joint and femur are very good, but the tibia and fibula show evidence of posterior bowing. Compare with figure at far left.

Faulty Leg, Knee, and Foot Positions: Analysis and Treatment

Postural Fault	Anatomical Position of Joints	Muscles in Shortened Position	Muscles in Lengthened Position	Treatment procedures, if indicated on the basis of tests for alignment and muscle length and strength tests
Hyperextended knee	Knee hyperextension Ankle plantar flexion	Quadriceps Soleus	Popliteus Hamstrings at knee	Instruct regarding overall postural correction with emphasis on avoiding knee hyperextension. In hemiplegics, short-leg brace with right-angle stop.
Flexed knee	Knee flexion Ankle dorsiflexion	Popliteus Hamstrings at knee	Quadriceps Soleus	Stretch knee flexors, if tight. Overall postural correction. Knee flexion may be secondary to hip flexor shortness. Check length of hip flexors; stretch if short.
Medially rotated femur (often associated with pronation of foot, see below.)	Hip joint medial rotation	Hip medial rotators	Hip lateral rotators	Stretch hip medial rotators, if tight. Strengthen hip lateral rotators, if weak. Young children should *avoid* sitting in reverse-tailor fashion ("W" position). (See below for correction of any accompanying pronation.)
Knock-knees (Genu valgum)	Hip joint adduction Knee joint abduction	Fascia lata Lateral knee joint structures	Medial knee joint structures	Inner wedge on heels, if feet are pronated. Stretch fascia lata, if indicated.
Postural bowlegs (Genu varum)	Hip joint medial rotation Knee joint hyperextension Foot pronation	Hip medial rotators Quadriceps Foot everters	Hip lateral rotators Popliteus Tibialis posterior and long toe flexors	Exercises for overall correction of foot, knee, and hip positions. *Avoid* knee hyperextension. Strengthen hip lateral rotators. Inner wedges on heels to correct foot pronation.
		Stand with feet straight ahead and about two inches apart. Relax the knees into an "easy" position, i.e., neither stiff nor bent. Tighten the muscles which lift the arches of the feet, rolling the weight *slightly* toward the outer borders of the feet. Tighten the buttocks muscles to rotate the legs slightly outward (until knee-caps face directly forward).		
Pronation	Foot eversion	Peroneals and toe extensors	Tibialis posterior and long toe flexors	Inner wedges on heels. (Usually 1/8 inch on wide heels, and 1/16 inch on medium heels.) Overall correction of posture of feet and knees. Exercises to strengthen the inverters. Instructions in proper standing and walking.
Supination	Foot inversion	Tibials	Peroneals	Outer wedge on heels. Exercise for peroneals.
Hammer toes and low metatarsal arch	Hyperext. m.p. joints, flex. i.p. joints	Toe extensors	Lumbricales	Stretch m.p. joints by flexion; stretch i.p. joints by extension. Strengthen lumricales by m.p. joint flexion. Metatarsal pad or bar.

Procedure for Postural Examination

Postural examination consists of three essential parts: 1) examination of the alignment in standing; 2) tests for flexibility and muscle length; and 3) tests for muscle strength.

EQUIPMENT

The equipment used (see p. 309) consists of the following:

Posture Boards. These are plywood boards on which foot prints have been drawn. Foot prints may be painted on the floor of the examining room, but the posture boards have the advantage of being portable.

Plumb line. The plumb line is suspended from an overhead bar, and the plumb bob is hung in line with the point on the posture board which indicates the standard base point. (See p. 312.)

Folding Ruler with Spirit Level. This is used to measure the difference in level of posterior iliac spines. It may be used also to detect any differences in shoulder level. A background with squares (as shown in many of the photographs) is a more practical aid in detecting differences in shoulder level.

Set of Six Blocks, four inches by 10 inches of the following thicknesses: $\frac{1}{8}$, $\frac{1}{4}$, $\frac{3}{8}$, $\frac{1}{2}$, $\frac{7}{8}$, and 1 inch. These are used for the purpose of determining the amount of lift needed to level the pelvis laterally. This means is preferred to the use of leg length measurement for this purpose (see discussion p. 310).

Marking Pencil. The pencil is used for marking the spinous processes in order to observe the position of the spine in cases of lateral deviation.

Tape Measure. This may be used in taking leg-length measurements, and for measuring the limitation of forward bending in reaching finger tips toward toes.

Chart for Recording Examination Findings. (See p. 313.)

Appropriate Clothing, such as a two-piece bathing suit for girls or trunks for boys, should be worn by subjects for a postural examination. Postural examination of school children is unsatisfactory when attempts are made to examine children clothed in ordinary gym suits.

In hospital clinics, gowns or other suitable garb should be provided.

ALIGNMENT IN STANDING

The subject stands on the posture boards with his feet in the position indicated by the foot prints. *Anterior View.* Observe the position of the feet, knees, and legs. Toe positions, appearance of the longitudinal arch, alignment in regard to pronation or supination of the foot, rotation of the femur as indicated by the position of the patella, knock-knees, or bowlegs should all be noted. Any rotation of the head, or abnormal appearance of the ribs should also be noted. Findings are recorded on the chart under "Segmental Alignment."

Lateral View. With the plumb line hung in line with a point just in front of the outer ankle bone, the relationship of the body as a whole to the plumb-line is noted and recorded under "Plumb Alignment." It should be observed from both right and left sides for the purpose of detecting rotation faults. Such descriptions as the following may be used in recording findings: "Body anterior from ankles up", "pelvis and head anterior", "good except lordosis", "upper trunk and head posterior".

Segmental alignment faults may be noted with or without the plumb line. Observe whether the knees are in good alignment, hyperextended, or flexed; note the position of the pelvis as seen from side view; whether the anteroposterior curves of the spine are normal or exaggerated; head position, forward or tilted up or down; chest position, whether normal, depressed, or elevated; and the contour of the abdominal wall. Findings are recorded on the chart under "Segmental Alignment."

Posterior View. With the plumb line hung in line with a point midway between the heels, the relationship of the body or parts of the body to the plumb line are expressed as good, or as deviations toward the right or left, and are so recorded on the chart. *In examining scoliosis patients it is especially important to observe the relationship of the overall posture to the plumb line.* Suspending a plumb line in line with the seventh cervical vertebra, or in line with the buttocks (as is frequently done) may be useful in ascertaining the curvature of the spine itself, but does not reveal the extent to which the spine may be compensating for a lateral shift of the pelvis or other postural faults that contribute to lateral pelvic tilt and associated spinal deviations.

From the standpoint of segmental alignment, one should observe the feet, noting the alignment of the tendo calcaneous, postural adduction or abduction of the hips, relative height of the posterior iliac spines, lateral pelvic tilt, lateral deviations of the spine, position of the shoulders and of the scapulae. In scoliosis patients, it is important to observe any asymmetrical segmental faults that may give rise to lateral pelvic tilts and accompanying deviations of the spine. For example, a lateral pelvic tilt may occur as a result of one foot being pronated or one knee being habitually flexed, allowing a dropping of the pelvis on that side in standing. Rotation of the spine or thorax, as seen in scoliosis cases, are observed with the patient bending forward. When

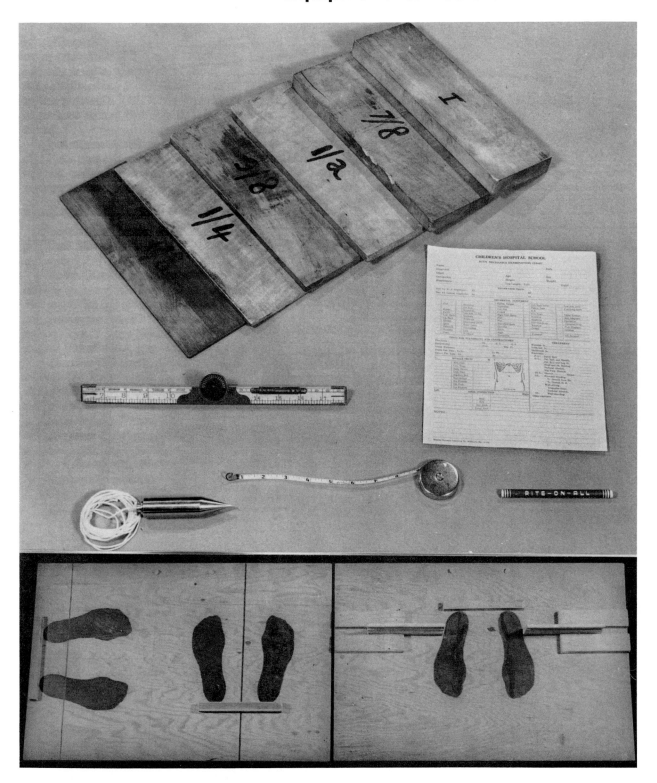

Equipment Used in Postural Examinations. Set of six blocks, ruler with spirit level, plumb line, tape measure, marking pencil, chart, and posture boards.

forward bending from a standing position, rotation of the bodies of the vertebrae is seen as a fullness or prominence on the side of convexity in a structural (fixed) curve of the spine. However, in a functional curve, there may be no evidence of rotation in forward bending. This is especially true if the functional curve is associated with hip abductor imbalance resulting in lateral pelvic tilt, or of anterior abdominal muscle imbalance because the effects of these imbalances on the spine are minimized in forward bending. Findings are recorded under "Segmental Alignment."

TESTS FOR FLEXIBILITY AND MUSCLE LENGTH

In this group of tests are included the tests for muscle length as described in Chapters 4, 5, and 6. Findings are recorded on the chart in the space provided. Forward bending is recorded as "Normal", "Limited", or "Normal +" with the number of inches from, or beyond, the toes recorded. (See p. 234 regarding normal for various ages in this test.) On the chart, "Bk" indicates back, "H.S.", Hamstrings, and "G.S.", Gastroc-soleus. An "X" is used to indicate which of these muscles is tight in case of limitation.

Forward bending may be checked in the standing or sitting position, but the authors consider the test in sitting more indicative of flexibility. If flexibility is normal in sitting and limited in standing, there is usually some rotation or lateral tilt of the pelvis resulting in rotation of the lumbar spine, which in turn restricts the flexion in the standing position.

Findings in regard to the arm overhead elevation tests are recorded as normal or limited, and, if limited, as slight, moderate, or marked. Hip flexor and fascia lata tightness are described in like manner.

Trunk extension is the movement of backward bending, and is done in the standing position to differentiate the flexibility test of the back from the strength test of back muscles as done in the prone position. Normally the back should arch in the lumbar region. If hyperextension is limited, the patient may try to simulate the backward bending by flexing his knees and leaning backward. Knees should be kept straight during the test.

Lateral flexion movements are used to test for lateral flexibility of the trunk. The length of the left lateral trunk muscles permits range of motion for trunk-bending toward the right, and vice versa. In other words, if flexibility of the trunk toward the right is limited it should be interpreted as some muscle tightness of the left lateral trunk muscles unless, of course, there is the element of limited spinal motion due to ligamentous or joint tightness.

Among other things, the variation among individuals in length of torso and in space between the ribs and iliac crest make for differences in flexibility. It is impractical to try to measure the degree of lateral flexion. Range of motion is considered to be normal for the individual when the rib cage and iliac crest are closely approximated in side bending. Most people can bring the fingertips to about the level of the knee when bending directly sideways.

MUSCLE STRENGTH TESTS

Muscle tests essential in postural examinations are described in Chapters 4, 5, and 6. They include tests of the "upper", "lower", and oblique abdominals, lateral trunk flexors, back extensors, middle and lower Trapezius, Serratus anterior, Gluteus medius, Gluteus maximus, Hamstrings, hip flexors, Soleus, and toe flexors.

In problems of anteroposterior deviations in postural alignment, it is especially important to test the abdominal muscles, back muscles, the hip flexors and extensors, and the Soleus. In problems of lateral deviation of the spine or lateral tilt of the pelvis it is especially important to test the oblique abdominal muscles, lateral trunk flexors and the Gluteus medius.

Lateral and anteroposterior deviations of alignment are frequently found in combination, and most of the above tests are then included in the examination.

LEG LENGTH MEASUREMENTS

So-called "actual leg length" is a measurement of length from the anterior-superior spine of the ilium to the medial malleolus. Obviously such a measurement is not an absolutely accurate determination of leg length because the points of measurement are from a landmark on the pelvis to one on the leg. Since it is impossible to palpate a point on the femur under the anterior superior spine, it is necessary to use the landmark on the pelvis. It becomes necessary, therefore, to fix the alignment of the pelvis in relation to the trunk and legs before taking measurements to insure the same relationship of both extremities to the pelvis. Pelvic rotation or lateral tilt will change the relationship of the pelvis to the extremities enough to make a considerable difference in measurement. To obtain as much accuracy as possible, the patient lies supine on a table with trunk, pelvis, and legs in straight alignment and legs close together. The distance from the anterior superior spine to the umbilicus is measured on right and left to check against lateral pelvic tilt or rotation. If there is a difference in measurements, the pelvis is leveled and any rotation corrected so far as possible before leg length measurements are taken.

"Apparent leg length" is a measurement from the umbilicus to the medial malleolus. This type of measurement is more often a source of confusion than an aid in determining differences in length for the purpose of applying a lift to correct pelvic tilt. The confusion arises from the reversal of the picture between standing and lying, which occurs when the pelvic tilt is due to muscle imbalance rather than actual leg length difference.

In standing, a fault in alignment will result when a weak muscle fails to provide adequate support for weight-bearing.

A weakness of the right Gluteus medius allows the pelvis to deviate toward the right and ride slightly upward giving the appearance of a *longer* right leg. If the postural fault has been of long standing there is usually an associated imbalance in the lateral trunk muscles in which the right laterals are stronger than the left. (See p. 291.)

In lying, a fault in alignment will more often result from the pull of a strong muscle. In the supine position an individual who has the type of imbalance described above, (i.e., a weak right medius and strong right laterals) will tend to lie with the pelvis higher on the right, pulled up by the stronger lateral abdominal muscles. This position in turn draws the right leg up so that it appears *shorter* than the left.

It is recommended that the need for an elevation on a shoe be determined by measurements in the standing rather than lying position. The boards of various thicknesses (see p. 309) are used for this purpose.

INTERPRETATION OF TEST FINDINGS

In the usual case of faulty posture, the pattern of faulty body mechanics as determined by the alignment test will be confirmed by the muscle tests if both procedures have been accurate. At times, there may be an apparent discrepancy in test findings. The inconsistency may be based on such things as the following: The effects of an old injury or disease may have altered the alignment pattern particularly as related to handedness patterns; effects of a recent illness or injury may have been superimposed on an established pattern of imbalance; or a child with a lateral curvature of the spine may be in a transition stage between a C-curve and an S-curve.

Except in flexible children, the postural faults seen at the time of examination will usually correspond with the habitual faults of the individual. With children it is necessary and advisable to do repeated tests of alignment, and to obtain information regarding the habitual posture from the parent and teacher who see them frequently. It is also advisable to keep photographic records of children's posture for a really worthwhile evaluation of postural changes in growing children.

It is of particular importance that girls between the ages of ten and fourteen have periodic examination of the spine because more spinal curvatures occur in girls than in boys, and they usually make their appearance between these ages.

Postural Examination

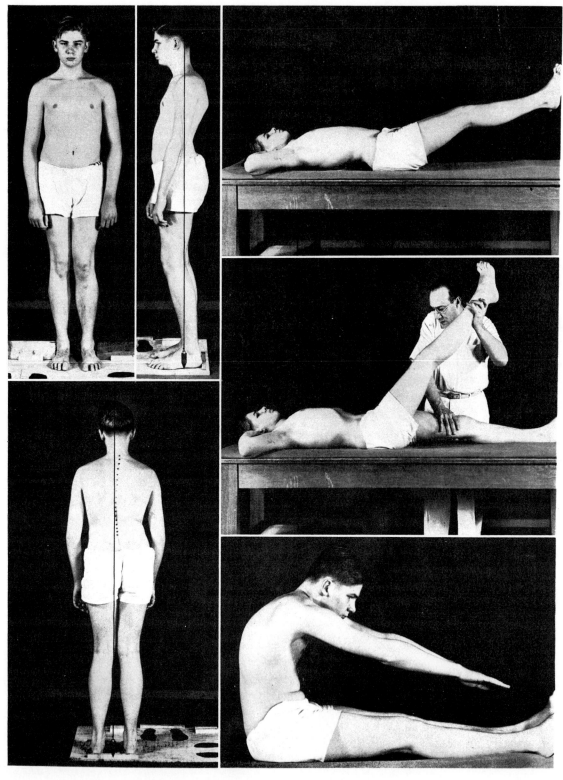

Photographs showing faulty alignment, limitation of motion in flexibility tests, and abdominal muscle weakness. Examination findings on this case are recorded on the Postural Examination Chart on the facing page.

POSTURAL EXAMINATION CHART

Name *D.L.* .... Cl. no. .... Doctor .... —

Diagnosis *Faulty posture* .... Date of 1st Ex. .... —

Onset .... Date of 2nd Ex. ... —

Occupation *High school student* .... Height — Weight —

Handedness *Right* .... Age *17* .... Sex *M* .... Leg length: Left — Right —

PLUMB ALIGNMENT

Side view: Lt. *Knee, pelvis, and head anterior* Rt. *(Same as from left)*

Back view: Deviated lt. .... Deviated rt. *Body from feet upward*

SEGMENTAL ALIGNMENT

Feet		X	Hammer toes		Hallux valgus		Low ant. arch		Ant. foot varus	
		L	Pronated >		Supinated		Flat long. arch		Pigeon toes	
Knees		B	Med. rotat. *R > L*		Lat. rotat.	B	Knock-knees *slight*			
			Hyperext. >	B	Flexed *L > R*		Bowlegs		Tibial torsion	
Pelvis		R	Leg in postural add.		Rotation	Ant.	Tilt	Ant.	Deviation	
Low back		X	Lordosis *marked*		Flat		Kyphosis		Operation	
Up. back		X	Kyphosis		Flat	B	Scap. abducted *R>L*		Scap. elevated	
Thorax		X	Depressed chest		Elevated chest		Rotation	Post.	Deviation *slight*	
Spine			Total curve	L	Lumbar – *Thoracic*		Thoracic	R	Cervical – *Thoracic*	
Abdomen		X	Protruding *slight*		Scars					
Shoulder			Low		High	B	Forward	B	Med. rotated	
Head		X	Forward		Torticollis		Tilt		Rotation	

TESTS FOR FLEXIBILITY AND MUSCLE LENGTH

Forward bending *Limited 7"* Bk. *Tight* H.S. *Tight* G.S. *Sl. tight*

Arm overhead elevation: Lt. *Slightly limited* Rt. *Normal length*

Hip flexors: Lt. *Marked tightness* Rt. *Marked tightness*

Tensor fas. lata: Lt. *Slight tightness* Rt. *Normal length*

Trunk extension: *Normal range*

Trunk lat. flex.: To lt. *Slightly limited* To rt. *Normal range*

MUSCLE STRENGTH TESTS

L		R
70	Mid. trapezius	70
(60)	Low. trapezius	60
100	Back extensors	100
100	Glut. medius	70
100	Glut. maximus	100
100	Hamstrings	100
100	Hip flexors	100
80	Tib. posterior	100
Weak	Toe flexors	Weak

R 90% TRUNK RAISING

Slight weakness

50% LEG LOWERING

SHOE CORRECTION

Left		Right
1/8"	(Wide Heel) Inner wedge (Narrow heel)	
3/16"	Level heel raise	
Medium, bar	Metatarsal support	Medium, bar
	Longitudinal support	

NOTES:

TREATMENT

Massage: ....

Infra-red: ....

Moist Heat: ....

Paraffin Bath: ....

Other: ....

Exercises:

Bk. Lying Pel. tilt and breath. ...X...

Pel. tilt and leg sl. ...X...

Head and sh. raising *(omit)*

Shoulder add. stretch ...X...

Straight leg-raise ...X...

Hip flex. stretch ...X...

Sd. Lying Stretch *left* tensor ...X...

Sitting Forward bending

To stretch low bk. ...X...

To stretch h. s. ...X...

Wall-sitting

Middle trapezius ...X...

Lower trapezius ...X...

Standing Foot and knee ex. ...X...

Wall-standing ...X...

Other Exercises: ....

Stretching toe extensors

Cross-sectional exercise for left

External oblique and right

Internal oblique

Support: ....

Good and Faulty Posture: Summary Chart

Good Posture	Part	Faulty Posture
In standing, the longitudinal arch has the shape of a half dome. Barefoot or in shoes without heels, the feet toe out slightly. In shoes with heels the feet are parallel. In walking with or without heels, the feet are parallel and the weight is transferred from the heel along the outer border to the ball of the foot. In running the feet are parallel or toe-in slightly. The weight is on the balls of the feet and toes because the heels do not come in contact with the ground.	Foot	Low longitudinal arch or flat foot. Low metatarsal arch, usually indicated by calluses under the ball of the foot. Weight borne on the inner side of the foot (pronation). "Ankle rolls in." Weight borne on the outer border of the foot (supination). "Ankle rolls out." Toeing out while walking, or while standing in shoes with heels ("outflared" or "slue-footed"). Toeing in while walking or standing ("pigeon-toed").
Toes should be straight, that is, neither curled downward nor bent upward. They should extend forward in line with the foot and not be squeezed together or overlap.	Toes	Toes bend up at the first joint and down at middle and end joints so that the weight rests on the tips of the toes (hammer toes). This fault is often associated with wearing shoes that are too short. Big toe slants inward toward the mid-line of the foot (hallus valgus). "Bunion." Ths fault is often associated with wearing shoes that are too narrow and pointed at the toes.
Legs are straight up and down. Knee caps face straight ahead when feet are in good position. Looking at the knees from the side, the knees are straight, i.e., neither bent forward nor "locked" backward.	Knees and Legs	Knees touch when feet are apart (knock-knee). Knees are apart when feet touch (bow-legs). Knee curves slightly backward (hyperextended knee). "Back-knee." Knee bends slightly forward, that is, it is not as straight as it should be (flexed knee). Knee-caps face slightly toward each other (medially rotated femurs). Knee-caps face slightly outward (laterally rotated femurs).
Ideally, the body weight is borne evenly on both feet and the hips are level. One side is not more prominent that the other as seen from front or back, nor is one hip more forward or backward than the other as seen from the side. The spine does not curve to the left or the right side. (A *slight* deviation to the left in right-handed individuals and to the right in left-handed individuals should not be considered abnormal, however. Also, since a tendency toward a *slightly* low right shoulder and *slightly* high right hip is frequently found in right-handed people, and vice versa for left-handed, such deviations should not be considered abnormal.)	Hips, Pelvis, and Spine Back view	One hip is higher than the other (lateral pelvic tilt). Sometimes it is not really much higher but appears so because a sideways sway of the body has made it more prominent. (Tailors and dressmakers often notice a lateral tilt because the hemline of skirts or length of trousers must be adjusted to the difference.) The hips are rotated so that one is farther forward than the other (clockwise or counter-clockwise rotation).

Good and Faulty Posture: Summary Chart

Good Posture	Part	Faulty Posture
The front of the pelvis and the thighs are in a straight line. The buttocks is not prominent in back but slopes slightly downward. The spine has four natural curves. In the neck and lower back the curve is forward, in the upper back and lowest part of the spine (sacral region) it is backward. The sacral curve is a fixed curve while the other three are flexible.	Spine and Pelvis Side view	The low back arches forward too much (lordosis). The pelvis tilts forward too much. The front of the thigh forms an angle with the pelvis when this tilt is present. The normal forward curve in the low back has straightened out. The pelvis tips backward and there is a slightly backward slant to the line of the pelvis in relation to the front of the hips (flat back). Increased backward curve in the upper back (kyphosis or round upper back). Increased forward curve in the neck. Almost always accompanied by round upper back and seen as a forward head. Lateral curve of the spine (scoliosis); toward one side (C-curve), toward both sides (S-curve).
In young children up to about the age of 10 the abdomen normally protrudes somewhat. In older children and adults it should be flat.	Abdomen	Entire abdomen protrudes. Lower part of the abdomen protrudes while the upper part is pulled in.
A good position of the chest is one in which it is slightly up and slightly forward (while the back remains in good alignment). The chest appears to be in a position about half way between that of a full inspiration and a forced expiration.	Chest	Depressed, or "hollow-chest" position. Lifted and held up too high, brought about by arching the back. Ribs more prominent on one side than on the other. Lower ribs flaring out or protruding.
Arms hang relaxed at the sides with palms of the hands facing toward the body. Elbows are slightly bent, so forearms hang slightly forward. Shoulders are level and neither one is more forward or backward than the other when seen from the side. Shoulder blades lie flat against the rib cage. They are neither too close together not too wide apart. In adults, a separation of about four inches is average.	Arms and Shoulders	Holding the arms stiffly in any position forward, backward, or out from the body. Arms turned so that palms of hands face backward. One shoulder higher than the other. Both shoulders hiked-up. One or both shoulders drooping forward or sloping. Shoulders rotated either clockwise or counterclockwise. Shoulder blades pulled back too hard. Shoulder blades too far apart. Shoulder blades too prominent, standing out from the rib cage ("winged scapulae").
Head is held erect in a position of good balance.	Head	Chin up too high. Head protruding forward. Head tilted or rotated to one side.

Terminology

	Nomina Anatomica, 3rd ed., 1966[32] Terminology used in this text:	Basle Nomina Anatomica, 1895 (BNA)
Vertebral column	Thoracolumbar	Dorsolumbar
Scapula	Lateral border Medial border Superior angle	Axillary border Vertebral border Medial angle
Forearm	Anterior surface Posterior surface	Ventral or volar surface Dorsal surface
Wrist	Flexor retinaculum Scaphoid (carpal) Trapezium	Transverse carpal ligament Navicular (carpal) Greater multangular
Ankle	Tendo calcaneous	Tendo Achilles
Nerves	Anterior interosseus nerve Dorsal root Ventral root Lateral pectoral nerve Medial pectoral nerve Vestibulocochlear nerve	Volar interosseus nerve Posterior root (of spinal nerve) Anterior root (of spinal nerve) Lateral anterior thoracic nerve Medial anterior thoracic nerve Auditory nerve
Muscles Upper extremity	Extensor digitorum Flexor digitorum superficialis	Extensor digitorum communis Flexor digitorum sublimis
Scapula	Levator scapulae	Levator anguli scapulae
Facial	Depressor anguli oris Depressor labii inferioris Levator anguli oris Nasalis, alar part Nasalis, transverse part Levator labii superioris Levator labii sup. alaeque nasi Zygomaticus minor Corrugator supercilii Epicranius, frontal belly	Triangularis Quadratus labii inferioris Caninus Dilator naris Depressor septi Quadratus labii superioris Corrugator Frontalis
Mastication	Pterygoidous lateralis Pterygoidous medialis	Pterygoideus externus Pterygoideus internus

Bibliography

References Cited in the Text

1. O'Connell AL, Gardner EB: *Understanding the Scientific Basis of Human Motion.* Baltimore, The Williams & Wilkins Co., 1972.
2. Haines PW: On muscles of full and of short action. *J. Anat.,* 69–B:20, 1934.
3. Mountcastle VB (editor): *Medical Physiology,* ed. 14. St. Louis, The C.V. Mosby Co., 1980.
4. Legg AT: Physical therapy in infantile paralysis. In Mock: *Principles and Practice of Physical Therapy,* Vol. II. Hagerstown, MD, W.F. Prior Co., Inc., 1932, p 45.
5. Lilienfeld AM, Jacobs M, Willis M: A study of the reproducibility of muscle testing and certain other aspects of muscle scoring. *Phys Ther Rev* 34:282, 1954.
6. Medical Research Council: *Aids to the Investigation of Peripheral Nerve Injuries.* War Memorandum No. 7, 2nd ed. revised. London, His Majesty's Stationery office, 1943.
7. Inman VT, Saunders JB de CM, Abbott LC: Observations on the function of the shoulder joint. *J Bone Joint Surg* 26:1, 1944.
8. Brodal A: *Neurological Anatomy: In Relation to Clinical Medicine,* ed. 3. New York, Oxford University Press, 1981.
9. Peele TL: *The Neuroanatomic Basis for Clinical Neurology,* ed. 3. New York, McGraw-Hill Book Co., 1977.
10. Keegan JJ, Garrett FD: The segmental distribution of the cutaneous nerves in the limbs of man. *Anat Rec* 102, 1948.
11. Goss CM (editor): *Gray's Anatomy of the Human Body,* ed. 28. Philadelphia, Lea & Febiger, 1966.
12. Eycleshymer AC, Shoemaker DM: *A Cross-Section Anatomy.* New York, D. Appleton and Co., 1923.
13. Brash JC: *Neuro-vascular Hila Muscles.* London, E. and S. Livingstone, Ltd., 1955.
14. Coyne JM, Kendall FP, Latimer RM, et al: Evaluation of brachial plexus injury. *J Am Phys Ther Assoc* 48:733, 1968.
15. Romanes GJ (editor); *Cunningham's Textbook of Anatomy,* ed. 10. London, Oxford University Press, 1964.
16. Anson BJ (editor): *Morris' Human Anatomy,* ed. 12. New York, McGraw Hill Book Co., 1966.
17. Spalteholz W: *Hand Atlas of Human Anatomy.* Vols. II and III, ed. 6. London, The J.B. Lippincott Co.
18. deJong RN: *The Neurologic Examination,* ed. 2 and 3. New York, Harper & Row Publishers, 1967.
19. Haymaker W, Woodhall B: *Peripheral Nerve Injuries,* ed. 2. Philadelphia, W.B. Saunders Co., 1953.
20. Foerster O, Bumke O: *Handbuch der Neurologie,* Vol. V. Breslau, 1936.
21. Schade JP: *The Peripheral Nervous System.* New York, Am. Elsevier Publishing Co., Inc., 1966.
22. Boileau JC, Basmajian JV: *Grant's Method of Anatomy,* 7th ed. Baltimore, Williams & Wilkins Co., 1965.
23. Crowe P, O'Connell A, Gardner E: An electromyographic study of the abdominal muscles and certain hip flexors during selected sit-ups. Part I. Abstract of paper presented to Research Section of the 78th Annual Convention of the Am. Assoc. for Health Physical Education and Recreation. (Mimeographed), 1961.
24. Nachemson A, Elfstron G: *Intravital Dynamic Pressure Measurements in Lumbar Discs.* Stockholm, Almqvista Wiksell, 1970.
25. Kendall HO, Kendall FP: Normal flexibility according to age groups. *J Bone Joint Surg* 33-A:690, 1948.
26. Steindler A: *Kinesiology of the Human Body under Normal and Pathological Conditions.* Springfield, Charles C Thomas, 1955.
27. Appleton AB, Hamilton WJ, Simon G: *Surface and Radiological Anatomy,* ed. 2. Baltimore, The Williams & Wilkins Co., 1946.
28. Morton DJ: *The Human Foot.* Columbia University Press, 1935.
29. Whitman R: *A Treatise on Orthopaedic Surgery.* Philadelphia, Lea & Febiger, 1919, p 660.
30. Sobotta-Figge: *Atlas of Human Anatomy,* Vol. 1. Munich, Urban and Schwarzenberg, 1974.
31. Ralston HJ: Mechanics of voluntary muscle. *Am J Phys Med* 32:166, 1953.
32. International Anatomical Nomenclature Committee: *Nomina Anatomica,* ed. 3. Amsterdam, Excerpta Medica Foundation, 1966.

Selected Readings

Abd-el-Malek S: The part played by the tongue in mastication. *J Anat* 89:250, 1955.

Ahlback S-D, Lindahl O: Sagittal mobility of the hip-joint. *Acta Orthop Scand* 34:310, 1964.

Allbrook D: Movements of the lumbar spinal column. *J Bone Joint Surg* 39-B:339, 1957.

Allsop KG: Potential hazards of abdominal exercises. *J Health, Phys Ed Recreation* 42:89, 1971.

Alston W, Carlson KE, Feldman DJ, et al: A quantitative study of muscle factors in chronic low back syndrome. *J Am Geriatrics Soc* 14:1041, 1966.

American Academy of Orthopoedic Surgeons: *Joint Motion, Method of Measuring and Recording.* Chicago, 29 East Madison St., 1965.

Andersson BJG, Jonsson B, Ortegren R: Myoelectric activity in individual lumbar erector spinae muscles in sitting: a study with surface and wire electrodes. *Scand J Rehab Med* 3 Suppl:91, 1974.

Andersson GBJ, Ortengren R, Herberts P: Quantitative electromyographic studies of back muscle activity related to posture and loading. *Ortho Clin North Am* 8:85, 1977.

Andersson GBJ, Ortengren R, Nachemson A, et al: Lumbar disc pressure and myoelectric back muscle activity during setting. *Scand J Rehab Med* 6:104, 1974.

Andersson GBJ, Ortengren R, Nachemson AL, et al: The sitting posture: an electromyographic and discometric study. *Orthop Clin North Am* 6:105, 1975.

Ardran GM, Kemp FH: A radiographic study of movements of the tongue in swallowing. *Dent Pract* 5:252, 1955.

Ardran GM, Kemp FH: The mechanism of the larynx. II, The epiglottes and closure of the larynx. *Br J Radiol* 40:372, 1967.

Arnold GE. Physiology and pathology of the cricothyroid muscle. *Laryngoscope* 71:687, 1961.

Asmussen E, Klausen K: Form and function of the erect human spine. *Clin Orthop* 25:55, 1962.

Atkinson M, Dramer P, Wyman SM, et al: The dynamics of swallowing. I. Normal phyryngeal mechanisms. *J Clin Invest* 36:581, 1957.

Backdahl M, Carlsöö S: Distribution of activity in muscles acting on the wrist. *Acta Morph Neer-Scand* 4:136, 1961.

Baker AB (editor): *Clinical Neurology,* Vol. IV, ed. 2. New York, Harper and Brothers, Hoeber Medical Division. 1962.

Basmajian JV: Electromyography of iliopsoas. *Anat Rec* 132:127, 1958.

Basmajian JV: Electromyography of two-joint muscles. *Anat Rec* 129:371, 1957.

Basmajian JV: *Grant's Method of Anatomy,* ed. 9. Baltimore, The Williams & Wilkins Co, 1975.

Basmajian JV: *Muscles Alive,* ed. 4. Baltimore, The Williams & Wilkins Co., 1978.

Basmajian JV: *Primary Anatomy,* ed. 5. Baltimore, The Williams & Wilkins Co., 1964.

Basmajian JV, Dutta CR: Electromyography of the pharyngeal constrictors and levator palati in man. *Anat Rec* 139:561, 1961.

Basmajian JV, Latif A: Integrated actions and functions of the chief flexors of the elbow. *J Bone Joint Surg* 39-A:1106, 1957.

Basmajian JV, Travill A: Electromyography of the pronator muscles in the forearm. *Anat Rec* 139:45, 1961.

Bender JA, Kaplan HM: The multiple angle testing method for the evaluation of muscle strength. *J Bone Joint Surg* 45 A:135, 1963.

Blackburn SE, Portney LG: Electromyographic activity of back musculature during Williams' flexion exercises. *Phys Ther* 61:878, 1981.

Blakely WR, Garety EJ, Smith DE: Section of the cricopharyngeus muscle for dysphagia. *Arch Surg* 96:745, 1968.

Blanton PL, Biggs NL, Perkins RC: Electromyographic analysis of the buccinator muscle. *J Dent Res* 49:389, 1970.

Bole CT, Lessler MA: Electromyography of the genioglossus muscles in man. *J Appl Physiol* 21:1695, 1968.

Bosma JF: Deglutition: pharyngeal stage. *Physiol Rev* 37:275, 1957.

Brand PW, Beach RB, Thompson DE: Relative tension and potential excursion of muscles in the forearm and hand. *J Hand Surg* 6:209, 1981.

Brantigan OC, Voshell AF: The mechanics of the ligaments and menisci of the knee joint. *J Bone Joint Surg* 23:44, 1941.

Brash, JC (editor): *Cunningham's Manual of Practical Anatomy*, Vol. I, ed. 11. New York, Oxford University Press, 1948.

Brunnstrom S: *Clinical Kinesiology*, ed. 3. Philadelphia, F.A. Davis Co., 1972.

Bunnell, S: *Surgery of the Hand*, ed. 4, revised by Joseph H. Boyes. Philadelphia, J.B. Lippincott Co., 1964.

Carmen DJ, Blanton PL, Biggs NL: Electromyographic Study of the anterolateral abdominal musculature utilizing indwelling electrodes. *Am J Phys Med* 51:113, 1972.

Campbell EJM: *The Respiratory Muscles and the Mechanics of Breathing*. Chicago, Year Book Publishers, Inc., 1958.

Campbell EJM, Agostini E, Davis JN: *The Respiratory Muscles: Mechanisms and Neural Control*, ed. 2. Philadelphia, W.B. Saunders Co., 1970.

Chusid JG: *Correlative Neuroanatomy and Functional Neurology*, ed. 15. Los Altos, CA, Lange Medical Publications, 1973.

Clayson SJ, Newman IM, Debevec DF, et al: Evaluation of mobility of hip and lumbar vertebrae of normal young women. *Arch Phys Med Rehabil* 43:1, 1962.

Cleall JF: Deglutition: A study of form and function. *Am J Orthodont* 51:566, 1965.

Close JR: *Motor Function in the Lower Extremity*. Springfield, IL, Charles C Thomas, 1964.

Close JR, Kidd CC: The functions of the muscles of the thumb, the index and long fingers. *J Bone Joint Surg* 51-A:1601, 1969.

Close RI: Dynamic properties of mammalian skeletal muscles. *Physiol Rev* 52:129, 1972.

Cole TM: Goniometry: the measurement of joint motion. In Krusen, Kottke, Elwood: *Handbook of Physical Medicine and Rehabilitation*, ed. 2, Philadelphia, W.B. Saunders Co., 1971.

Corbin KB, Harrison F: Proprioceptive components of cranial nerves. The spinal accessory nerve. *J Comp Neuro* 69:315, 1938.

Cunningham DP, Basmajian JB: Electromyography of genioglossus and geniohyoid muscles during deglutition. *Anat Rec* 165:401, 1969.

Currier DP: Maximal isometric tension of the elbow extensors at varied positions. *Phys Ther* 52:1265, 1972.

Currier DP: Positioning for knee strengthening exercises. *Phys Ther* 57:148, 1977.

Cyriax J: *Textbook of Orthopaedic Medicine*, ed. 7, Vol. 1, Diagnosis of soft tissue lesions. London, Bailliere Tindall, 1978.

Davis GG: *Applied Anatomy*, ed. 5. Philadelphia, J.B. Lippincott Co., 1918.

deJong RN: *The Neurological Examination*, ed. 4. New York, Harper & Row Publishers, 1979.

DeLuca CJ, Forrest WJ: Force analysis of individual muscles acting simultaneously on the shoulder joint during isometric abduction. *J Biomech* 6:385, 1973.

DeSousa OM, Berzin F, Berardi AC: Electromyographic study of the pectoralis major and latissimus dorsi during medial rotation of the arm. *Electromyography* 9:407, 1969.

DeSousa OM, Demoraes JL, Demoraes Vieira FL: Electromyographic study of the brachioradialis muscle. *Anat Rec* 139:125, 1961.

DeSousa, OM, Furlani J: Electromyographic study of the m. rectus abdominis. *Acta Anat* 88:281, 1974.

Duchenne GB: In Kaplan EB: *Physiology of Motion*. Philadel-phia, J.B. Lippincott Co., 1949.

Duvall EN: *Kinesiology: The Anatomy of Motion*. Englewood Cliffs, NJ, Prentice-Hall, Inc., 1959.

Eaton RG, Littler JW: A study of the basal joint of the thumb. *J Bone Joint Surg* 51-A:661, 1969.

Ekholm J, Arborelius U, Fahlcrantz A, et al: Activation of abdominal muscles during some physiotherapeutic exercises. *Scand J Rehab Med* 11:75, 1979.

Elftman H: Biomechanics of muscle. *J Bone Joint Surg* 48-A:363, 1966.

Elliott HC: *Textbook of Neuroanatomy*, ed. 2. Philadelphia, J.B. Lippincott Co., 1969.

Faaborg-Andersen K: Electromyographic investigation of intrinsic laryngeal muscles in humans. *Acta Physiol Scand* 41:Suppl 140–11, 1957.

Farfan HF: *Mechanical Disorders of the Low Back*. Philadelphia, Lea & Febiger, 1973.

Farfan HF: Muscular mechanism of the lumbar spine and the position of power and efficiency. *Orthop Clin North Am* 6:135, 1975.

Fenn WO, Rahn H: *Handbook of Physiology*, Section 3, Respiration, Vol. 1. Washington, DC, Am. Physiol. Society, 1964.

Fischer FJ, Houtz SJ: Evaluation of the function of the gluteus maximus muscle. *Am J Phys Med* 47:182, 1968.

Flint MM: Lumbar posture: A study of Roentgenographic measurement and the influence of flexibility and strength. *Res Quart* 34:15, 1963.

Flint MM: An electromyographic comparison of the function of the iliacus and the rectus abdominis muscles. *J Am Phys Therap Assoc* 45:248, 1965.

Flint MM: Abdominal muscle involvement during performance of various forms of sit-up exercise. *Am J Phys Med* 44:224, 1965.

Flint MM, Gudgell J: Electromyographic study of abdominal muscular activity during exercise. *Res Quart* 36:29, 1965.

Floyd WF, Silver PHS: Electromyographic study of patterns of activity of the anterior abdominal wall muscles in man. *J Anat* 84:132, 1950.

Floyd WF, Silver PHS: The function of the erectores spinae muscles in certain movements and postures in man. *J Physiol* 129:184, 1955.

Frankel VH, Nordin M: *Basic Biomechanics of the Skeletal System*. Philadelphia, Lea & Febiger, 1980.

Fujiwara M, Basmajian JV: Electromyographic study of two-joint muscles. *Am J Phys Med* 54:234, 1975.

Furlani J: Electromygraphic study of the m. biceps brachii in movements of the glenohumeral joint. *Acta Anat* 96:270, 1976.

Gardiner MD: *The Principles of Exercise Therapy*. London, G. Bell & Sons, Ltd., 1956.

Gardner E, Gray DJ, O'Rahilly R: *Anatomy*, ed. 4. Philadelphia, W.B. Saunders Co., 1975.

Girardin Y: EMG action potentials of rectus abdominis muscle during two types of abdominal exercises. In Cerquigleni S, Venerando A, Wartenweiler J: *Biomechanics III*, Baltimore, University Park Press, 1973.

Godfrey KE, Kindig LE, Windell EJ: Electromyographic study of duration of muscle activity in sit-up variations. *Arch Phys Med Rehabil* 58:132, 1977.

Goss CM (editor): *Gray's Anatomy of the Human Body*, ed. 29. Philadelphia, Lea & Febiger, 1973.

Gowitzke BA, Milner MM: *Understanding the Scientific Basis of Human Motion*, ed. 2. Baltimore, The Williams & Wilkins Co., 1980.

Gracovetsky S, Farfan HF, Lamy C: The mechanism of the lumbar spine. *Spine* 6:249, 1981.

Gray ER: The role of leg muscles in variations of the arches in normal and flat feet. *J Am Phys Therap Assoc* 49:1084, 1969.

Grieve GP: The sacro-iliac joint. *Physiotherapy* 62:384, 1976.

Grieve DW, Arnott AW: The production of torque during axial rotation of the trunk. *J Anat* 107:147, 1970.

Grinker RR, Sahs AL: *Neurology*, ed. 6. Springfield, Charles C Thomas, 1966.

Gutin B, Lipetz S: Electromyographic investigation of rectus abdominis in abdominal exercises. *Res Quart* 42:256, 1971.

Guyton AC: *Textbook of Medical Physiology*, ed. 3. Philadelphia, W.B. Saunders Co., 1966.

Haffajee D, Moritz U, Svantesson G: Isometric knee extension as a function of joint angle, muscle length and motor unit activity. *Acta Orthop Scand* 43:138, 1972.

Halpern A, Bleck E: Sit-up exercise: an electromyographic study. *Clin Orthop Related Res* 145:172, 1979.

Harvey VP, Scott GD: Reliability of a measure of forward flexibility and its relation to physical dimensions of college women. *Res Quart* 38:28, 1965.

Harvey VP, Scott GD: An investigation of the curl-down test as a measure of abdominal strength. *Res Quart* 38:22, 1965.

Hasue M, Fujiwara M, Kikuchi S: A new method of quantitative measurement of abdominal and back muscle strength. *Spine* 5:143, 1980.

Haymaker W: *Bing's Local Diagnosis in Neurological Diseases*, ed. 15. St. Louis, The C.V. Mosby Co., 1969.

Hicks JH. The three weight-bearing mechanisms of the foot. In Evans FG: *Biomechanical Studies of the Musculo-Skeletal System*. Springfield, IL, Charles C Thomas, 1961.

Hirano M, Koike Y, von Leden H: The sterno-hyoid muscle during phonation. *Acta Otolaryngol* 64:500, 1967.

Hirano M, Koike Y, Joyner J: Style of Phonation. *Arch Otolaryngol* 89:902, 1969.

Hollinshead WH, Jenkins DB: *Functional Anatomy of the Limbs and Back*, ed. 5. Philadelphia, W.B. Saunders Co., 1981.

Houtz SJ, Lebow MJ, Beyer FR: Effect of posture on strength of the knee flexor and extensor muscles. *J Appl Physiol* 11:475, 1957.

Houtz SJ: Influence of gravitational forces on function of lower extremity muscles. *J Appl Physiol* 19:999, 1964.

Ingelmark BE, Lindstrom J: Asymmetries of the lower extremities and pelvis and their relations to lumbar scoliosis. *Acta Morph Neerl Scand* 5:221, 1963.

Johnson JTH, Kendall HO: Isolated paralysis of the serratus anterior muscle. *J. Bone Joint Surg* 37-A:567, 1955. Also reprinted in *Orthop Appl J* 18:201, 1964.

Johnson JTH, Kendall HO: Localized shoulder girdle paralysis of unknown etiology. *Clin Orthop* 20:151, 1961.

Jones FW: *The Principles of Anatomy*, ed. 2. Baltimore, The Williams & Wilkins Co., 1942.

Jonsson B, Olofsson BM, Steffner LCh: Function of the teres major, latissimus dorsi and pectoralis major muscles. *Acta Morph Neerl Scand* 9:275, 1972.

Kaplan EB: *Functional and Surgical Anatomy of the Hand*, ed. 2. Philadelphia, J.B. Lippincott Co., 1965.

Keagy RD, Brumlik J, Bergan JJ: Direct electromyography of the psoas major muscle in man. *J Bone Joint Surg* 48-A: 1377, 1966.

Kendall HO, Kendall FP: Developing and maintaining good posture. *J Am Phys Ther Assoc* 48:319, 1968.

Kendall FP: A criticism of current tests and exercises for physical fitness. *J Am Phys Ther Assoc* 45:187, 1965.

Kendall HO, Kendall FP: Posture, Flexibility, and Abdominal Muscle Tests. Leaflet, Baltimore, Waverly Press, 1963. Revised, 1964.

Kendall HO: Watch those T.V. exercises. *T.V. Guide* II-31:5, 1963.

Kendall FP: Range of motion. *The Correlation of Physiology with Therapeutic Exercise*, New York, Am Phys Ther Assoc, 1956.

Kendall HO, Kendall FP, Boynton DA: *Posture and Pain.* Baltimore, The Williams & Wilkins Co., 1952. Reprinted by Robert E. Krieger Publishing Co., Melbourne, FL, 1971.

Kendall HO, Kendall FP: Functional muscle testing. *Physical Medicine and General Practice* Chapt. XII. New York, Paul B. Hoeber, Inc., 1952.

Kendall HO, Kendall FP: Orthopedic and physical therapy objectives in poliomyelitis treatment. *Physiother Rev* 27:159, 1947.

Kendall HO, Kendall FP: Assisted in writing *Physical Therapy for Lower Extremity Amputees.* War Department Technical Manual TM-8-293:14–42 and 58–65, Washington, D.C., U.S. Government Printing Office, 1946.

Kendall HO, Kendall FP: Unpublished report on the Posture Survey at U.S. Military Academy, West Point, 1945.

Kendall HO, Kendall FP: The role of abdominal exercise in a program of physical fitness. *J Health Phys Ed* 480, 1943.

Kendall HO, Kendall FP: Gluteus medius and its relation to body mechanics. *Physiother Rev* 21:131, 1941.

Kendall HO, Kendall FP: *Care During the Recovery Period of Paralytic Poliomyelitis.* U.S. Public Health Bulletin No. 242, Washington, D.C., U.S. Government Printing Office, 1939.

Kendall HO: Some interesting observations about the after care of infantile paralysis patients. *J Excep Child* 3:107, 1937.

Kendall HO, Kendall FP: Study and Treatment of Muscle Imbalance in Cases of Low Back and Sciatic Pain. Pamphlet. Privately printed, Baltimore, 1936.

Kendall PH, Jenkins JM: Exercises for backache. *Physiotherapy* 54:158, 1968.

Klopsteg PE, Wilson PD, et al: *Human Limbs and Their Substitutes.* New York, McGraw-Hill Book Co., 1954.

Klousen K, Rasmussen B: On the location of the line of gravity in relation to L5 in standing. *Acta Physiol Scand* 72:45, 1968.

Kotby MN: Electromyography of the laryngeal muscles. *Electroenceph Clin Neurophysiol* 26:341, 1969.

Kramer P, Atkinson M, Wyman SM, et al: The dynamics of swallowing. *J Clin Invest* 36:589, 1957.

Kraus H: Effects of lordosis on the stress in the lumbar spine. *Clin Orthop* 117:56, 1976.

Landsmeer JMF: The anatomy of the dorsal aponeurosis of the human finger and its functional significance. *Anat Rec* 104:31, 1949.

Last RJ: Innervation of the limbs. *J Bone Joint Surg* 31-B:452, 1949.

Lindstrom A, Zachrisson M: Physical therapy for low back pain and sciatica. *Scand J Rehabil Med* 2:37, 1970.

Lieb FJ, Perry J: Quadriceps function. *J Bone Joint Surg* 53-A:749, 1971.

Lindahl O: Determination of the sagittal mobility of the lumbar spine. *Acta Orthop Scand* 37:241, 1966.

Lindahl O, Movin A: The mechanics of extension of the knee-joint. *Acta Orthop Scand* 38:226, 1967.

Lipetz S, Gutin B: Electromyographic study of four abdominal exercises. *Med Sci Sports* 2:35, 1970.

Lockhart RD, Hamilton GF, Fyfe FW: *Anatomy of the Human Body.* Philadelphia, J.B. Lippincott Co., 1959.

Loebl WY: Measurement of spinal posture and range of spinal movement. *Ann Phys Med* 9:103, 1967.

Long C: Intrinsic-extrinsic muscle control of the fingers. *J Bone Joint Surg* 50-A:973, 1968.

Loptata M, Evanich MJ, Lourenco RV: The electromyogram of the diaphragm in the investigation of human regulation of ventilation. *Chest* 70 Suppl:162S, 1976.

Low JL: The reliability of joint measurement. *Physiotherapy* 62:227, 1976.

Mann R, Inman VT: Phasic activity of intrinsic muscles of the foot. *J Bone Joint Surg* 46-A:469, 1964.

Manter JT: Variations of the interosseous muscles of the human foot. *Anat Rec* 93:117, 1945.

McKenzie J: The development of the sternomastoid and trapezius muscles. *Contributions to Embryology, No. 258*, 37:121. Washington, DC, Carnegie Institution, 1962.

Michelle AA: *Iliopsoas.* Springfield, IL, Charles C Thomas, 1962.

Moore KL: *Clinically Oriented Anatomy.* Baltimore, Williams & Wilkins, 1980.

Moore ML: The measurement of joint motion. Part II: The technique of goniometry. *Phys Ther Rev* 29:256, 1949.

Moore ML: Clinical assessment of joint motion. In Licht S: *Therapeutic Exercise*, ed 2. Baltimore, Waverly Press, Inc., 1965.

Morris JM, Benner G, Lucas DB: An electromyographic study of the intrinsic muscles of the back in man. *J Anat* 96:509, 1962.

Murphey DL, Blanton PL, Biggs NL: Electromyographic investigation of flexion and hyperextension of the knee in normal adults. *Am J Phys Med* 50:80, 1971.

Nachemson A: Electromyographic studies on the vertebral portion of the psoas muscle. *Acta Orthop Scand* 37:177, 1966.

Nachemson A: Physiotherapy for low back pain patients. *Scand J Rehab Med* 1:85, 1969.

Nachemson A: Towards a better understanding of low back pain: A review of the mechanics of the lumbar disc. *Rheumatol Rehab* 14:129, 1975.

Nachemson A: A critical look at the treatment for low back pain. *Scand J Rehab Med* 11:143, 1979.

Nachemson A, Lindh M: Measurement of abdominal and back muscle strength with and without low back pain. *Scand J Rehab Med* 1:60, 1969.

Ouaknine G, Nathan H: Anastomotic connections between the eleventh nerve and the posterior root of the first cervical nerve in humans. *J Neurosurg* 38:189, 1973.

Paré EB, Schwartz JM, Stern JT: Electromyographic and anatomical study of the human tensor fasciae latae muscle. In *Proceedings of the 4th Congress of the International Society of Electrophysiological Kinesiology.* Boston, Published by the organizing committee, 1979.

Partridge MJ, Walters CE: Participation of the abdominal muscles in various movements of the trunk in man. *Phys Ther Rev* 39:791, 1959.

Patton NJ, Mortensen OA: A study of some mechanical factors affecting reciprocal activity in one-joint muscles. *Anat Rec* 166:360, 1970.

Pearson AA: The spinal accessory nerve in human embryos. *J Comp Neurol* 68:243, 1938.

Pearson AA, Sauter RW, Herrin GR: The accessory nerve and its relation to the upper spinal nerves. *J Anat* 114-A:371, 1964.

Pennal GF, Conn GS, McDonald G, et al: Motion studies of the lumbar spine. *J Bone Joint Surg* 54-B:442, 1972.

Pressman JJ, Kelemen G: Physiology of the larynx. *Physiol Rev* 35:506, 1955.

Provins KA: Maximum force exerted about the elbow and shoulder joints on each side separately and simultaneously. *J Appl Physiol* 7:393, 1955.

Quiring DP: In Warfel JH: *The Head, Neck and Trunk,* ed. 3. Philadelphia, Lea & Febiger, 1967.

Ralston HJ, Todd FN, Inman VT: Comparison of electrical activity and duration of tension in the human rectus femoris muscle. *Electromyog Clin Neurophysiol* 16:271, 1976.

Ramsey GH, Watson JS, Gramiak R, et al: Cinefluorographic analysis of the mechanism of swallowing. *Radiology* 64:498, 1955.

Romanes GJ (editor): *Cunningham's Textbook of Anatomy,* ed. 11. London, Oxford University Press, 1972.

Rustad WH, Morrison LF: Revised anatomy of the recurrent laryngeal nerves. *Laryngoscope* 62:237, 1952.

Salter N, Darcus HD: The effect of the degree of elbow flexion on the maximum torques developed in pronation and supination of the right hand. *J Anat* 86-B:197, 1952.

Saunders JB deCM, Davis C, Miller ER: The mechanism of deglutition. *Ann Otol Rhinol Laryngol* 60:897, 1951.

Schewing LE, Pauly JE: An electromyographic study of some muscles acting on the upper extremity of man. *Anat Rec* 135:239, 1959.

Scudder GN: Torque curves produced at the knee during isometric and isokinetic exercise. *Arch Phys Med Rehabil* 61:68, 1980.

Sharp JT, Draz W, Danon J, et al: Respiratory muscle function and the use of respiratory muscle electromyography in the evaluation of respiratory regulation. *Chest* 70 Suppl:150S, 1976.

Sharrard WJW: The segmental innervation of the lower limb muscles in man. *Ann Ray Col Surg* 35:106, 1964.

Shelton RL, Bosma JF, Sheets BV: Tongue, hyoid and larynx displacement in swallow and phonation. *J Appl Physiol* 15:283, 1960.

Shevlin MG, Lehmann JF, Lucci JA: Electromyographic study of the function of some muscles crossing the glenohumeral joint. *Arch Phys Med* 50:264, 1969.

Silbiger M, Pikielney R, Douner M: Neuromuscular disorders affecting the pharynx. *Invest Radiol* 2:442, 1967.

Smith JW: Muscular control of the arches of the foot in standing: an electromyographical assessment. *J Anat* 88-B:152, 1954.

Soderberg GL: Exercises for the abdominal muscles. *J Health, Phys Ed Recreation* 37:67, 1966.

Stokes IAF, Abery JM: Influence of the hamstring muscles on lumbar spine curvature in sitting. *Spine* 5:6, 1980.

Straus WL, Howell AB: The spinal accessory nerve and its musculature. *Quart Rev Biol* 11:387, 1936.

Sunderland S: *Nerves and Nerve Injuries,* ed. 2. Edinburgh, Churchill Livingstone, 1978.

Suzuki N: An electromyographic study of the role of muscles in arch support of the normal and flat foot. *Nagoya Med J* 17:57, 1972.

Tavores AS: L'Innervation Des Muscles Pectoraux. Acta Anat 21:132, 1954.

Travill AA: Electromyographic study of the extensor apparatus of the forearm. *Anat Rec* 144:373, 1962.

Truex RC, Carpenter MB (editors): *Strong and Elwyn's Human Neuroanatomy,* ed. 6. Baltimore, The Williams & Wilkins Co., 1969.

Vogel PH: The innervation of the larynx of man and dog. *Am J Anat* 90:427, 1952.

Walters CE, Partridge MJ: Electromyographic study of the differential action of the abdominal muscles during exercise. *Am J Phys Med* 36:259, 1957.

Warwick R, Williams PL (editors): *Gray's Anatomy,* British ed. 36. Philadelphia, W.B. Saunders Co., 1980.

Weathersby HT, Sutton LR, Erusen UL: The kinesiology of muscles of the thumb: an electromyographic study. *Arch Phys Med Rehabil* 44:321, 1963.

Wells KF: *Kinesiology,* ed. 4. Philadelphia, W.B. Saunders Co., 1966.

White A, Panjabi M: *Clinical Biomechanics of the Spine.* Philadelphia, J.B. Lippincott Co., 1978.

Wiles P: Movements of the lumbar vertebrae during flexion and extension. *Proc Roy Soc Med* 28:647, 1935.

Wilkie DR: The mechanical properties of muscle. *Br Med Bull* 12:177, 1956.

Williams M, Lissner HR: *Biomechanics of Human Motion.* Philadelphia, WB Saunders Co., 1962.

Williams M, Stutzman L: Strength variation through the range of joint motion. *Phys Ther Rev* 39:145, 1959.

Williams PC: *The Lumbosacral Spine.* New York, McGraw-Hill Book Co., The Blakiston Division, 1965.

Wolf S: Normative data on low back mobility and activity levels. *Am J Phys Med* 58:217, 1979.

Woodburne RT: *Essentials of Human Anatomy,* ed. 5. New York, Oxford University Press, 1973.

Wright WG: *Muscle Function.* New York, Paul B. Hoeber, Inc., 1928.

Zemlin WR: *Speech and Hearing Science.* Englewood Cliffs, NJ, Prentice-Hall, Inc., 1968.

Index

A

Abdominal muscles, 189–193
  action of, during, arched-back sit-up, 197–210
    curled-trunk sit-up, 199–205
    leg raising or lowering, 213
    trunk raising, 187, 195, 196, 198
  anterior, strength of, 13, 14
  exercises for, 211, 220, 289, 301
    concerns regarding, 186
  fixation by, 6, 7, 258, 260
  function of, in standing, 275
  grading of, 209, 215, 222, 225, 226
  imbalance of, 224
    deviations of umbilicus in relation to, 224
  lateral, 222
  oblique, 226
  origins and insertions of, 189, 193
  tests for strength of, 208, 214
Abduction, 21 (see also specific joint)
  forefoot, 27
  muscles acting in, 123–125, 181, 182
Abductor digiti minimi, 69
  hallucis, 131
  pollicis brevis, 62, 66 (Note)
    longus, 68
Abductors of hip, 166–169
  scapula, 118, 119–121
  shoulder, 98–101
"Acquired" weakness, 14
Acromioclavicular joint, 22
Action, head and neck muscles, 257
Actively insufficient, 5
Adaptive shortness, 270
Adduction, 21 (see also specific joint)
  deformity of hip, 177
  forefoot, 27
  horizontal, of shoulder, 23
  muscles acting in, 123–125, 180, 181
Adductor brevis, 176, 177
  hallucis, 131
  longus, 176, 177
  magnus, 176, 177
  pollicis, 61
Adductors, hip, 176–178
  scapula, 112–117
  shoulder, tests for length of, 109
Alignment, faulty, 271, 272, 279, 281–287
  ideal, 194, 278, 280, 290
Analysis and treatment charts, 288, 296, 301, 307
Anatomical position, 19
Anconeus, triceps and, 96, 97
Ankle, and foot plantar flexors, 146
  joint, 26
Anterior, deltoid, 100, 101
  and lateral neck flexors, 254
  neck flexors, 258
Aponeurosis, abdominal, 189–193
Arc of motion, 11–13
Arch, longitudinal, 135, 141, 143
  transverse, 133, 137
Arcuate line, 189, 193
Arm movements used in testing abdominal muscles, 224
  pain, 106
Articularis genus, 158
Axes, 21
Axis, mechanical of femur, 178

B

Back extensors, 188, 228–230
  and hip extensors, 231

Bell's palsy, 251
Bibliography, 317–320
Biceps brachii, 90, 94, 95
  femoris, 155
Bowlegs, postural, 304, 305
  structural, 303
Brachial plexus, 53, 55
  lesion, case history of, 38–41
Brachialis, 94
Brachioradialis, 92
Buccinator, 243

C

Calcaneous, 137, 144, 146
Calcaneovalgus, 137
Calcaneovarus, 137, 141
Carpometacarpal joints, of fingers, 24
  of thumb, 24, 25
Cases, recorded examinations of, 40, 42–46, 252, 253, 313
Cavus, 133, 137, 144
Center of gravity, 19
Cervical plexus, 53, 54
Charts, action of head and neck muscles, 257
  analysis, of muscle imbalance, lower extremity, 183
    upper extremity, 127
    and treatment, 288, 296, 301, 307
  cranial nerve and muscle, 250
  deglutition, muscles of, 266, 267
  joint measurement, lower extremity, 30
    upper extremity, 29
  key to muscle grading, 12
  muscle, lower extremity, 182
    upper, 126
  muscles listed according to spinal segment innervation and
      grouped according to joint action, lower extremity, 180,
      181
    scapular, 123
    upper extremity, 124, 125
  origins and insertions, head and neck muscles, 256
  neck and back extensors, 228
  postural examination, 313
  respiratory muscle, 262
  spinal nerve and motor point, lower extremity, 36
    upper extremity, 34
  spinal nerve and muscle, lower extremity, 37
    upper extremity, 35
  spinal segment supply to muscles, neck and upper extremity,
      48, 49
    trunk and lower extremity, 50, 51
  spinal segment supply to nerves, neck and upper extremity, 58
    trunk and lower extremity, 58
  summary, good and faulty posture, 314, 315
  use of in differential diagnosis, 38
Circumduction, 22 (see also specific joints.)
Clawhand, 75, 76
Clockwise rotation, 21, 28, 190, 191
Compilation, spinal segment, to muscles, 49, 51
  to nerves, 58
Contracture, 9 (see also specific muscles)
Coracobrachialis, 93
Coronal axis, 21
  plane, 20
Corrugator supercilii, 238
Counterpressure, 6 (see also hip and shoulder rotators)
Counter-clockwise rotation, 21, 28, 190, 191
Cranial nerve and muscle chart, 250–253
  nerves and facial muscles, drawings, 236, 237
Curled-trunk sit-up, 197, 199–204, 236, 237
Curve, lateral of spine, 170, 192, 230, 260, 295–297, 308, 309
Cutaneous distribution, sensory areas of, 33, 35, 37

D

Definitions and descriptions of terms, 194–197
Deformities, caused by, 9
(see "weakness" and "contracture" on pages listed below)
  elbow, 92–97
  fingers, 72–83
  foot, 131–146
  forearm, 88–91
  hip, 165–179
  knee, 156, 157, 159
  neck, 258–261
  scapula, 106, 112–122
  shoulder, 98–105, 107–111
  thumb, 61–68
  wrist, 84–87
Deglutition, muscles of, 266, 267
Deltoid, 99–101
  anterior, 100, 101
  middle, 99
  posterior, 100, 101
  support of, 8
Depression, of scapula, 22, 123
Depressor anguli oris, 244
  labii inferioris, 242
    superioris, 242
  septi and transverse portion nasalis, 239
Dermatomes, 33, 35, 37, 250
Deviation of umbilicus, 224
Diaphragm, 263, 264
Digastric, 255, 266
Dorsal interossei, of foot, 136, 137
  of hand, 74–76
Dorsiflexion of ankle, 26
  muscles acting in, 138–141, 181
Drop-foot, 138–141

E

Ear muscles, 237, 250
Elbow flexors, 95
Elevation of scapula, 22, 123
Epicranius, 238
Equinovalgus, 137
Equinovarus, 137, 142
Equinus, 137, 146, 177
Erector spinae, 228–230
Eversion, 27
  muscles acting in, 182
Exercise programs, 272
Exercises, 211, 220, 289, 297, 301, 307
Expiration, muscles of, 262, 265
Extension, 21 (see also specific joints)
  muscles acting in, 123–125, 180, 181
  of thumb, in radial nerve lesion, 65 (Note)
Extensor carpi radialis longus and brevis, 86
  carpi ulnaris, 87
  digiti minimi, 78
  digitorum, 78, 79
    longus and brevis, 138, 139
  hallucis longus and brevis, 140
  indicis, 78
  pollicis brevis, 67
    longus, 66
External oblique, 190, 192, 198, 216, 218, 219, 222, 226, 284
  in relation to posture, 218, 219
Eye muscles, 246, 247

F

Facial muscles, drawings of, 236, 237
  paralysis of, 251
  tests of, 236–248
Fascia lata, 175, 297, 307
Faulty posture, analysis and treatment, 288, 296, 301, 307 (see also posture)
Feet, effect of holding down during trunk raising, 205, 206
  posture of, 276, 277, 302

Fixation

Fixation, 6 (see also specific muscles)
  muscles, 10
Flexibility, forward bending tests for, 232, 233, 310
  normal according to age level, 234 (see also tests for muscle length and range of motion)
Flexion, 21 (see also specific joints)
  deformity, hip, 177
    knee, 146, 156, 157
  lateral, 21, 28
  muscles acting in, 124, 125, 180, 181
Flexor carpi radialis, 84
  ulnaris, 85
  digiti minimi, 71
  digitorium brevis, 133
    longus and quadratus plantae, 135
    profundus, 82, 83
    superficialis, 80, 81
  hallucis brevis, 132
    longus, 134
  pollicis brevis, 65, 66 (Note)
    longus, 64
Flexors, elbow, 94, 95, 124
  neck, 257, 258
Forward bending test for length of posterior muscles, 232–234
  head, 28, 117, 258, 281, 284, 285, 300
Frontalis, (see Epicranius)
Foot, boards, 308, 309
  deformities, 131–146
  position of, in standing and walking, 276, 277
  pronation of, 302
  supination of, 302

G

Gastrocnemius, 145–147
  limp, 142, 146
  posture associated with weakness of, 147
Gemellus inferior and superior, 172
Gliding movement, 21
Gluteus maximus, 174, 175
  medius, 169
    limp, 170
    weakness, effect on posture, 170
  minimus, 168
Gracilis, 156, 176, 177
Grading, 8, 10–13
  abdominal muscles, anterior, leg lowering, 215
    trunk raising, 209
    lateral, 222
    oblique, 226
  back extensors, 230
  completion of anti-gravity arc of motion in, 11
  extremity muscles, 8, 10–13
  facial muscles, 11, 252, 253
  gravity factor in, 11
  gravity-lessened position in, 11–13
  in test of young children, 15
  key to, 12
  Lovett's system of, 10
  test position in relation to, 7
Gravity, center of, 19
  lessened position, 11–13
  line of, 19
  plumb line in relation to, 18

H

Hallux valgus, 131
Hammer toes, 132, 134, 137
Hamstrings and gracilis, 156
  lateral, 155
  medial, 154
  tests for length of, 148–153
Handedness, effect on posture, 294
  on muscle balance, 291, 292
Head, forward, 28, 117, 258, 281, 284, 285, 300
  and neck muscles, 228, 229, 256, 257
  rotation of, 21, 28

Hip, abduction, true and apparent, 10, 223
    abductors, 166–169
        weakness of, effect on posture, 170
    adduction deformity, 177
    adductors, 176–178
    extensors, 154, 155, 174, 175
    flexors, action of, during trunk raising, 199–204
        paralysis of, 196
        test for length of, 160–163
            for strength of, 164, 165
        stretching, 289
    joint, 25, 26
        position of, in standing, 275
    lateral rotators, 172, 173
    medial rotators, 171
Hyperextension, 22, 28, 195
    of hip, 284
        knee, 26, 146, 147, 156, 157, 167, 303

I

Ideal alignment, 194, 278, 280, 290
Iliacus, 164, 202, 280
Iliopsoas, 164
    in relation to posture, 275, 281, 284
    tests for length of, 160–163
        for strength of, 164
Iliotibial tract, 175, 290, 296
Infrahyoid muscles, 248, 255, 258, 267
Infraspinatus, 110
Innervation (see nerves)
Insertions (see origins and insertions)
Inspiration, muscles of, 262–265
Intercostals, 264, 265, 301
Internal oblique, 191, 192, 219, 222, 226
Interossei, of foot, 136, 137
    of hand, 72, 73, 75, 76
        tests for length of, 76
Interphalangeal joints, of fingers, 24
    thumb, 25
    toes, 27
Inversion, 27
    muscles acting in, 182
Isolation of muscles in testing, 4

J

Joint, acromioclavicular, 22
    action, muscles grouped according to, 123–125, 180, 181
    ankle, 26
    calcaneocuboid, 275, 278
    carpometacarpal, of fingers, 24
        of thumb, 25
    elbow, 23
    hip, 25, 26
    interphalangeal, of fingers, 24
        of thumb, 25
        of toes, 27
    knee, 26
    metacarpophalangeal, of fingers, 24
        of thumb, 25
    metatarsophalangeal, 27
    movements, 21, 22
        of scapula and shoulder girdle, 22
    radioulnar, 23
    shoulder, 22, 23
    subtalar, 26, 27
    transverse tarsal, 26, 27
    types of, 21
    wrist, 24
Joint measurement charts, 29, 30
Joints of vertebral column, 27, 28

K

Key to muscle grading, 12
Knee, anteromedial instability of, 156, 179
    bent position during sit-ups, 201–204, 206, 210, 212
        temporary use of, 212
    flexion of, 146, 156, 157, 177, 179, 303
    hyperextension of, 26, 146, 147, 156, 159, 281, 284, 285, 303, 306
    joint, 26
    lateral instability of, 156, 167
    position in standing, 275
Knock-knee, 156, 167, 171, 173, 303–305
    postural, 305
    structural, 303, 304
Kyphosis, 281, 287, 300
    -lordosis posture, 194, 281

L

Larynx, 250, 267
Lateral, abdominal muscles, 222
    hamstrings, 155
    neck flexion, muscles acting in, 257
    rotators of hip, 172, 173
        of shoulder, 110, 111
    trunk flexors and hip abductors, 223
    trunk raising, 222, 223
Latissimus dorsi, 102, 103, 108, 222, 230
Leg, length measurements, 310, 311
    lowering test, 214, 215
    raising or lowering with lumbar spine flexed, 216
        with lumbar spine hyperextended, 217
Levator anguli oris, 240
    labii superioris, 242
    alaeque nasi, 242
    palpebrae superioris, 246
    scapulae, 102, 112, 113, 123
Leverage, 8
Limp, gastrocnemius, 142, 146
    gluteus medius, 170
Linea alba, 189–193
Line of gravity, 19
Longitudinal axis, 21
Lordosis, 194, 281–283, 286–288
Lovett systems of grading, 10
Lower trapezius, 116
Lumbar plexus, 53, 56
Lumbricales of foot, 136, 137
    of hand, 74–76
    test for length of, 76

M

Masseter, 245
Mechanical axis of femur, 178
Medial hamstrings, 154
    rotators of hips, 171
        of shoulder, 108
Mentalis, 244
Metacarpophalangeal joints, 25
Metatarsal arch, 133, 137, 279, 307
Metatarsophalangeal joints, 27
Middle deltoid, 99
    trapezius, 114
Midsagittal plane, 20
Mobility, excessive, 272
    limited, 271, 272
Movements, joint, 21–28
    of spine and hip joints during trunk raising, 200, 201
Muscle, contracture (see specific muscles)
    grading, 10–13
    imbalance, chart for analysis of, 127, 183
    shortness, 9, 270 (see also specific muscles)
    testing, 3
        fundamental components of, 3
        of individual muscles, 3, 4, 6

Muscle, continued
  tests, for length, 76, 102–104, 106, 109, 148–153, 160–163, 166,
    232–234
    for strength (see specific muscles)
    suggested order of, 15
  weakness, 9, 270 (see specific muscles)
Muscles, abdominal, 189–193
  back, 228–230
  extremity, lower, 129–183
    upper, 59–127
  facial, 236, 237, 250
    tests, 238–248
  grouped according to joint action, 123–125, 180, 181
  multi-joint, 5, 7
  neck, 228, 229, 254, 256, 257, 261
  of deglutition, 266, 267
  one-joint, 4, 5, 7
  "safeguard" action of 5, 156

                        N

Nasalis, alar portion, 239
  transverse portion, 239
"Natural" weakness, 15
Neck extensors, test for, 261
  and back extensors, 228
  flexors, test for, 258, 260
  muscles, action of, 257
    origins and insertions of, 256
Nerves, cranial, 236, 237, 250
  spinal, neck and upper extremity, 34, 35
    trunk and lower extremity, 36, 37 (see also specific muscles)
Neutral position of pelvis, 25
Normal, flexibility according to age, 234
  strength in anterior neck and abdominal muscles, 13, 14
  subjects, used for tests, 6
  use of word, in grading, 13, 14

                        O

Oblique trunk raising, 226
Obliquus externus abdominis (external oblique), 190–192, 198,
    216, 218, 219, 222, 226, 284
  inferior, 246
  internus abdominis (internal oblique), 190–192, 226
  superior, 246
Obturator externus, 172
  internus, 172
Opponens digiti minimi, 70
  pollicis, 63
Opposition, 25
Orbicularis oculi, 248
  oris, 243
Order of muscle tests, 15
Origins and insertions, abdominal muscles, 189–193
  back muscles, 228
  deglutition, muscles of, 266, 267
  extremity muscles, lower, 131–179
    upper, 61–118
  facial, 238–247
  head and neck muscles, 228, 256
Overstretch, 270

                        P

Pain, in relation to faulty posture, 187, 271, 288, 296
  arm, 106
Palate, 250, 266
Palmar interossei, 73, 75, 76
  first, 66 (Note)
Palmaris longus and brevis, 77
Parentheses, use of in recording grades, 8, 12
Passive insufficiency, 9

Patient, position of, for tests, 6 (see also specific muscles)
Pectineus, 176
Pectoralis, major, 104, 105, 108
  nerves to, 52
  minor, 106, 123
Pelvic tilt, 21, 25
  anterior, 25, 194, 281–283
  lateral, 25, 194, 291–295
  posterior, 25, 194, 284, 285
Pelvis, anterior deviation of, 191, 219, 237
  neutral position of, 25, 194, 278, 280
  rotation of, 21, 190–192
Percentage grades, 10–14
Peripheral nerves, cutaneous distribution of, 33, 35, 37, 250
Peroneus longus and brevis, 143
  tertius, 138, 139
Pharynx, 250, 267
Piriformis, 172
Planes, 20
  lines of reference representing, 274, 278, 290
Plantar flexion of ankle, 26
  muscles acting in, 145, 181
Plantaris, 145
Platysma, 242, 258
Plexus, nerve, 53–57
  brachial, 55
  cervical, 54
  lumbar, 56
  sacral, 57
Plumb line, 19, 278, 279, 308, 309
  reference points for, 278
Popliteus, 157, 307
Posterior deltoid, 100, 101
Postural bowlegs, 304, 305
  knock-knees, 305
Postural examination chart, 313
  equipment for, 308, 309
  findings, interpretation of, 311
  procedure for, 308
Posture, 269–315
  associated with weakness of gastrocnemius and soleus, 147
    of hip abductors, 170
    of hip flexors, 284, 285
    of transversus abdominis, 193
  boards, 308, 309
  external oblique in relation to, 218, 219
  faulty alignment, flat back, 194, 285, 288
  lordosis, 194, 283, 288
  "military", 282
  ideal alignment, 194, 278
  iliopsoas in relation to, 275, 281, 284
  of children, 286, 287
  of feet and knees, 302–306
  of head and neck, 300
  of shoulders and scapula, 298, 299
  of trunk, 194 (see also 278–297)
  standard, 273–277
  summary charts of, 314, 315
  sway-back, 194, 284, 288
Pressure, 8 (see also specific muscles)
Prime mover, 3
Procerus, 240
Pronation, of foot, 27, 302
  of forearm, 23, 24
  muscles acting in, 124, 125, 180, 181
Pronator quadratus, 88, 89
  teres, 88
Psoas major, 164
  minor, 164, 166
Pterygoideus lateralis, 245
  medialis, 245

                        Q

Quadratus femoris, 172
  lumborum, 222, 227, 230
Quadriceps femoris, 158, 159

R

Radial nerve lesion, 66 (Note)
Radioulnar joint, 23, 24
Range of motion, 7, 8, 11, 22–30
Recording muscle grades on abdominal muscle charts, 225
    on cranial nerve and muscle charts, 252, 253
    on spinal nerve and muscle charts, 40, 42–46
Rectus abdominis, 189, 198, 216, 222, 226
    femoris, shortness of, 159, 161, 162, 174
    inferior, 246, 247
    lateralis, 247
    medialis, 247
    superior, 246, 247
Resistance, 8
Respiration, muscles of, 262–265
Respiratory muscle chart, 262
Rhomboids, major and minor, 112, 113, 115, 123
Risorius, 241
Rotation, 21, 28
    clockwise and counter-clockwise, 21, 28, 190, 191, 279
    muscles acting in, 123–125, 180, 181, 257

S

Sacral plexus, 53, 57
Sacrotuberous ligament, 174
"Safeguard," 5, 6, 156
Sagittal axis, 21
    plane, 20
Sartorius, 165, 179
Scaleni, 256, 257, 258
Scapula, movements of, 123
Scoliosis, 192, 230, 260, 295–297, 308, 309, 311
Semimembranosus and semitendinosus, 154
Sensory, areas of cutaneous distribution, 33, 35, 37
    dermatome areas, 33, 35, 37
Serratus anterior, 118–123
    paralysis or weakness of, 120–123
Shoe corrections, 288, 307
Shortness, 9 (see also specific muscles)
Shoulder adductors, test for length of, 103, 104
    joint, 22, 23
    lateral rotators, 110, 111
    medial rotators, 108
Sit-up, arched-back, 197, 210
    curled-trunk, 197, 199–204
    exercises, 196
        indications and contraindications, 206, 207
Slack, putting a muscle on a, 5, 206
Soleus, 144–147
    posture associated with weakness of, 147
Spinal, nerve and motor point charts, 34, 36, 38
    nerve and muscle charts, 33–37
        examinations recorded on, 40, 42–46
        use of in differential diagnosis, 38
    segment to muscles, 47–52
        to nerves, 58
Spine
    movements of, 27, 28
    normal curves of, 194
Stabilization, 6 (see also fixation under specific muscles)
    of pelvis, 168, 223, 231
Sternoclavicular joint, 22
Sternocleidomastoid, 258–260
Strength, average, 13
    "normal" for age, 13, 14
    return of, 3
Stretch-weakness, 270, 273
Subclavius, nerve to, 35, 55
Subjects, normal, used for tests, 5
Subscapularis, 108
Substitution, 9, 10
    anterior deltoid, 101
    elbow flexors, 94
    gastrocnemius, 146
    hamstrings, 156
    hip abductors, 10, 168

    adductors, 177
    flexors, 165
    quadriceps, 159
    soleus, 144
    wrist extensors, 87
        flexors, 85
Subtalar joint, 26, 27
Summary charts of good and faulty posture, 314, 315
Supination, foot, 27
    forearm, 23, 24
    muscles acting in, 125, 181
Supinator, 90, 91
    and biceps, 90
Support, 6, 286, 288, 297
Suprahyoid muscles, 248, 255, 258, 266
Supraspinatus, 98, 99
Sway-back posture, 194, 284, 288
Synergistic action, 5

T

Tautness, 9
Temporalis, 245
Tendocalcaneous, 290, 291, 308
Tensor fasciae latae, 165–167
Teres major, 107, 108
    minor and infraspinatus, 110
Terminology, 316
Test, 7, 8 (see also specific muscles)
    movement, 7, 8, 13
    position, 7, 8, 13
Testing, 3
    differentiation of muscles in, 4, 5
    fundamental components of, 3
    in cases of relaxed joint, 8
    "isolation" of muscles in, 4
    leverage in, 8
    of individual muscles, 3, 4
    of infants and young children, 15
    of multi-joint muscles, 4, 5, 7
    of one-joint muscles, 4, 5, 7
    useful in, 3
Tests, abdominal, leg lowering, 214
    trunk raising, 208
    back, 230
    extremities (see specific muscle)
    facial, 238–248
    for muscle length, 76, 102–104, 106, 109, 148–153, 160–163, 232–234
    neck, 258–261
    respiratory, 264, 265
    suggested order of, 6, 15
    terms used in descriptions of, 6–9
Therapeutic exercises, 211, 220, 272, 289, 297, 301, 307
Tibialis anterior, 141
    posterior, 142
Tightness, 9
Tilt of pelvis, 25
    of scapula, 22, 123
Toe extensors, 137–140
    flexors, 14, 132–135
Tongue muscles, 266
Torticollis, 117, 260, 261
Transverse axis, 21
    plane, 20
Transverse tarsal joints, 26, 27
Transversus abdominis, 191, 193
Trapezius, 112, 114–117, 121–123
    lower, 116
    middle, 114
        error in testing, 115
    paralysis of, 122
    upper, 117, 257, 261
Treatment, 211, 220, 272, 289, 297, 301, 307
Trendelenberg sign, 291
Triceps and anconeus, 96, 97
Trunk, curl, 196, 200, 201, 203, 209
    raising forward with lumbar spine hyperextended, 210
        with spine flexed, 209
        obliquely forward, 226
        sideways, 222

U

Umbilicus, deviations of, 224
Upper trapezius, 117, 257, 261

V

Valgus, forefoot, 131, 137
  equino-, 134, 135, 137, 143
  genu, 303
  hallux, 131
Varus, forefoot, 137
  equino-, 143
  genu, 303
Vertebral column, 27, 28, 189–192, 194
  flexibility of, 232–234
  movement of, during sit-ups, 200, 201

W

Wall-sitting exercise, 301
  -standing exercise, 301
"Weak back," 188
Weakness, 3, 7, 9 (see also specific muscles)
  "acquired," 14, 15
  distinguished from restriction of range of motion, 7
  due to, 3, 273
  effect on posture, 147, 170, 193, 218, 219 (see also chapter 8.)
  "natural," 15
  permitting a deformity, 9
  stretch-, 270, 273
Wrist joint, 24

Z

Zero position, 19
Zygomaticus major, 241
  minor, 242